INFRARED SPECTROSCOPY
FUNDAMENTALS AND APPLICATIONS

Kruger Brentt
Publishers

INFRARED SPECTROSCOPY
FUNDAMENTALS AND APPLICATIONS

VOLUME-1

Lawrence Baker
Adam Kecskemeti

Kruger Brentt
Publishers

2025

Kruger Brentt Publishers UK. LTD.
Company Number 9728962

Regd. Office: 68 St Margarets Road, Edgware, Middlesex HA8 9UU

ISBN: 978-1-78715-359-2 (Vol.1)

For information on all our publications visit our website at http://krugerbrentt.com/

PREFACE

"Infrared Spectroscopy: Fundamentals and Applications" serves as a comprehensive guide to understanding the principles, techniques, and diverse applications of infrared spectroscopy in various fields of science and technology. This book aims to provide researchers, students, and professionals with a thorough understanding of the theoretical foundations, instrumentation, and practical considerations associated with infrared spectroscopy.

Infrared spectroscopy is a powerful analytical technique used to study the chemical composition, structure, and properties of materials by analyzing their interaction with infrared radiation. From molecular identification and chemical analysis to materials characterization and quality control, infrared spectroscopy plays a vital role in a wide range of disciplines, including chemistry, biology, pharmacy, environmental science, materials science, and forensic science.

As editors of this volume, we have endeavored to compile a diverse array of perspectives, methodologies, and case studies from leading experts and practitioners in the field of infrared spectroscopy. Through a blend of theoretical insights, practical guidance, and real-world examples, this book covers a wide range of topics, including infrared spectroscopic techniques, instrumentation, data analysis, and applications in various scientific and industrial settings.

Our overarching goal in presenting this book is to provide readers with a comprehensive and up-to-date resource that fosters a deeper understanding of the principles and applications of infrared spectroscopy. By offering insights into the latest advancements, best practices, and emerging trends in the field, we aim to inspire innovation, collaboration, and excellence in the use of infrared spectroscopy across different domains of research and industry.

We extend our sincere gratitude to all the contributors who have generously shared their expertise, experiences, and insights in the field of infrared spectroscopy. It is our hope that this book will serve as a valuable reference and guide for researchers, students, and professionals as they explore the diverse capabilities and applications of infrared spectroscopy in their respective fields of study and practice.

Lawrence Baker | Adam Kecskemeti

CONTENTS

SECTION-2: POLYMERS AND BIOPOLYMERS

VOLUME-2

SECTION-3: MATERIALS TECHNOLOGY

INTRODUCTION TO INFRARED SPECTROSCOPY

1. INTRODUCTION: AN OVERVIEW

1.1 Short history of the technique

Sir William Herschel is credited with the discovery of infrared light in the year 1800 [1]. Herschel was studying the energy levels linked to the wavelengths of light in the visible spectrum. The sunlight was passed through a prism, resulting in the familiar visible spectrum of colors found in a rainbow. This spectrum ranges from blue to red and is characterized by corresponding wavelengths or frequencies [2, 3] (refer to Fig.1).Sir William Herschel discovered infrared radiation in 1800 [1]. Herschel was studying the energy levels linked to the wavelengths of light within the visible spectrum. By passing sunlight through a prism, the familiar visual spectrum of colors, known as the rainbow, was observed. This spectrum ranges from blue to red and is characterized by corresponding wavelengths or frequencies [2, 3] (refer to Fig.1).

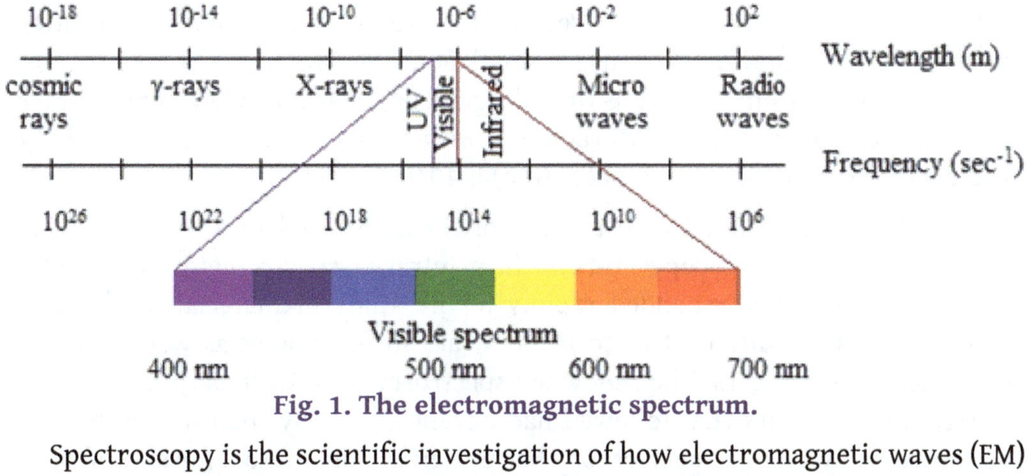

Fig. 1. The electromagnetic spectrum.

Spectroscopy is the scientific investigation of how electromagnetic waves (EM) and matter interact. The wavelengths of the colors are directly related to the energy levels of the rainbow colors. By systematically scanning the visible spectrum from blue to red and measuring the corresponding temperatures, Herschel observed a progressive increase in temperature. Herschel therefore opted to measure the temperature just below the red part, assuming that the temperature rise would cease beyond the visible spectrum. However, he was astonished to discover that the temperature was much greater. He referred to the rays that were lower in frequency than red rays as “non colorific rays” or invisible rays, which were later named "infrared rays" or IR light. The human eye cannot perceive this light. The human eye typically detects electromagnetic radiation with wavelengths ranging from 390 to 750 nm. The infrared spectrum begins at a wavelength of 0.75 nanometers. A nanometer (nm) is equivalent to 10-9 meters (m). The Infrared spectrum is categorized into three regions: Near Infrared (NIRS), Mid Infrared (MIRS), and Far Infrared (FIRS) [4-6].

1.2 The three Infra red regions of interest in the electromagnetic spectrum

In terms of wavelengths the three regions in micrometers (μm) are the following:

1. NIRS, (0.7 μm to 2,5 μm)
2. MIRS (2,5 μm to 25 μm)
3. FIRS (25 μm to300 μm).

In terms of wavenumbers the three regions in cm^{-1} are:

1. (NIRS), 14000-4000 cm^{-1}
2. (MIRS), 4000-400 cm^{-1}
3. (FIRS), 400-10 cm^{-1}

The first region (NIRS) allows the study of overtones and harmonic or combination vibrations. The MIRS region is to study the fundamental vibrations and the rotation-vibration structure of small molecules, whereas the FIRS region is for the low heavy atom vibrations (metal-ligand or the lattice vibrations).Infrared (IR) light is electromagnetic (EM) radiation with a wavelength longer than that of visible light: ≤0.7μm. One micrometer (μm) is 10^{-6}m.

Experiments continued with the use of these infrared rays in spectroscopy called, Infrared Spectroscopy and the first infrared spectrometer was built in 1835. IR Spectroscopy expanded rapidly in the study of materials and for the chemical characterization of materials that are in our planet as well as beyond the planets and the stars. The renowned spectroscopists, Hertzberg, Coblenz and Angstrom in the years that followed had advanced greatly the cause of Infrared spectroscopy. By 1900 IR spectroscopy became an important tool for identification and characterization of chemical compounds and materials. For example, the carboxylic acids, R-COOH, show two characteristic bands at 1700 cm^{-1} and near 3500 cm^{-1}, which correspond to the C=O and O-H stretching vibrations of the carboxyl group, -COOH. Ketones, R-CO-R absorb at 1730-40cm^{-1}. Saturated carboxylic acids absorb at 1710 cm^{-1}, whereas saturated/aromatic carboxylic acids absorb at 1680-1690 cm^{-1} and carboxylic salts or metal carboxylates absorb at 1550-1610 cm^{-1}. By 1950 IR spectroscopy was applied to more complicated molecules such as proteins by Elliot and Ambrose [2]. These later studies showed that IR spectroscopy could also be used to study biological molecules, such as proteins, DNA and membranes and could be used in bioSciences, in general [2-8].

Physicochemical techniques, particularly infrared spectroscopy approaches, are non-invasive and have the potential to gather information about molecular structure and characterization of various materials at different levels. Spectroscopic techniques, which rely on the interaction between light and matter, have been extensively employed to investigate materials in both in vivo and ex vivo or in vitro settings. Infrared spectroscopy can yield data on various substances, including isolated materials, biomaterials like biopolymers, and biological materials such as connective tissues, single cells, and biological fluids, to name a few examples. A wide range of information can be gained from a single experiment using extremely little samples. Infrared spectroscopy is a technique that examines the energy levels of molecules in wavenumbers (cm-1) within the electromagnetic spectrum. It does this by analyzing the vibrations of the molecules, which are also expressed in wavelengths (μm). Infrared spectroscopy is the examination of how matter interacts with light radiation as waves pass through it. The waves possess an electromagnetic nature and engage with the polarity of the chemical bonds of the molecules [3]. If a molecule lacks polarity, meaning it does not possess a dipole moment, it will not exhibit any infrared interaction and so will not generate an infrared spectrum.

1.3 Vibrational degrees of freedom

Chemical bonds are the forces responsible for holding the atoms together in a molecule. The chemical link in a diatomic molecule, such as hydrochloric acid (H-Cl), occurs between hydrogen (H) and chlorine (Cl). The chemical forces responsible for the bonding between these two atoms are regarded as analogous to the forces exerted by springs without mass. In a Cartesian coordinate system, each mass in a molecule is defined by three coordinates, namely x, y, and z, which determine its position in space. Consequently, the molecule possesses three distinct degrees of freedom for movement. The total number of degrees of freedom for all the atoms in a molecule is directly proportional to the number of atoms in the molecule, with a factor of 3. By removing the degrees of freedom related to translation and rotation from the total 3N degrees of freedom, we obtain 3N-6 internal movements for a non-linear molecule and 3N-5 for a linear molecule. In a linear molecule like H-Cl, the rotational motion around the bond axis does not affect the energy of the molecule. The oscillations occurring within an object's interior structure are referred to as the normal modes of vibration.

Thus, in the example of H-Cl we have one vibration,(3x2)-5=1, i.e. only one vibration along the H-Cl axis or along the chemical bond of the molecule. For a non linear molecule as H_2O we have (3x3)-6=3 vibrations, the two vibrations along the chemical bonds O-H symmetrical (v_s) and antisymmetrical (vas) O-H bonds and the bending vibration (δ) of changing the angle H-O-H of the two bonds [3,4]. In this way we can interpret the IR-spectra of small inorganic compounds, such as, SO_2, CO_2 and NH_3 quite reasonably. For the more complicated organic molecules the IR spectrum will give more vibrations as calculated from the 3N-6 vibrations, Since the number of atoms in the molecule increases, however the spectrum is interpreted on the basis of characteristic bands.

2. THEORY

2.1 Interaction of light waves with molecules

The interaction of light and molecules forms the basis of IR spectroscopy. Here it will be given a short description of the Electromagnetic Radiation, the energy levels of a molecule and the way the Electromagnetic Radiation interacts with molecules and their structure [5, 6].

2.2 Electromagnetic radiation

The EM radiation is a combination of periodically changing or oscillating electric field (EF) and magnetic field (MF) oscillating at the same frequency, but perpendicular to the electrical field [7] (see Fig.2).

The wavelength is represented by λ [6], which is the wavelength, the distance between two positions in the same phase and frequency (ν) is the number of oscillations per unit time of the EM wave per sec or vibrations/unit time. The wavenumber is the number of waves/unit length [7]. It can be easily seen [3] that c is given by equation 1:

$$c = \lambda\nu \qquad (1)$$

where, c is the velocity of light of EM waves, or light waves, which is a constant for a medium in which the waves are propagating, c=3x 10^8m/s

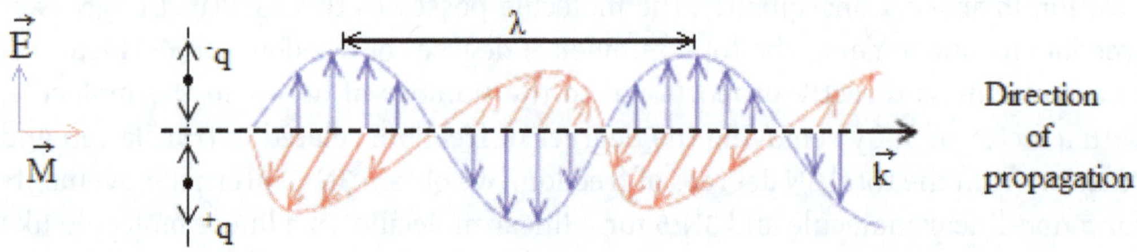

Fig. 2. An Illustration of Electromagnetic Radiation can be imagined as a self-propagating transverse oscillating wave of electric and magnetic fields. This diagram shows a plane linearly polarized wave propagating from left to right. The electric field is in a vertical plane (E) blue and the magnetic field in a horizontal plane (M) red

The wavelength (λ) is inversely proportional to the frequency, 1/ν. The Energy in quantum terms [8]: is given by Planck's equation:

$$E = h\nu \qquad (2)$$

Which was deduced later also by Einstein, where, E is the energy of the photon of frequency ν and h is Max Planck's constant [8], h=6.62606896x 10^{-34} Js or h =4.13566733x 10^{-15} ev. Wave number and frequency are related by the equation

$$\nu = c\tilde{\nu} \qquad (3)$$

The EM spectrum can be divided as we have seen into several regions differing in frequency or wavelength. The relationship between the frequency (ν) the wavelength (λ) and the speed of light(c) is given below:

$$\nu = \frac{c}{\lambda} \quad \nu = \frac{E}{h} \quad E = \frac{hc}{\lambda} \qquad (4)$$

The frequency in wavenumbers is given by the equation:

$$\tilde{\nu} = \frac{1}{2\pi c}\sqrt{\frac{k}{\mu}} \ (\mathrm{cm}^{-1}) \qquad (5)$$

Where, k=bond spring constant,μ= reduced mass, c=velocity of light (cm/sec),

μ is the reduced mass of the AB bond system of masses and m= mass of the atoms, mA=mass of A and mB= mass of B. The isotope effect can also be calculated using the reduced mass and substituting the isotopic mass in the equation of the frequency in wavenumbers.

Example, the H-Cl molecule

$$\mu = \frac{m_H m_{Cl}}{m_H + m_{Cl}}$$

m_H and m_{Cl} are the atomic masses of H and Cl atoms.

2.3 Energy of a molecule

The term "atom" was derived by Democritus [9] from the Greek word α-τέμνω, which means "indivisible" or "unable to be further divided" in Greek. It was first hypothesized that the atom is the fundamental unit of matter, possessing distinct properties, and serving as the fundamental component of all substances in the universe. Molecules are formed by combinations of atoms.

The energy of a molecule is the sum of 4 types of energies [3]:

$$E = E_{ele} + E_{vib} + E_{rot} + E_{tra} + E_{nuc} \quad (6)$$

E_{ele}: is the electronic energy of all the electrons of the molecule

E_{vbr}: is the vibrational energy of the molecule, i.e., the sum of the vibrations of the atoms in the molecule

E_{rot}: is the rotational energy of the molecule, which can rotate along the three axes, x,y,z

E_{tra}: is the translational energy of the molecule, which is due to the movement of the molecule as a whole along the three cartesian axes, x, y, z .

E_{nuc}: is the nuclear energy

Energy level electronic transitions (see Figs 3A, 3B):

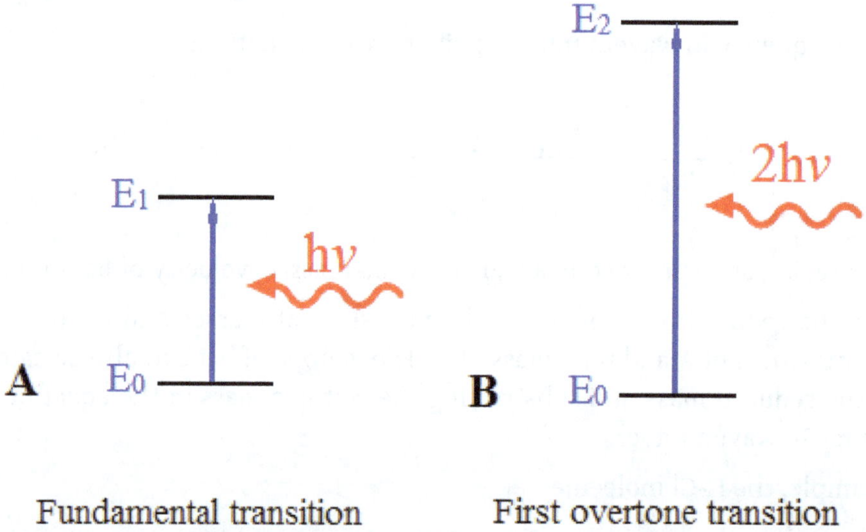

Fig. 3. A: increasing the energy level from E0 to E1 with the wave energy hv, which results in the fundamental transition, B: increasing the energy level from E0 to E2 leads to the first overtone transition or first harmonic.

3. INFRARED SPECTROSCOPY TECHNIQUES

We provide two types of IR spectrophotometers: the classical and the Fourier. Enhance spectrophotometers using the interferometer

3.1 The classical IR spectrometers [3, 4]

The main elements of the standard IR classical instrumentation consist of 4 parts (see Fig.4)

1. A light source of irradiation
2. A dispersing element, diffraction grating or a prism
3. A detector

4. Optical system of mirrors
5. SchematiCs of a two-beam absorption spectrometer are shown in. Fig. 4.

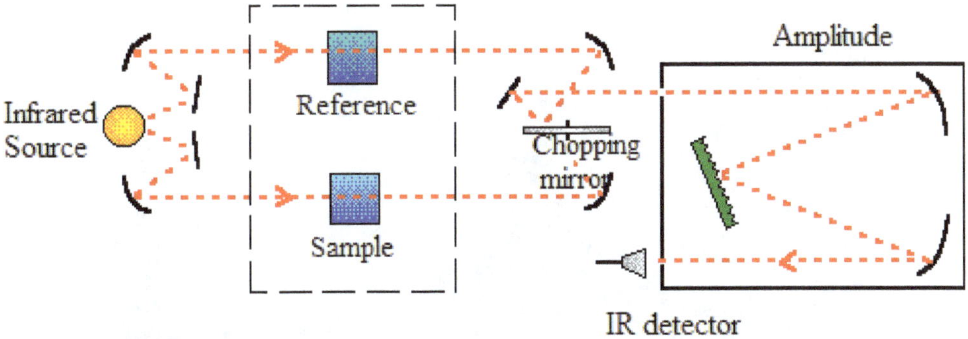

Fig. 4. A schematic diagram of the classical dispersive IR spectrophotometer.

The infrared light emitted by the source is reflected off a flat mirror and then passes through both the sample and reference monochromator. The beams are reflected by a revolving mirror, which alternates the passage of the sample and reference beams to the dispersing element and ultimately to the detector, resulting in the spectrum (see to Figure 4). As the beams alternate, the mirror undergoes a gradual rotation, allowing various frequencies of infrared radiation to reach the detector.

3.2 Fourier Transform IR spectrometers

The advent of high-performance Fourier Transform Infrared Spectroscopy (FT-IR) using a Michelson Interferometer [10] led to the development of contemporary spectrometers [7]. Both classical and current IR spectrometers provide identical information, with the key distinction being the utilization of a Michelson interferometer. This interferometer enables all frequencies to simultaneously reach the detector, rather than one frequency at a time.

In the 1870's A.A. Michelson [11] was measuring light and its speed with great precision(3) and reported the speed of light with the greatest precision to be 299,940 km/s and for this he was awarded the Nobel Prize in 1907. However, even though the experiments in interferometry by Michelson and Morley [12] were performed in 1887 the interferograms obtained with this spectrometer were very complex and could not be analyzed at that time because the mathematical formulae of Jean Baptiste Fourier series in 1882 could not be solved [13]. We had to wait until the invention of Lasers and the high performance of electronic computers in order to solve the mathematical formulae of Fourier to transform a number of points into waves and finally into the spectra [14]

The addition, of the lasers to the Michelson interferometer provided an accurate method (see Figs. 5A & 5B) of monitoring displacements of a moving mirror in the interferometer with a high performance computer, which allowed the complex interferogram to be analyzed and to be converted via Fourier transform to give spectra.

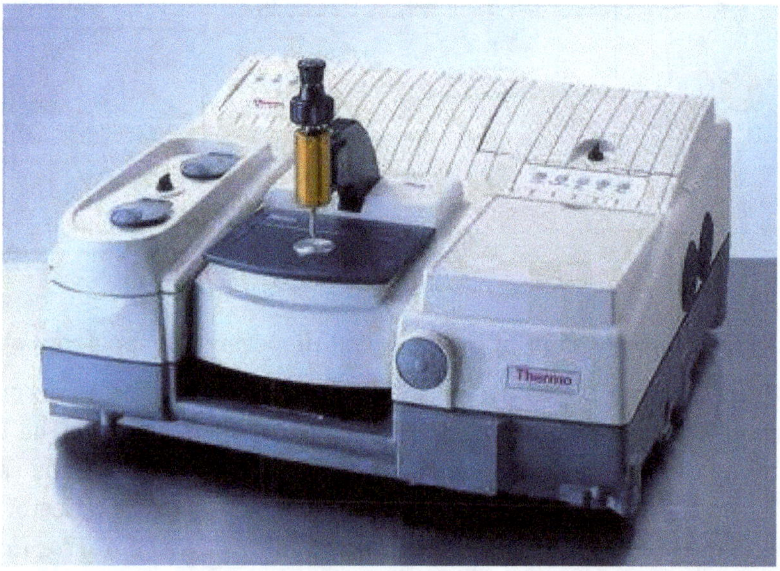

Fig. 5 A. Michelson FT-IR Spectrometer has the following main parts.

1. Light source
2. Beam splitter (half silvered mirror)
3. Translating mirror
4. Detector
5. Optical System (fixed mirror)

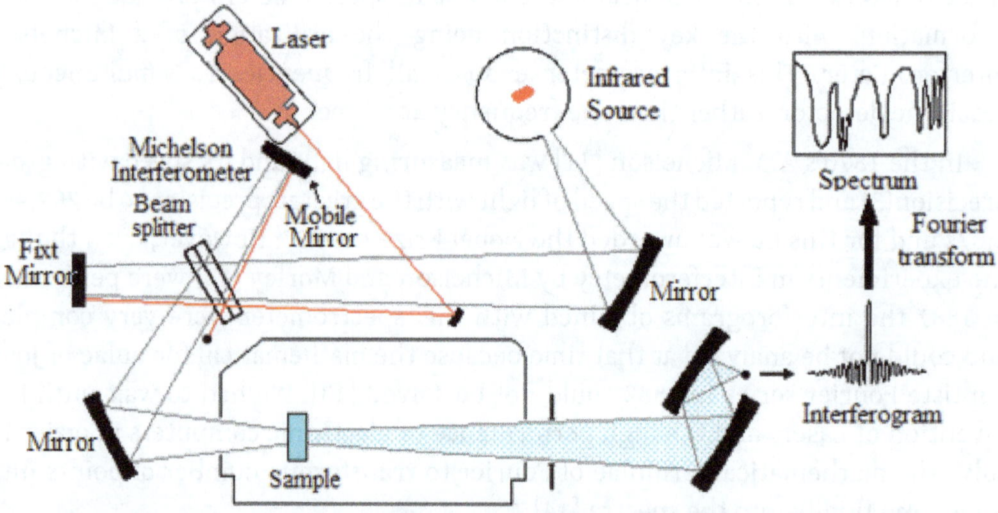

Fig. 5B. Schematic illustration of a modern FTIR Spectrophotometer.

Infrared spectroscopy experienced significant advancements following World War II and particularly after 1950, thanks to enhancements in instrumentation and electronics. These improvements positioned the technique as a focal point in chemical research and later, in the 1980s, in the field of biosciences. The introduction of new sample handling techniques, such as the attenuated total reflection method (ATR) and the interferometer, further contributed to its prominence [13]. The Fourier Transform is a mathematical technique used to transform a function from the time domain to the frequency domain.IR spectrophotometry is extensively employed in both research and industry as a standard method and a dependable tool for quality control, analysis of molecular structure, and kinetics [14-16] in the field of biosciences (refer to Fig. 6). Presented here is the spectrum of an intricate substance, namely an atheromatic plaque, along with its interpretation.

Currently, contemporary methodologies known as FT-methods are employed in practical applications. The non-Fourier Transform (FT) methods refer to the traditional approaches of dispersing light using a prism or a diffraction grating in the field of infrared (IR) spectroscopy. The FT approach provides a more accurate determination of absorption spectra. To get the IR spectra today, it is recommended to employ a Michelson interferometer [17]. The FT- method has the advantage of continuously detecting a wide range of radiation (known as the multiplex or Fellget advantage). Additionally, a larger portion of the source radiation is able to pass through the instrument due to the circular aperture (referred to as the Jacquinot advantage), as opposed to the narrow slit used in classical instruments with prisms or diffraction gratings.

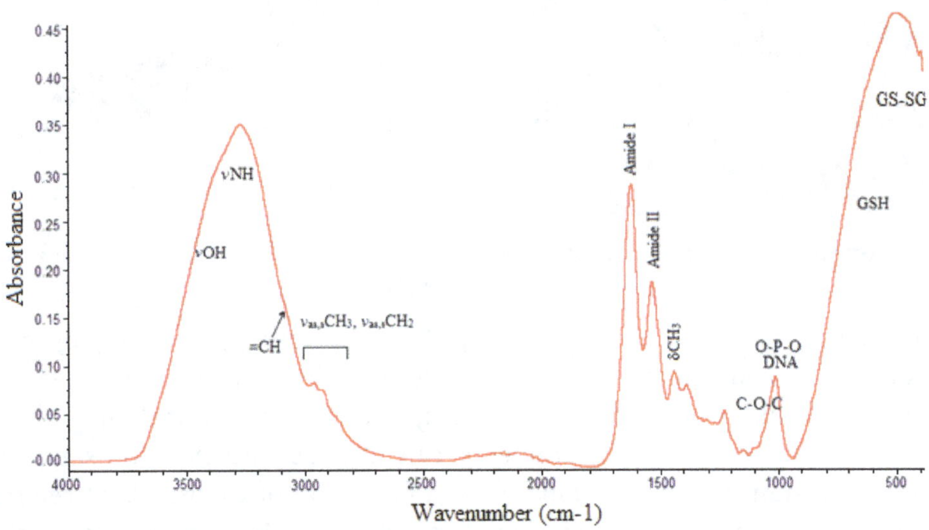

Fig. 6. FT-IR spectrum of a coronary atheromatic plaque is shown with the characteristic absorption bands of proteins, amide bands, O-P-O of DNA or phospholipids, disulfide groups, etc.

3.3 Micro-FT-IR spectrometers

By incorporating a reflecting microscope into the IR spectrometer, it becomes possible to get IR spectra of tiny molecules, crystals, and tissue cells. This enables the use of IR spectroscopy to biological systems, including connective tissues, blood samples, and bones, in the field of pathology in medicine [15, 26-27]. Figure 7 displays the microscopic visualization of malignant breast tissues along with its corresponding spectrum.

4. APPLICATIONS

Infrared spectroscopy is employed in the fields of chemistry and industry to identify and characterize substances. IR spectroscopy is employed to characterize compounds since the IR spectrum of a molecule serves as its unique "fingerprint" [16, 17]. Infrared spectroscopy is a non-invasive technique that is valuable for examining the secondary structure of complex systems, including biological molecules such as proteins, DNA, and membranes, without causing any damage. In recent years, infrared spectroscopy has been employed in medical sciences to analyze and differentiate between healthy and diseased human tissues.

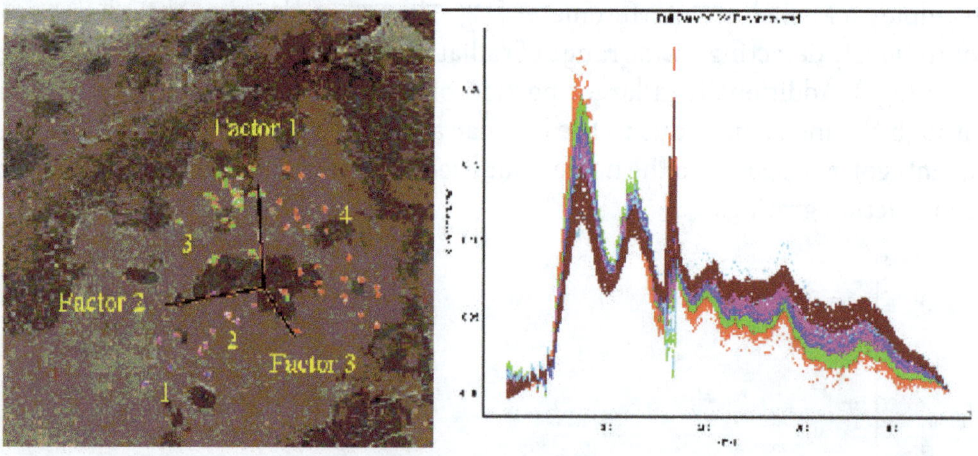

Fig. 7. Breast tissue: a 3-axis diagram and the mean spectral components are shown [25].

In both research and industry, IR spectroscopy is employed for measurement and quality control purposes. The instruments have been miniaturized and made portable for easy transportation, including for use in field trials. Accurate measurements can also be obtained for samples in solution. Substances can be compared to a collection of thousands of reference spectra [18]. Examples of specialized applications of IR spectroscopy include: In polymer manufacturing, IR spectroscopy has proven to be particularly effective in quantifying the degree of polymerization [18]. Infrared spectroscopy is a valuable technique for the identification, characterization, and verification of compounds. Given that the IR spectrum serves as the distinctive characteristic of a substance. Hence, infrared (IR) spectroscopy has a forensic function and is employed for the analysis of many compounds, including alcohol, narcotics, fibers, blood, and paints [19-28]. The book contains multiple parts that provide numerous examples of these uses.

5. REFERENCES

1. W. Herschel, Phil. Trans.R.Soc.London, 90, 284 (1800)
2. Elliot and E. Ambrose, Nature, Structure of Synthetic Polypeptides 165, 921 (1950); D.L.Woernley, Infrared Absorption Curves for Normal and Neoplastic Tissues and Related Biological Substances, Current Research, Vol. 12, , 1950 , 516p
3. T. Theophanides, In Greek, National Technical University of Athens, Chapter in "Properties of Materials", NTUA, Athens (1990); 67p
4. J. Anastasopoulou and Th. Theophanides, Chemistry and Symmetry", In Greek National Technical University of Athens, NTUA, (1997), 94p
5. G.Herzberg, Atomic spectra and atomic structure, Dover Books, New York,Academic press, 1969, 472 p
6. Maas, J.H. van der (1972) Basic Infrared Spectroscopy.2nd edition. London: Heyden & Son Ltd. 105p
7. Colthup, N.B., Daly, L.H., and Wiberley, S.E.(1990).Introduction to Infrared and Raman Spectroscopy.Third Edition. London: Academic press Ltd, 547 p.
8. Fowles, G.R. (1975).Introduction to Modern OptiCs. Second Edition. New York: Dover publications Inc., 336 p
9. Democritos, Avdera, Thrace, Greece, 460-370 BC
10. Hecht, E. OptiCs .Fourth edition. San Francisco: Pearson Education Inc. (2002
11. A.A. Michelson, Studies in OptiCs, University of Chicago, Press, Chicago (1962), 208 p
12. A.A Michelson and Morley, "on the Relative Motion of the Earth and the luminiferous Ether" Am. J. of Science, 333-335(1887); F.Gires and P.Toumois," L'interférométrie utilizable pour la compression lumineuse module en fréquence "Comptes Rendus de l'Académie des Sciences de Paris, 258, 6112-6115(1964)
13. Jean Baptiste Joseph Fourier, Oeuvres de Fourier, (1888); Idem Annals de Chimie et de Physique, 27, Paris, Annals of Chemistry and Physics, (1824) 236-281p

14. S. Tolansky, An Introduction to Interferometry, William Clowes and Sons Ltd. (1966), 253 p

15. J. Anastassopoulou, E. Boukaki, C. Conti, P. Ferraris, E.Giorgini, C. Rubini, S. Sabbatini,

16. T. Theophanides, G. Tosi, Microimaging FT-IR spectroscopy on pathological breast tissues, Vibrational Spectroscopy, 51 (2009)270-275

17. Melissa A. Page and W. Tandy Grubbs, J. Educ., 76(5), p.666 (1999)

18. Modern Spectroscopy, 2nd Edition, J.Michael Hollas,ISBN: 471-93076-8.

19. Wikipedia, the free encyclopedia. Infrared spectroscopy http://en.wikipedia.org (July 28, 2007).

20. Mount Holyoke College, South Hadley, Massachusetts. Forensic applications of IR

21. http://www.mtholyoke.edu (July 28, 2007

22. T. Theophanides, Infrared and Raman Spectra of Biological Molecules, NATO Advanced Study Institute, D. Reidel Publishing Co. Dodrecht, 1978,372p.

23. T. Theophanides, C. Sandorfy) Spectroscopy of Biological Molecules, NATO Advanced Study Institute, D. Reidel Publishing Co. Dodrecht, 1984 , 646p

24. T. Theophanides Fourier Transform Infrared Spectroscopy, D. Reidel Publishing Co. Dodrecht, 1984.

25. T. Theophanides, Inorganic Bioactivators, NATO Advanced Study Institute, D. Reidel Publishing Co. Dodrecht, 1989,415p

26. G. Vergoten and T. Theophanides, Biomolecular Structure and DynamiCs: Recent experimental and Theoretical advances, NATO Advanced Study Institute, Kluwer Academic Publishers, The Netherlands, 1997, 327p

27. C. Conti, P. Ferraris, E. Giorgini, C. Rubini, S. Sabbatini, G. Tosi, J. Anastassopoulou, P. Arapantoni, E. Boukaki, S FT-IR, T. Theophanides, C. Valavanis, FT-IR Microimaging Spectroscopy:Discrimination between healthy and neoplastic human colon tissues , J. Mol Struc. 881 (2008) 46-51.

28. M. Petra, J. Anastassopoulou, T. Theologis & T. Theophanides, Synchrotron micro-FT- IR spectroscopic evaluation of normal paediatric human bone, J. Mol Structure, 78 (2005) 101

29. P. Kolovou and J. Anastassopoulou, "Synchrotron FT-IR spectroscopy of human bones. The effect of aging". Brilliant Light in Life and Material Sciences, Eds. V. Tsakanov and H. Wiedemann, Springer, 2007 267-272p.

30. T. Theophanides, J. Anastassopoulou and N. Fotopoulos, Fifth International Conference on the Spectroscopy of Biological Molecules, Kluwer Academic Publishers, Dodrecht, 1991,409p

SECTION- 1
MINERALS AND GLASSES

1

Chapter

UTILIZING INFRARED SPECTROSCOPY FOR THE IDENTIFICATION OF NOVEL AMORPHOUS PHASES

A CASE STUDY ON THE CARBONATO COMPLEX FORMED VIA MECHANOCHEMICAL PROCESSING

1. INTRODUCTION: AN OVERVIEW

1.1 Mechanochemistry and high-energy milling

Since the late 19th century, mechanochemistry has emerged as a significant field in chemistry, focusing on the study of chemical changes caused by mechanical stress. It has gained importance in both fundamental research and practical applications, as demonstrated by the pioneering work of M. Carey Lea and the original definition by F. W. Ostwald (Takacs, 2004; Boldyrev & Tkačova, 2000). While the basic principles of mechanochemistry are still being thoroughly investigated, mechanical alloying, a powder metallurgy technique that involves ball milling particles under high-energy impact conditions, gained commercial success as early as 1966. It was utilized to create enhanced nickel- and iron-based alloys for the aerospace industry (Suryanarayana et al., 2001). The topic of mechanochemical processes is constantly advancing in numerous disciplines, such as ceramics processing, mineral processing, catalysis, pharmaceutics, and more, in addition to metallurgy. The primary method for applying mechanical stress to a material is by ball-particle collisions in a milling device, due to its simplicity and technological feasibility. This technique is commonly known as "high-energy milling". This method differs from classical "wet ball-milling" in that it typically involves milling a powder or mixture of powders without the presence of liquid. In these conditions, a greater amount of kinetic energy from a moving ball in a grinding bowl is transferred to the powder particles during collisions. This is why it is referred to as "high-energy" milling. Due to the practicality of performing chemical reactions by high-energy milling, a commonly used term in scientific literature is "mechanochemical synthesis". Various types of milling machines, such as shaker, planetary, horizontal, and attrition mill, are employed to perform mechanochemical processes (Lu & Lai, 1998). A widely utilized tool, especially for scientific investigations, is the planetary ball mill (Fig. 1a). Figure 1b depicts a schematic representation of the movement of the ball within a grinding bowl of a planetary mill. The distinct ball motion is a consequence of two types of rotations: i) the grinding bowl rotates around its center, and ii) the supporting disc, to which the bowls are connected, also rotates. These two rotations have opposite directions (see to Fig. 1b). In a rotational geometry, the forces exerted on the milling balls cause them to move in a periodic manner, as shown by the arrows in Figure 1b. During this movement, under specific conditions, the balls become detached from the internal surface of the bowl and collide with the powder particles on the opposite side. The mathematical model, obtained from the simplified ball movement, exhibited good agreement with the experimental measurements of power consumption during milling, as reported by Burgio et al. (1991) and Iasonna & Magini (1996). Furthermore, the periodic motion was validated using numerical simulations conducted by Watanabe et al. in 1995a, as well as high-speed video camera recordings performed by Le Brun et al. in 1993. The intense energy generated during collisions between balls and powder results in several effects on the solid material. These effects include the formation of numerous defects in the crystal structure, the transformation of the structure into an amorphous state or the complete loss of its regular arrangement, the deformation of particles through both plastic and elastic processes, the reduction of particle size to the nanometer range, the enlargement of the powder's specific surface area, the occurrence of polymorphic transitions, and even the initiation of chemical reactions (Fig. 1c). These modifications lead to unique characteristics of the powder. Mechanochemical reactions, occurring during milling without the need for external thermal energy, are distinct from conventional synthesis methods that rely on thermally driven reactions. This characteristic makes the method intriguing and noteworthy. Comprehending mechanochemical processes and their underlying mechanisms is challenging due to their intricate nature. Furthermore, the analysis of the high-energy milling process has shown that, in addition to generating local heat through intense impacts, there is a significant rise in pressure at the point of contact between two milling balls. This pressure can reach levels as high as several gigapascals (GPa). Both the increase in temperature and pressure occur within tenths of microseconds, which is the anticipated period of a collision. This demonstrates the non-equilibrium nature of the mechanochemical process (Maurice & Courtney, 1990). During high-energy collisions, powder particles experience a combination of hydrostatic and shear stress components. This adds complexity to the overall understanding, even in seemingly straightforward situations such polymorphic phase transitions. For instance, it has been demonstrated that traditional thermodynamic phase diagrams are not applicable to polymorphic phase transitions that occur during high-energy milling (Lin & Nadiv, 1979). Introducing the shear component into the calculations significantly modifies the classical hydrostatic-pressure-temperature (p-T) phase diagram. For example, in the case of a polymorphic transition between litharge and massicot forms of PbO, a two-phase field region emerges in the phase diagram. This indicates that the two polymorphs coexist, rather than a distinct transition line as seen in the conventional PbO p-T diagram. This phenomenon can be used to explain why two different forms of a substance can coexist when milling is done for a long time under stable conditions. This has been reported in previous studies by Lin and Nadiv in 1979, as well as by Iguchi and Senna in 1985. The impact of shear stress and localized temperature increase on more intricate mechanochemical reactions is currently the focus of many debates.

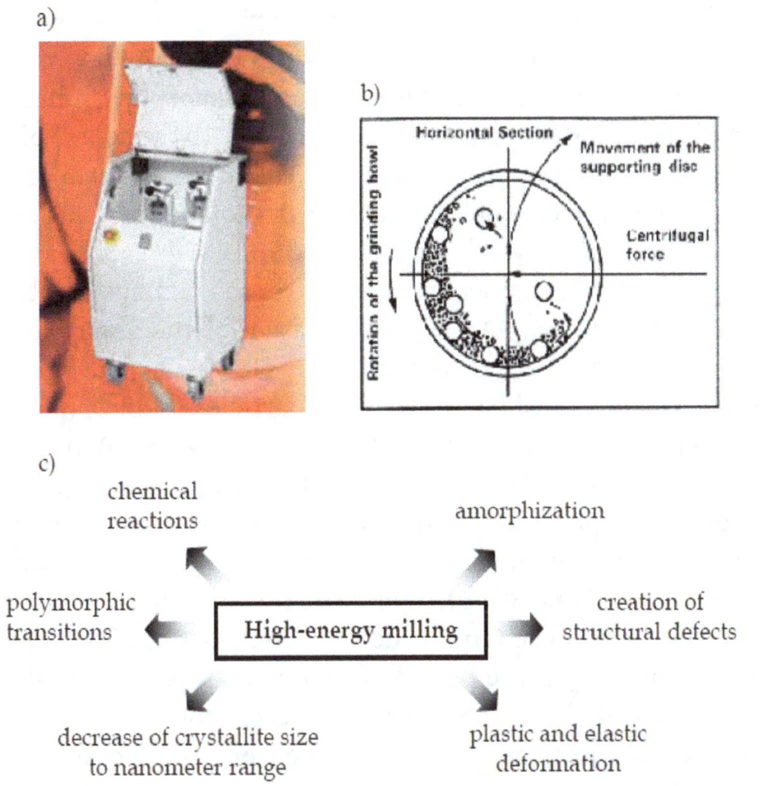

Fig. 1. a) Laboratory-scale planetary mill Fritsch Pulverisette 4, b) schematic representation of the movement of milling balls in a planetary mill (from Suryanarayana, 2001) and c) characteristic phenomena taking place in the solids as a result of high-energy collisions.

1.2 Mechanochemical synthesis of complex ceramic oxides and underlying reaction mechanisms

Mechanochemical synthesis (or high-energy milling assisted synthesis) has been found particularly useful for the synthesis of ceramic oxides with complex chemical composition, ranging from ferroelectric, magnetic and multiferroic oxides to oxides exhibiting semiconducting and catalytic properties. For an overview of the research activity in this field the reader should consult Kong et al. (2008) and Sopicka-Lizer (2010).

Whereas, in general, extensive literature data can be found on the mechanochemical synthesis of complex oxides, only limited studies are devoted to the understanding of mechanochemical reaction mechanisms. Primarily driven by the need to enrich our fundamental knowledge of mechanochemistry, the studies of reaction mechanisms have also been found to be essential in order to efficiently design a mechanochemical process, which includes the selection of milling parameters, milling regime, etc. (Rojac et al., 2010).

One of the main difficulties in analyzing the complex mechanisms of mechanochemical reactions is the identification of amorphous phases, which are metastable and appear often transitional with respect to the course of the reaction. To illustrate an example, we present in Fig. 2 the mechanochemical synthesis of $KNbO_3$ from a powder mixture of K_2CO_3 and NB_2O_5 (Rojac et al., 2009). In the first 90 hours of milling, the initial crystalline K_2CO_3 and NB_2O_5 (Fig. 2a, 0 h) are transformed into an amorphous phase, characterized by two broad "humps" centred at around 29° and 54° 2-theta (Fig. 2a, 90 h). The formation of the amorphous phase was confirmed by transmission electron microscopy (TEM), i.e., an amorphous matrix was observed with embedded nanocrystalline particles of NB_2O_5 (Fig. 2b), which is consistent with the X-ray diffraction (XRD) pattern (Fig. 2a, 90 h). Further milling from 90 to 350 hours resulted in the crystallization from the amorphous phase; this is evident from the appearance of new peaks after 150 and 350 hours of milling, which were assigned to various potassium niobate phases with different K/Nb molar ratio (Fig. 2a, 150 and 350 h). Therefore, the amorphous phase represents a transitional phase of the reaction. In addition, comparison of the 90-hours milled K_2CO_3–NB_2O_5 mixture (Fig. 2a, 90 h) with the 90-hours separately milled NB_2O_5 (Fig. 2a, NB_2O_5 90 h), revealed a much larger degree of amophization of NB_2O_5 when co-milled with K_2CO_3; note the considerably weaker NB_2O_5 peaks and higher XRD background in the case of the mixture as compared to separately milled NB_2O_5. This suggests that the amorphization of NB_2O_5 is not a consequence of the high-energy impacts only, but has its origin in the mechanochemical interaction with the carbonate. It should be emphasized that this is not an isolated case; examples involving transitional amorphous phases can also be found during mechanical alloying of mixture of metals (El-Eskandarany et al., 1997). Finally, a nucleation-and-growth mechanism from amorphous phase was recently proposed as a general concept to explain the mechanochemical synthesis of a variety of complex oxides, such as $Pb(Zr_{0.52}Ti_{0.48})O_3$, $Pb(Mg_{1/3}NB_{2/3})O_3$, $Pb(Zn_{1/3}NB_{2/3})O_3$, etc. (Wang et al., 2000a, 2000b; Kuscer et al., 2006). In order to understand mechanochemical reactions, it is thus indispensable to analyze more closely the transitional amorphous phase.

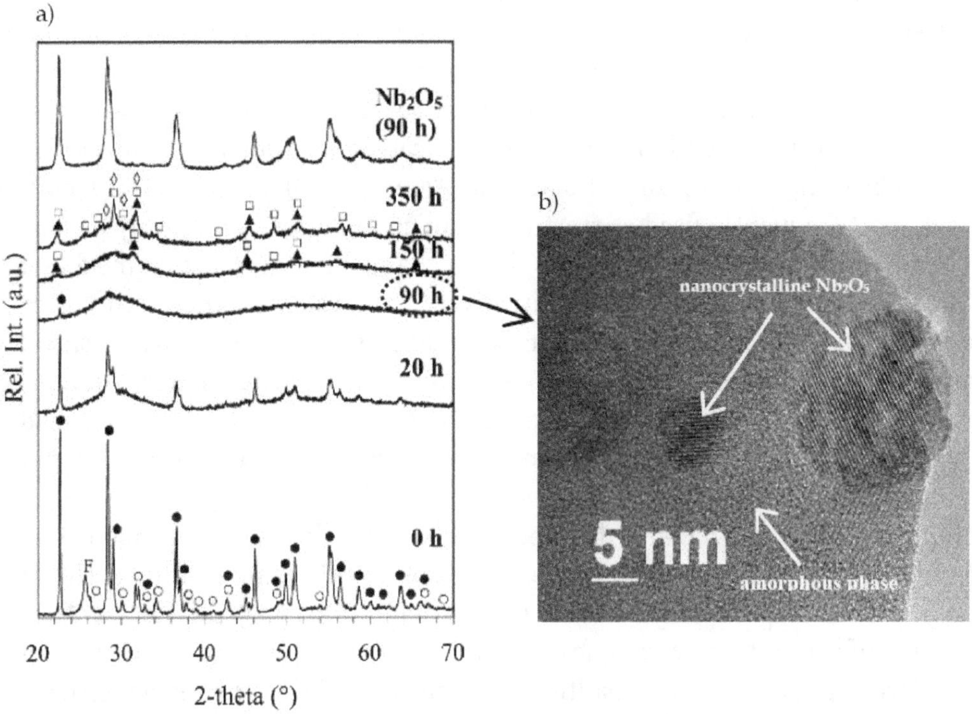

Fig. 2. a) XRD patterns of K_2CO_3–NB_2O_5 powder mixture after high-energy milling for 20, 90, 150 and 350 hours. The non-milled mixture is denoted as "0 h". The pattern of the 90-hours- separately-milled NB_2O_5 is added for comparison. In order to prevent adsorption of water during XRD measurements, a polymeric foil was used to cover the non-milled powder mixture. b) TEM image of the K_2CO_3–NB_2O_5 powder mixture after high-energy milling for 90 hours. Notations: K_2CO_3 (○, PDF 71-1466), NB_2O_5 (●, PDF 30-0873), KNbO3 (▲, PDF 71-0946), K6Nb10.88O30 (□ , PDF 87-1856), K8Nb18O49 (◊, PDF 31-1065), polymeric foil (F); "h" denotes milling hours (from Rojac et al., 2009).

It is clear from the above considerations that the most often used and widely reported XRD analysis becomes insufficient to provide Detailed information about amorphous phases. The benefits of in-depth studies of mechanochemical reaction mechanisms by selection of appropriate analytical tools, able to provide data on a short-range (local) structural scale, such as nuclear magnetic resonance (NMR), X-ray photoelectron spectroscopy (XPS), electron paramagnetic resonance (EPR) spectroscopy, infrared spectroscopy (IR), Raman spectroscopy, etc., were demonstrated by the pioneering work of Senna, Watanabe and co- workers (Watanabe et al., 1996, 1997; Senna, 1997). In those cases, the synthesis of selected complex oxide systems have been studied from starting mixtures comprising typically hydroxide and oxide compounds; extensive data on these studies can be found in Avvakumov et al. (2001).

Mechanochemical processing has recently provided important improvements in the synthesis of ceramic materials in the family of alkaline niobates tantalates, a rich group of materials exhibiting wide applicability; this includes $KTaO_3$ and $(K,Na,Li)(Nb,Ta)O_3$ (KNLNT), which are considered as promising materials for dielectric (microwave) and piezoelectric applications, respectively (Glinsek et al., 2011; Tchernychova et al., 2011; Rojac et al., 2008a, 2010). Since alkali carbonates are the most frequently used as starting alkali compounds, it naturally became of interest to understand in more details the mechanochemical reaction mechanisms in which carbonate ions (CO_3^{2}–) are involved. The results of these studies carry important practical consequences. For example, in the case of the synthesis of the complex KNLNT solid solution, it was demonstrated that the identification of the reaction mechanism during mechanochemical processing is a key step leading to highly homogeneous KNLNT ceramics with excellent piezoelectric response. After identifying an intermediate amorphous carbonato complex, to which the present chapter is particularly devoted, it was found that a homogeneous KNLNT can only be obtained by providing the formation of this complex during the high-energy milling step. In other words, milling conditions that did not lead to the formation of the carbonato complex, e.g., milling in the "friction" mode instead of the "friction+impact" mode, resulted into considerable Ta-inhomogeneities and, consequently, to a reduced piezoelectric response (Rojac et al., 2010).

In this chapter we present an overview of the studies of reaction mechanisms in systems comprising CO_3^{2}– ions. The chapter Aims primarily at showing the importance of combining various analytical methods, including quantitative XRD analysis, thermal analysis and infrared spectroscopy, to obtain an overall picture of a complex reaction mechanism, such as the one encountered during mechanochemical processing. The first part of the chapter is devoted to the synthesis of $NaNbO_3$ from a mixture of NA_2CO_3 and NB_2O_5. After demonstrating the feasibility of synthesizing $NaNbO_3$ directly by high-energy milling, we show systematically how a mechanism can be revealed by a built-up of data from various analytical methods. The focus is to gain insight into the amorphous phase, which represents a transitional phase in the synthesis of $NaNbO_3$. In the second part of the chapter we will extent the studies to other systems based on sodium carbonate, i.e., NA_2CO_3–M_2O_5 (M = V, Nb, Ta). The transition-metal oxides were selected through the 5th group of the periodic table to allow systematic comparisons and propose potentially a general reaction mechanism.

2. MECHANOCHEMICAL REACTION MECHANISM IN THE NA_2CO_3–NB_2O5 SYSTEM STUDIED BY A COMBINATION OF QUANTITATIVE X-RAY DIFFRACTION, THERMAL AND INFRARED SPECTROSCOPY ANALYSIS

2.1 Quantitative X-ray diffraction analysis

The mechanochemical synthesis of $NaNbO_3$ from a NA_2CO_3–NB_2O_5 mixture was followed by XRD analysis. Fig. 3 shows the XRD patterns of the NA_2CO_3–NB_2O_5 mixture after selected milling times. The pattern of the non-milled mixture (Fig. 3, 0 h), which is a homogenized mixture of NA_2CO_3 and NB_2O_5 powders just before mechanochemical treatment, can be fully indexed with the initial monoclinic NA_2CO_3 and orthorhombic NB_2O_5 (Fig. 3, 0 h). The first 5 hours of high-energy milling are characterized by broaden peaks of the two reagents together with reduced peak intensity (Fig. 3, 5 h). After 40 hours of milling NA_2CO_3 was not observed anymore in the mixture, whereas traces of the newly formed $NaNbO_3$ were first detected (Fig. 3, 40 h). Further milling from 40 to 400 h leaded to a progressive disappearance of NB_2O_5 from the mixture at the expense of the growing $NaNbO_3$. Note the long milling time, i.e., 400 hours, needed to obtain the final $NaNbO_3$ free of any reagents (Fig. 3, 400 h). The low rate of the reaction between NA_2CO_3 and NB_2O_5 resulted from the mild milling conditions, which were applied intentionally in order to enable a careful analysis of the individual reaction stages. It should be noted, however, that more intensive milling, resulting into $NaNbO_3$ after 32 hours of milling, did not change qualitatively the course of the reaction (for details see Rojac et al., 2008b). The results of the XRD analysis from Fig. 3 confirm the mechanochemical formation of $NaNbO_3$ according to the following reaction:

$$Na_2CO_3 + Nb_2O_5 \rightarrow 2NaNbO_3 + CO_2 \quad (1)$$

In order to obtain a more quantitative picture of the mechanochemical reaction, we performed a quantitative XRD phase analysis using the Rietveld refinement method. In addition to the amount of the crystalline phases, i.e., NA_2CO_3, NB_2O_5 and $NaNbO_3$, we determined also the contribution from the XRD background, which we denoted as "XRD- amorphous" phase. This was done using an internal standard method; details of the method can be found in Kuscer et al. (2006) and Rojac et al. (2008b).

The results of the refinement analysis in terms of the amounts of NA_2CO_3, NB_2O_5, $NaNbO_3$ and XRD-amorphous phase as a function of milling time are shown in Fig. 4. The amounts of both NA_2CO_3 and NB_2O_5 decrease with milling time (Fig. 4a). While NB_2O_5 persists in the mixture up to 280 hours (Fig. 4a, closed rectangular), NA_2CO_3 is no longer detected after 20 hours of milling (Fig. 4a, open rectangular). The amount of the XRD-amorphous phase rapidly increases in the initial part of the reaction, reaching a maximum of 91% after 110 hours of milling, after which it decreases with further milling. Note the constant amount of the XRD-amorphous phase after reaching 600 hours of milling. The formation of $NaNbO_3$ follows a sigmoidal trend: at the beginning of the reaction the formation rate is low, after which it increases and slows down again in the final part of the reaction (Fig. 4b, open circles). Similarly like the XRD-amorphous phase, no differences in the amount of $NaNbO_3$ are observed with milling from 600 to 700 hours, suggesting a constant $NaNbO_3$ -to- amorphous-phase mass ratio upon prolonged milling.

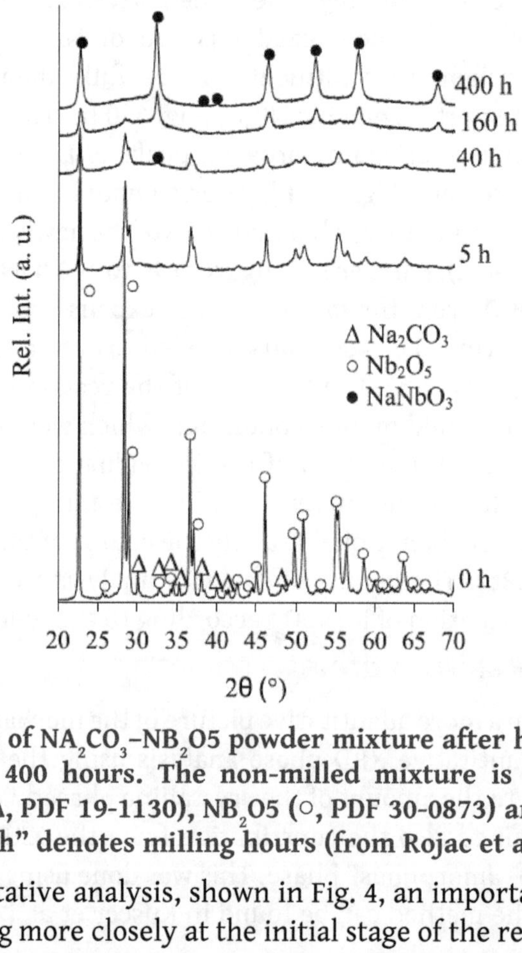

Fig. 3. XRD patterns of NA_2CO_3–NB_2O5 powder mixture after high-energy milling for 5, 40, 160 and 400 hours. The non-milled mixture is denoted as "0 h". Notations: NA_2CO_3 (Δ, PDF 19-1130), NB_2O5 (○, PDF 30-0873) and NaNbO3 (•, PDF 33-1270); "h" denotes milling hours (from Rojac et al., 2008b).

From the quantitative analysis, shown in Fig. 4, an important observation can be derived by looking more closely at the initial stage of the reaction. An enlarged view of this part of the reaction is shown as inset in Fig. 4b. Here, we can see that in the initial 20 hours of milling, during which no $NaNbO_3$ was detected, a large amount, i.e., 73%, of the amorphous phase was formed. Only subsequently, i.e., after 40 hours of milling, $NaNbO_3$ was firstly detected.

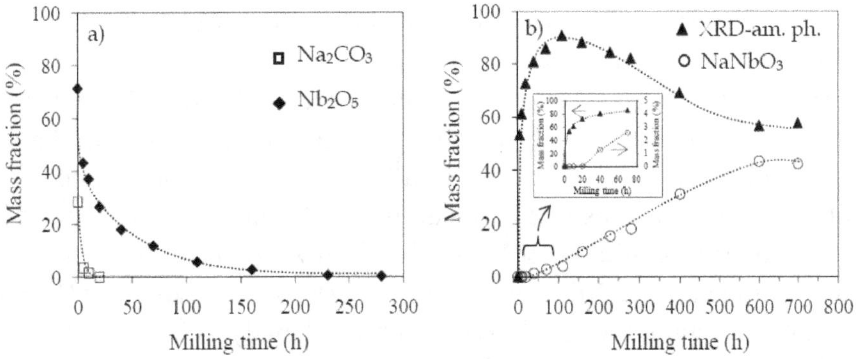

Fig. 4. Fractions of crystalline phases (NA_2CO_3, NB_2O_5 and $NaNbO_3$) and XRD-amorphous phase, determined by Rietveld refinement analysis, as a function of milling time. a) NA_2CO_3 and NB_2O5, b) $NaNbO_3$ and XRD-amorphous phase. The inset of b) shows an enlarged view of the curves in the initial 80 hours of milling. The lines are drawn as a guide for the eye (from Rojac et al., 2008b).

From this simple observation we can infer that $NaNbO_3$ is not formed directly, like assumed by equation 1, but through an intermediate amorphous phase. The transitional nature of the amorphous phase is further confirmed by the maximum in its amount after 110 hours of milling. Moreover, literature data go in favour of our conclusions. In fact, based on studies of the kinetics, the sigmoidal trend, like that observed in the case of $NaNbO_3$ (Fig. 4b, open circles), is characteristic for multistep mechanochemical processes, such as the amorphization of a mixture of metals, where the phase transformation requires two or more impacts on the same powder fraction. In contrast, continuously decelerating processes, described by asymptotic kinetics, are typical for the amorphization of single-phase compounds, such as intermetallics, where the structure is already altered after the first impact (Delogu & Cocco, 2000; Cocco et al., 2000; Delogu et al., 2004). Therefore, independently of the analysis on the XRD-amorphous phase, the sigmoidal-like trend in the formation of $NaNbO_3$ (Fig. 4b, open circles) suggests that the niobate is formed via a transitional phase.

In addition to the XRD-amorphous phase, we shall look at the changes induced in the NA_2CO_3 in the initial part of milling. Fig. 5 compares the XRD patterns of the NA_2CO_3–NB_2O_5 mixture in the first 40 hours of milling (Fig. 5a) with the XRD patterns of NA_2CO_3 (Fig. 5b), which was high-energy milled alone, without NB_2O_5, with exactly the same milling conditions as the mixture. While the peaks of NA_2CO_3 when milled together with NB_2O_5 completely disappeared after 20 hours of milling (see open triangles in Fig. 5a), this is clearly not the case even after 40 hours if NA_2CO_3 was milled alone (see Fig. 5b). The broader peaks of NA_2CO_3 after 40 hours of separate milling (Fig. 5b, 40 h) are most probably a consequence of reduced crystallite size and increase in microstrains due to creation of structural disorder. The disappearance of the original crystalline NA_2CO_3 from the mixture, suggesting amorphization, is therefore an effect triggered by the presemce of NB_2O_5 rather than a pure effect of the high-energy collisions. In relation to this mechanochemical interaction between NA_2CO_3 and NB_2O_5, a question that arises at this point is whether this interaction resulted into the carbonate decomposition. This is also relevant with respect to the nature of the amorphous phase. Obviously, further information could be obtained by following the decomposition of the carbonate during milling. This can be done using thermogravimetric (TG) analysis; the results of TG coupled with differential thermal analysis (DTA) and evolved-gas analysis (EGA) are presented in the following section.

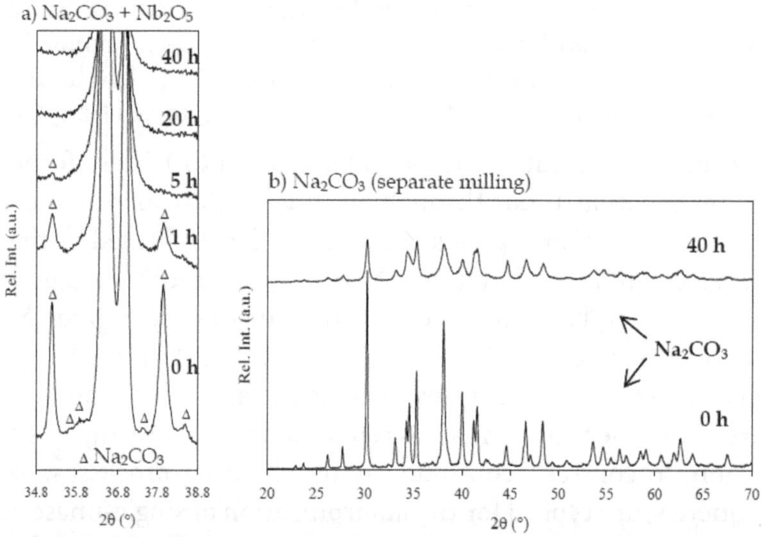

Fig. 5. XRD patterns of a) NA_2CO_3–NB_2O5 mixture and b) NA_2CO_3 after high-energy milling for up to 40 hours. The pattern in a) shows a narrow 2-theta region, i.e., from 34.8 to 38.8°, to highlight the changes upon milling in the peaks corresponding to NA_2CO_3. Note that all the peaks on the patterns of non-milled and 40-hours-separately-milled NA_2CO_3 in b) are indexed with monoclinic NA_2CO_3. Notation: NA_2CO_3 (Δ, PDF 19-1130); "h" denotes milling hours (from Rojac et al., 2006).

2.2 Thermal analysis

In order to explore the origin of the reaction-induced amorphization and/or possible decomposition of NA_2CO_3 (Fig. 5) we were further focused on the initial part of milling, i.e., results are presented for the samples treated in the first 40 hours of milling.

Fig. 6 presents the thermogravimetric (TG), derivative thermogravimetric (DTG), differential thermal analysis (DTA) and evolved-gas analysis (EGA) curves of the NA_2CO_3–NB_2O_5 powder mixture in the first 40 hours of high-energy milling. The non- milled NA_2CO_3–NB_2O_5 mixture looses mass in several steps in a broad temperature range from 400 °C to 800 °C (Fig. 6a and b, 0 h). The total mass loss of this mixture upon annealing to 900 °C amounts to 11.7%, which agrees well with the theoretical mass loss of 11.8%, calculated according to equation 1 for the complete decomposition of NA_2CO_3 in an equimolar mixture with NB_2O_5. The carbonate decomposition is further confirmed by EGA, which shows a release of CO_2 in the temperature range 400-800 °C (Fig. 6d, 0 h, full line). Note also that the DTG peaks (Fig. 6b, 0 h) coincide with the EGA(CO_2) peaks (Fig. 6d, 0 h, full line), showing that the measured mass loss in this sample is indeed entirely related to the decomposition of NA_2CO_3, which is triggered by the reaction with NB_2O_5, like represented by equation 1.

High-energy milling resulted into several changes in the thermal behaviour of the NA_2CO_3–NB_2O_5 mixture. Firstly, by inspecting the TG curves, a mass loss appears in the milled samples in the temperature range 25-300 °C, which was not observed prior milling (Fig. 6a, compare milled samples with the non-milled). According to the DTA curves (Fig. 6c), these mass losses between room temperature and 300 °C are accompanied by endothermic heat effects, which first manifest as a sharp endothermic peak at around 100 °C (Fig. 6c, 1 h), progressively evolving with milling into a broader endothermic peak, which expands from 80 °C to 250 °C (see for example Fig. 6c, 40 h). According to EGA(H_2O), the mass losses in this low temperature range correspond to the removal of H_2O (Fig. 6d, milled samples, dashed lines). The amounts of H_2O removed from the samples milled for 0, 1, 5, 20, 40 hours, as determined from the TG curves (Fig. 6a, milled samples, 25-300 °C), are 0%, 2.5%, 4.0%, 4.8% and 5.1%, respectively. This suggests gradual adsorption of H_2O on the powder with increasing milling time; taking into account that the milling was performed in open Air and also considering the hygroscopic nature of NA_2CO_3, the adsorption of H_2O is not surprising. We note that the H_2O removal from the samples milled for longer periods, i.e., 5, 20 and 40 hours, takes place at temperatures higher than 100 °C (Fig. 6d, dashed lines), which might suggest water chemisorption rather than physical adsorption.

In addition to water adsorption, high-energy milling induced considerable changes in the thermal decomposition of the carbonate. This is best seen by inspecting the DTG and EGA (CO_2) curves of the milled samples (Fig. 6b and 6d, milled samples). Firstly, it should be noted that in the temperature range between 350 °C and 500 °C the DTG peaks of the milled mixtures (Fig. 6b, milled samples) coincide with those of EGA(CO_2) (Fig. 6d, milled samples, full lines), which means that the mass loss in this temperature range is related to the CO_2 removal, i.e., to the carbonate decomposition. For the sake of discussion, we consider in the following only the EGA(CO_2) curves (Fig. 6d, full lines). In contrast to the carbonate decomposition in the non-milled mixture (Fig. 6d, 0 h, 400-800 °C), occurring in several steps and in a broad temperature range, which is characteristic for a physical mixture of NA_2CO_3 and NB_2O_5 particles (Jenko, 2006), the mixture milled for only 1 hour releases CO_2 in a much narrower temperature range, i.e., 400-500 °C (Fig. 6d, 1 h). We attribute this effect to the smaller particle size after 1 hour of milling, which is known to decrease considerably the decomposition temperature of NA_2CO_3 in the NA_2CO_3–NB_2O_5 mixture due to reduced diffusion paths (Jenko, 2006). In comparison with the 1-hour milled sample, upon milling for 5 hours only small changes are observed in the shape of the EGA(CO_2) peak (Fig. 6d, 5 h, 400-500 °C). After 20 hours of milling a new, weak EGA(CO_2) peak appears at 370 °C (Fig. 6d, 20 h), suggesting two-step carbonate decomposition; this peak then shifts to 400 °C upon 40 hours of milling (Fig. 6d, 40 h). Note that after 40 hours of milling the intense EGA(CO_2) peak at 420 °C becomes sharper in comparison with shorter milling times, i.e., 1, 5 and 20 hours, indicating a more uniform decomposition of the carbonate.

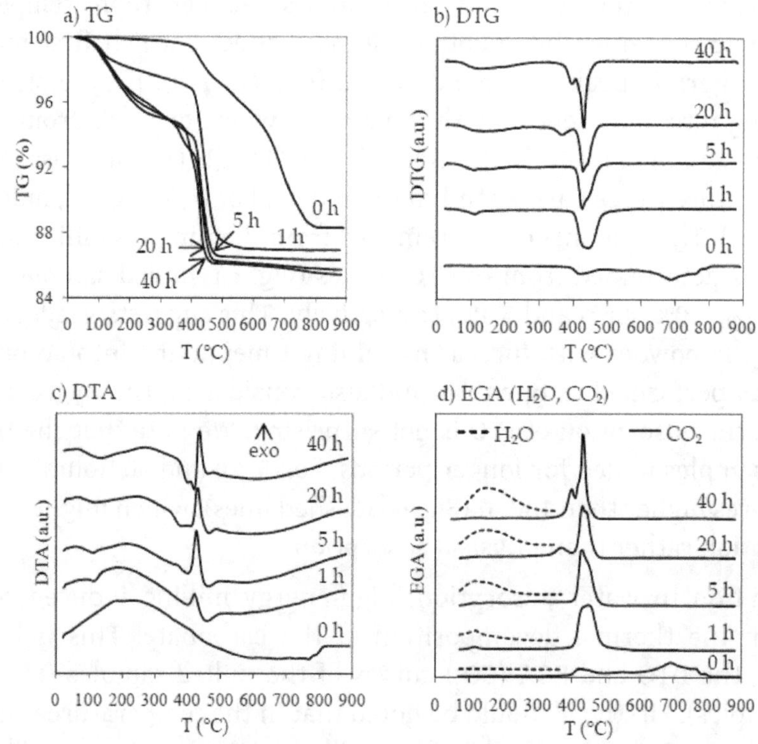

Fig. 6. a) TG, b) DTG, c) DTA and d) EGA(H_2O, CO_2) curves of the NA_2CO_3–NB_2O_5 powder mixture after high-energy milling for 1, 5, 20 and 40 hours. The non-milled mixture is denoted as "0 h". Since the main EGA(H_2O) signal was observed in the temperature range 25-350 °C the data are plotted accordingly. "h" denotes milling hours (from Rojac et al., 2006).

According to DTA, the decomposition of the carbonate in the milled samples is accompanied by an exothermic heat effect (Fig. 6c, milled samples). This is seen from the sharp and intense exothermic peaks appearing in all the milled samples in the temperature range where the CO_2 is released, i.e., 400-500 °C (compare Fig. 6c with Fig. 6d).

To summarize, the DTA and EGA(CO_2) analyses on the milled samples (Fig. 6c and d, milled samples) suggest a rather defined carbonate decomposition occurring in a narrow temperature range, which is not typical for a physical mixture of NA_2CO_3 and NB_2O_5 (compare 0 h with milled samples in Fig. 6c and 6d; see also Jenko, 2006); this indicates a change in the chemical state of the carbonate upon milling and formation of a new phase.

According to the mass loss related to the CO_2 release, which can be separated from the loss of H_2O by combining EGA and TG curves, we can calculate the amount of the residual carbonate in the mixture, i.e., the amount of the carbonate that did not decompose during high-energy milling. The total CO_2 loss from the sample milled for 40 hours is 9.6%, corresponding to 85.0% of residual carbonate. Therefore, in the first 40 hours of milling, a minor amount of the carbonate decomposed, whereas the major part, according to XRD analysis (Fig. 5a), became amorphous. As mentioned in the previous section, the NA_2CO_3 amorphization is stimulated by the mechanochemical interaction with NB_2O_5. This observation, together with the characteristic changes in the decomposition of the carbonate upon milling (Fig. 6), indicates a formation of a new carbonate compound. As a next step, it seems reasonable to explore the symmetry of the CO_3^{2-} ions, which was done using infrared spectroscopy.

2.3 Infrared spectroscopy analysis

The IR spectra of the NA_2CO_3–NB_2O_5 mixture before and after milling for various periods are shown in Fig. 7a. The two separate graphs in Fig. 7a show two different wavenumber regions, i.e., 950–1150 cm^{-1} and 1280–1880 cm^{-1}. The spectrum of the non-milled mixture is composed of a weak band at 1775 cm^{-1} and a strong one at 1445 cm^{-1}; no bands are observed in the lower wavenumber region between 950 and 1150 cm^{-1} (Fig. 7a, 0 h). Based on the literature data, the spectrum of the non-milled mixture can be entirely indexed with vibrational bands of the CO_3^{2-} ions, present in the initial NA_2CO_3 (Harris & Salje, 1992; Gatehouse et al., 1958). This is consistent with the fact that NB_2O_5, which is also a part of the mixture, did not show any IR bands in the two examined wavenumber regions (the IR spectrum of NB_2O_5 is not shown).

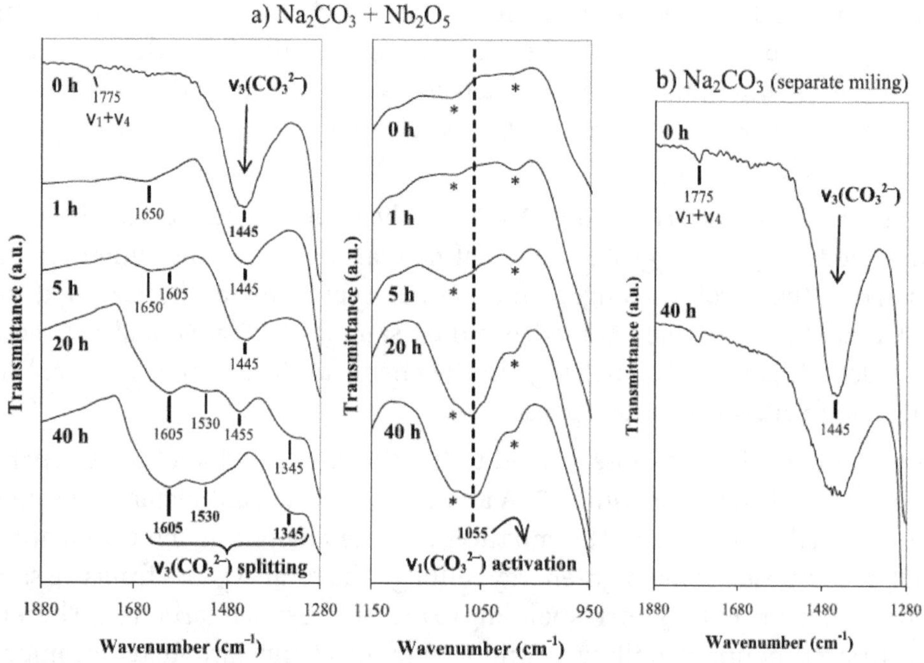

Fig. 7. FT-IR spectra of a) NA_2CO_3–NB_2O5 powder mixture after high-energy milling for 1, 5, 20 and 40 hours and (b) NA_2CO_3 subjected to separate high-energy milling for 40 hours. The non-milled powders are denoted as "0 h". Note that, in contrast to the NA_2CO_3–NB_2O5 mixture (a), no splitting of n_3($CO_3$2–) is observed in the case of the separately milled NA_2CO_3 (b). Notation: * Nujol, for bands assignment refer to Table 1; "h" denotes milling hours. (from Rojac et al., 2006).

The IR vibrations of the free CO_3^{2-} ion having D_{3h} point group symmetry are listed in Table 1. The CO_3^{2-} ion possesses two stretching and two bending vibrational modes. The symmetrical C–O stretching vibration, denoted as n_1, is IR-inactive, while the n_2, n_3 and n_4 are IR-active. According to Harris & Salje (1992), and Table 1, the strongest band of the non- milled sample at 1445 cm^{-1} (Fig. 7a, 0 h) belongs to the assymetrical C–O stretching vibration of CO_3^{2-} (n_3), while the weak band at 1775 cm^{-1} can be assigned to the combinational band of the type n_1+n_4. No bands are observed in the 950–1150 cm^{-1} region (Fig. 7a, 0 h), consistent with absence of the IR-inactive n_1 vibration. With the exception of some differences in the position, the bands of the non-milled mixture, which belong to NA_2CO_3, are consistent with vibrations characteristic for the free CO_3^{2-} ion with D_{3h} symmetry. This is in agreement with the literature data and was explained as being a consequence of the small effect of the crystal field of Na+ ions on the symmetry of the CO_3^{2-} in the NA_2CO_3 structure. This is somewhat different, for example, in Li_2CO_3, where a stronger interaction between crystal lattice and CO_3^{2-} ions leads to lowered CO_3^{2-} symmetry and, consequently, to a more complex IR spectrum (Buijs & Schutte, 1961; Brooker & Bates, 1971).

Table 1. Fundamental IR vibrations of carbonate ($CO_3$2–) ion with D3h symmetry. n2, n3 and n4 are IR-active vibrations, while n1 is IR-inactive (Gatehouse et al., 1958; Nakamoto, 1997).

Type of vibration	Notation	Wavenumber (cm^{-1})
C–O symmetrical stretching	n_1 (A_1')	1063
Out-of-plane CO_3^{2-} bending	n_2 (A_2")	879
C–O asymmetrical stretching	n_3 (E')	1415
In-plane CO_3^{2-} bending	n_4 (E')	680

Upon milling the NA_2CO_3–NB_2O_5 mixture, considerable changes can be observed in the IR spectra (Fig. 7a, milled samples). After 1 hour of milling a new weak band appears at 1650 cm^{-1}. The position of this band coincides with one of the strongest HCO_3– bands typical for alkaline hydrogencarbonates (Watters, 2005). This is in agreement with the simultaneous loss of H_2O and CO_2 upon annealing this sample (Fig. 6d, 1 h), which is characteristic for the hydrogencarbonate decomposition. Furthermore, we should not eliminate the possibility of having the in-plane bending vibration of H_2O, which also appears near 1650 cm^{-1} (Venyaminov & Prendergast, 1997).

By further milling from 1 hour to 40 hours related and simultaneous trends can be noted: i) the $n_3(CO_3^{2-})$ vibration shifts from 1445 cm^{-1} (Fig. 7a, 1 and 5 h) to 1455 cm^{-1} (Fig. 7a, 20 h) and decreases in intensity until it completely disappears after 40 h of milling, ii) the n_3 vibration is gradually replaced by new absorption bands appearing at 1605, 1530 and 1345 cm^{-1} (Fig. 7a, 40 h), and iii) a new band arises during milling, located at 1055 cm^{-1}, which belongs to the symmetrical C–O stretching vibration of the CO_3^{2-} ions (n_1) (Fig. 7a, see region 950–1150 cm^{-1}). We can conclude from these results that milling induced a splitting of n_3 and activation of n_1 vibrations, suggesting a change of the CO_3^{2-} symmetry from the original D_{3h}. We shall come back to this point after examining the fundamental relation between symmetry and IR vibrations of the carbonate ion.

An extensive review on the IR spectroscopic identification of different species arising from the reactive adsorption of CO_2 on metal oxide surfaces can be found in Busca & Lorenzelli, 1982. In principle, the carbonate ion is a highly versatile ligand, which gives rise not only to simple mono- or bidentate structures, but also to a number of more complicated bidentate bridged structures. Some examples of CO_3^{2-} coordinated configurations are schematically illustrated in Fig. 8.

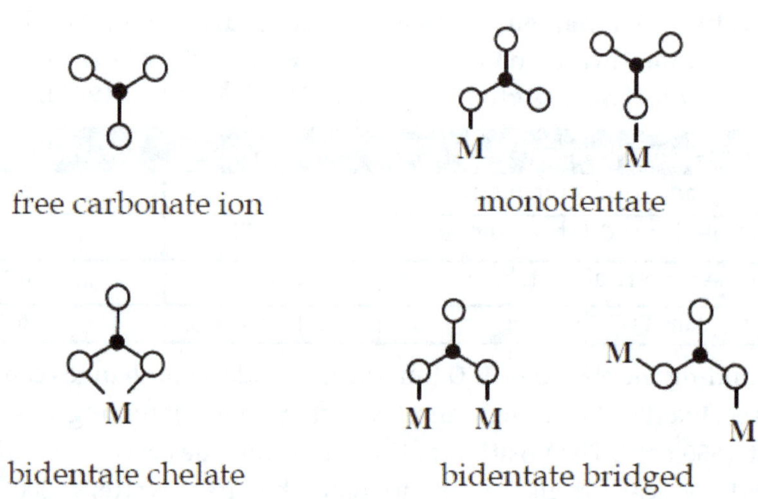

Fig. 8. Schematic view of free (non-coordinated) and various types of coordinated $CO_3$2– ions.

When the CO_3^{2-} ion is bound, through one or more of its oxygens, to a metal cation (denoted as "M" in Fig. 8), its point group symmetry is lowered. It is well known from the literature that the lowering of the CO_3^{2-} symmetry, resulting from the coordination of the carbonate ion in a carbonato complex, causes the following changes in the IR vibrational modes of the free carbonate ion (Gatehouse et al., 1958; Hester & Grossman, 1966; Brintzinger & Hester, 1966; Goldsmith & Ross, 1967; Jolivet et al., 1980; Busca & Lorenzelli, 1982; Nakamoto, 1997):

1. Activation of IR-inactive n_1 vibration
2. Shift of n_2 vibration
3. Splitting of n_3 vibration
4. Splitting of n_4 vibration

The most characteristic of the above IR spectroscopic changes upon CO_3^{2-} coordination is the infrared activation of the n_1, i.e., the symmetrical C–O stretching vibration. This vibration, as mentioned earlier, is IR-inactive for the free carbonate ion, but also for most alkali, alkaline-earth and heavy-metal carbonates; it appears as a weak band only in certain carbonates of the aragonite type (Gatehouse et al., 1958). To derive the relation between symmetry and IR vibrations, we shall first look at the details of the n_1 vibration. According to the IR selection rule, which states that the vibration is IR-active if the dipole moment is changed during vibration, we can understand that there will be no net change in the dipole moment during symmetrical C–O stretching vibration (n_1) of the CO_3^{2-} ion with D_{3h} symmetry; this comes from the equivalence of the three C–O bonds, which is schematically illustrated in Fig. 9 (bottom-left quadrant). The equivalence of these three C–O bonds is lost upon coordination, so that typically the C–O bond coordinated to the metal cation becomes weaker, while the C–O bonds not involved in metal binding becomes stronger with respect to the C–O bond in the free, non-coordinated, CO_3^{2-} ion (Fig. 9, upper-right quadrant) (Fujita et al., 1962; Brintzinger & Hester, 1966). This in turn leads to lowered CO_3^{2-} symmetry, e.g., from D_{3h} to C_{2n}, and to the activation of the n_1 vibration (Fig. 9, bottom-right quadrant). In the case of monodentate coordination, also the Cs symmetry is possible and arises when the M–O–C bond is not collinear (Fig. 9, upper-right quadrant); same IR spectroscopic changes also apply for this case (Fujita et al., 1962; Nakamoto, 1997).

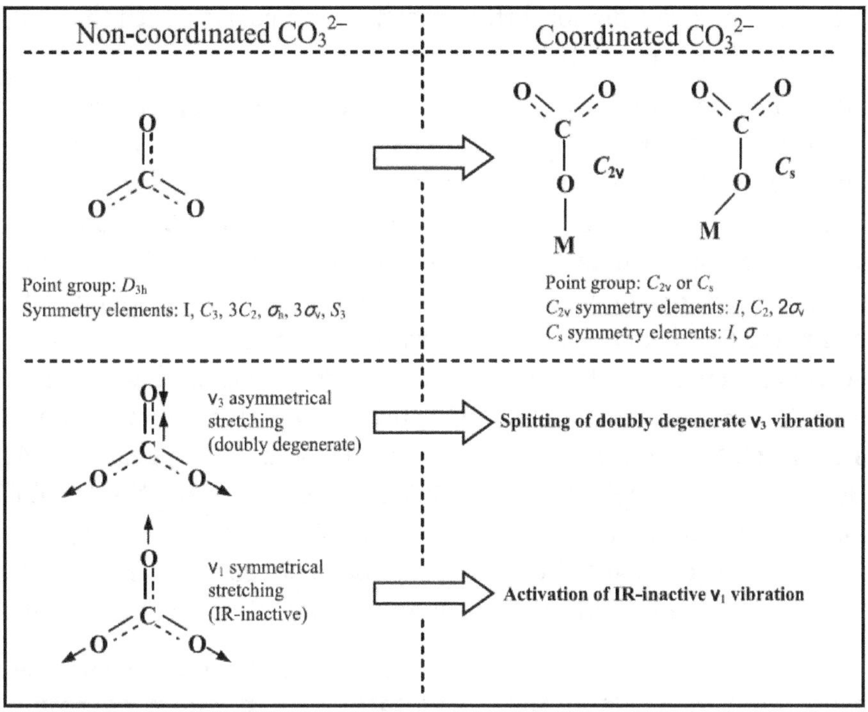

Fig. 9. Schematic representation of non-coordinated and coordinated CO_3^{2-} ion and the corresponding point group symmetry elements. The changes in the n_1 and n_3 IR vibrations of the CO_3^{2-} ion upon coordination are also shown. For simplicity, only monodentate coordination is presented. Notations: I – identity, Cn – n-fold axis of rotation, sh, sn – mirror planes perpendicular and parallel to the principal axis, respectively, Sn – n-fold rotation- reflection operation. The number preceding the symmetry operation symbol refers to number of such symmetry elements that the molecule possesses. For further details consult Nakamoto, 1997.

In parallel with the n_1 activation, also the splitting of the n_3 vibration occurs upon coordination. In the free CO_3^{2-} ion, the n_3 vibration is doubly degenerate (Fig. 9, bottom-left quadrant). Doubly degenerate vibrations occur only in molecules possessing an axis higher than twofold, which is the case of the D_{3h} symmetry, having a three-fold rotational axis (see Fig. 9, upper-left quadrant; C_3 – three-fold axis) (Nakamoto, 1997). The lowering of the symmetry of the carbonate ion from D_{3h} to either C_{2n} or Cs, which means loss of the equivalence of the three C–O bonds in the CO_3^{2-} and, therefore, loss of the three-fold rotational axis, leads to the separation (splitting) of the doubly degenerate vibrations (Fig. 9, bottom-right quadrant).

With respect to the changes upon coordination in the other two vibrational modes, i.e., n_2 and n_4, it should be noted that the splitting of the n_4 vibration has been studied to a lesser extent in comparison with the characteristic n_3 splitting. In addition, the shift of the n_2 vibration upon coordination is typically small and the values for complexes do not differ greatly in comparison with those of simple carbonates (Gatehouse et al., 1958).

Coming back to our case from Fig. 7a, we can interpret the n_3 splitting and n_1 activation of the CO_3^{2}– vibrations during milling of the NA_2CO_3–NB_2O_5 mixture as characteristic of lowered CO_3^{2}– symmetry, which is related to the mechanochemical formation of a carbonato complex. For comparison, we compiled in Table 2 the data of a number of carbonato complexes, with metals such as Cu and Co, provided from the literature. According to the notation of the coordinated C_{2n} symmetry, the n_3 vibration of the D_{3h} symmetry is now split into two components, which are denoted as n_1 and n_4 (second and third column in Table 2); the activated n_1 vibration becomes n_2 (fourth column in Table 2). The regions in which the carbonato complex absorption bands appear are 1623-1500 cm^{-1}, 1362-1265 cm^{-1} (n_3 splitting) and 1080-1026 cm^{-1} (n_1 activation) (Table 2). By comparing these data with the our case, we can observe that the n_3 split bands at 1605, 1530 and 1345 cm^{-1}, and the n_1 activated band at 1055 cm^{-1} from Fig. 7a (40 h) fall entirely within the wavenumber regions of carbonato complexes from Table 2.

Table 2. Some copper and cobalt carbonato complexes and the corresponding IR absorption bands related to CO_3^{2-} vibrations. n_1, n_2 and n_4 correspond to vibrations of CO_3^{2-} in the C_{2n} symmetry notation; according to this notation, the doubly degenerate n_3 vibration of the free CO_3^{2-} ion, which splits into two components, is denoted as n_1 and n_4, whereas the activated n_1 vibration is denoted as n_2 (data compiled from Gatehouse et al., 1958; Fujita et al., 1962; Jolivet et al., 1982; Healy & White, 1972).

Carbonato complex	$n_4(B_2)$ (cm^{-1})	$n_1(A_1)$ (cm^{-1})	$n_2(A_1)$ (cm^{-1})
$NA_2Cu(CO_3)2$	1500	1362	1058
$NA_2Cu(CO_3)2{\times}3H_2O$	1529	1326	1066, 1050
$K_3Co(CO_3)3{\times}3H_2O$	1527	1330	1080, 1037
$KCo(NH_3)2(CO_3)2$	1623, 1597	1265	1026
$Co(NH_3)6Co(CO_3)3$	1523	1285	1073, 1031
$Co(NH_3)4CO_3Cl$	1593	1265	1030
$Co(NH_3)4CO_3ClO4$	1602	1284	not reported

It is important to note that, in contrast to the NA_2CO_3–NB_2O_5 mixture, the n_3 vibration did not split when NA_2CO_3 was milled alone, i.e., without NB_2O_5. This is seen in Fig. 7b, which shows the IR spectra of NA_2CO_3 before and after separate milling. Except for the reduced intensity, which might be related to the decreased crystallite size and structural disordering induced by milling, the n_3 band at 1445 cm^{-1} is still present after 40 hours of separate milling. We emphasize that for this separate NA_2CO_3 milling the same milling conditions and same milling time, i.e., 40 hours, were applied as for the NA_2CO_3–NB_2O_5 mixture. Therefore, the lowering of the CO_3^{2-} symmetry and the corresponding coordination of the CO_3^{2-} ions can only be explained by the presemce of NB_2O_5 or, in other words, by the participation of Nb^{5+} as central cation.

The mechanochemical formation of the carbonato complex is further supported by XRD analysis. By comparing XRD and IR data, we find that the n_3 splitting and n_1 activation, which took place progressively from 5 to 40 hours of milling (Fig. 7a), coincide with the amorphization of NA_2CO_3 (Fig. 5a). For example, after 20 hours of milling, when the split n_3 bands are resolved for the first time and intense n_1 band appeared (Fig. 7a, 20 h), the XRD peaks of NA_2CO_3 could not be detected anymore (Fig. 5a, 20 h). From this comparison we can conclude that the amorphization of NA_2CO_3 is closely related to the formation of the complex. The conclusion seems reasonable if we consider that the formation of the complex requires a reconstruction, i.e., coordination, of the CO_3^{2-} ions; such reconstruction can eventually ruin the original NA_2CO_3 structure over the long range, make it undetectable to X-ray diffraction. The relation between amorphization and coordination is further supported by the fact that neither the amorphization of NA_2CO_3 nor the n_3 splitting were observed during separate milling of NA_2CO_3 (see Fig. 5b and Fig. 7b).

We finally note that after 40 hours of milling the powder mixture contains 81% of XRD- amorphous phase (inset of Fig. 4b). According to this large amount and based on the fact that we did not detect any new crystalline phase during 40 hours of milling we can conclude that the carbonato complex is amorphous or eventually nanocrystalline to an extent that is undetectable with X-ray diffraction methods. This example illustrates that enriched information on a local structural scale can only be achieved by appropriate selection of analytical tools. The amorphous carbonato complex has recently been confirmed using Raman and nuclear magnetic resonance (NMR) spectroscopies (Rojac et al., to be published).

Another important aspect to discuss is the possible role of water on of the formation of the complex. Jolivet et al. (1982) emphasized the influence of the water molecules on the n_3 splitting, which can be significant, depending on whether they can interact via hydrogen bonding with the carbonate group. This was demonstrated through various examples of lanthanide carbonates, where the hydrated forms showed different n_3 splitting with respect to their dehydrated analogues. As an example, the hydrated form of the $NA_2Cu(CO_3)2$ complex, that is $NA_2Cu(CO_3)2{\times}3H_2O$, showed larger Dn_3 splitting, i.e., 203 cm^{-1}, in comparison with its dehydrated form, i.e., 138 cm^{-1} (see also Table 2, first two examples).

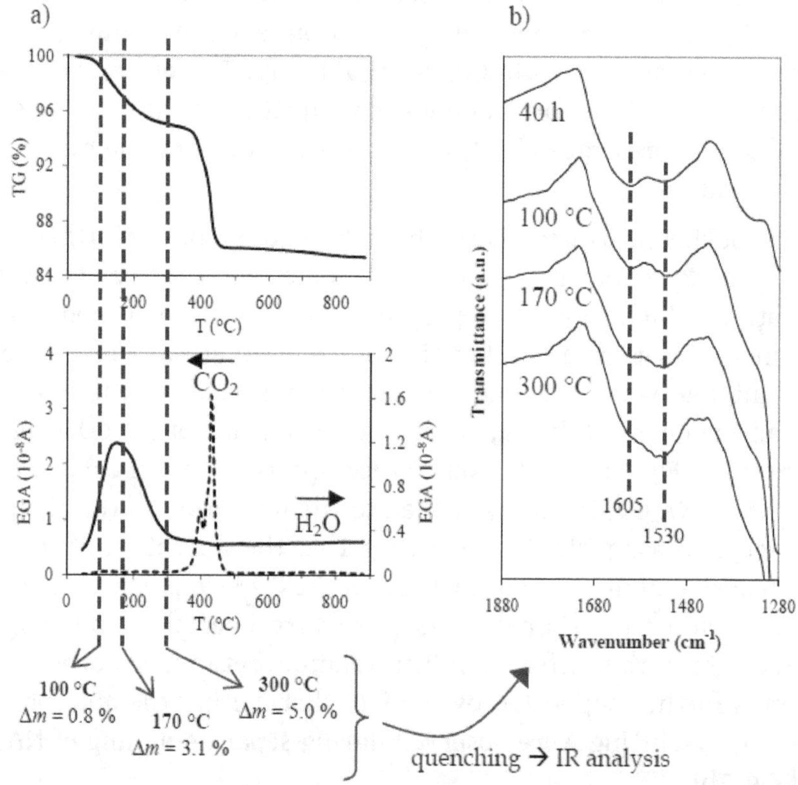

Fig. 10. a) TG and EGA(H_2O,CO_2) curves of the 40-hours high-energy milled NA_2CO^{3-}-NB_2O_5 powder mixture. The dashed lines on the graphs represent the temperatures at which the 40- hours milled sample was Air-quenched. b) FT-IR spectra of the 40-hours high-energy milled NA_2CO^{3-}-NB^2O^5 powder mixture Air-quenched from various temperatures (from Rojac et al., 2006).

In our case, the possible influence of water molecules on the carbonate ion should be considered. In fact, we showed in section 2.2 (Fig. 6) that an amount of water was introduced in the sample from the Air during the milling. Therefore, we have to examine more carefully the possible influence of water adsorption on the n_3 splitting. This was done by quenching the 40-hours-treated sample in Air from different temperatures so that controlled amounts of water were released; the quenched samples were then analyzed using IR spectroscopy. The IR spectra together with the TG and EGA curves are shown in Fig. 10. The mass losses after quenching at 100 °C, 170 °C and 300 °C were 0.8 %, 3.1 % and 5.0 %, respectively.

According to EGA, these mass losses correspond entirely to the removal of water (Fig. 10a, see dashed lines). From Fig. 10b we can see that with increasing amount of released H_2O the band at 1605 cm^{-1} gradually decreases at the expense of the band at 1530 cm^{-1}. Note that the intensity of the band at 1345 cm^{-1} decreases, too. The results confirm the influence of H_2O on the splitting of the n_3 vibration.

There are some cases of carbonato complexes, as pointed out by Jolivet et al. (1982), in which water molecules can even modify the coordination state of the carbonate ion. This is also the case of the $NA_2Cu(CO_3)2{\times}3H_2O$ complex, which contains both bidentate chelate and bridged carbonate ions, whereas its dehydrated form is exclusively a bridged structured (see also Fig. 8). Concerning our carbonato complex, it would be interesting to get more information about the actual influence of H_2O on the CO_3^{2-} coordination. Since apparently the H_2O has an active role in the mechanochemical formation of the complex, namely, it affects the n_3 splitting, and taking into account that this complex represents an intermediate phase from which the $NaNbO_3$ is formed, it would also be interesting to find out whether milling in humid-free conditions will affect the formation of the niobate. These questions will be left for further investigations.

3. MECHANOCHEMICAL REACTION RATE IN NA2CO3–M_2O_5 (M = V, NB, TA) POWDER MIXTURES

3.1 Quantitative X-ray diffraction, infrared spectroscopy and thermogravimetric analysis

In-depth study of reaction mechanism limited to one system is often insufficient if fundamental characteristics governing certain type of mechanochemical reaction are to be determined. Following the results from the previous section, in which we identified an amorphous carbonato complex as a transitional stage of the reaction between NA_2CO_3 and NB_2O_5, it is the next step to find out i) whether this mechanism is general for mechanochemical reactions involving CO_3^{2-} ions and ii) which parameters control the decomposition of the carbonato complex. The latter is particularly important as the decomposition of the complex is a necessary step for the formation of the final binary oxide.

In order to study systematically the mechanochemical interaction between CO_3^{2-} ions and various metal cations, which could possibly lead to the formation of the carbonato complex, we explored the reactions involving NA_2CO_3, as one reaction counterpart, and various 5th group transition-metal oxides, including V_2O_5, NB_2O_5 and TA_2O5. The Aim of the study was to determine the influence of the transition-metal oxide on i) the mechanochemical decomposition of NA_2CO_3 and ii) the rate of formation of the target binary oxides, i.e., $NaVO_3$, $NaTaO_3$ and $NaNbO_3$.

The mechanochemical formation of $NaMO_3$ (M = V, Nb, Ta) from respective NA_2CO_3–M_2O_5 (M = V, Nb, Ta) mixtures was followed by quantitative X-ray diffraction phase analysis using Rietveld refinement method. The fractions of $NaMO_3$ (M = V, Nb, Ta) as a function of milling time are shown in Fig. 11. The rate of formation of the final oxides follows the order $NaVO_3$>$NaTaO_3$>$NaNbO_3$. Note that the vanadate was formed within 4 hours, while the tantalate and niobate required much longer milling times to be the only crystalline phase detected in the mixtures, i.e., 72 and 150 hours, respectively. The results show that the type of the transition-metal oxide plays an important role in the formation kinetics of $NaMO_3$ (M = V, Nb, Ta).

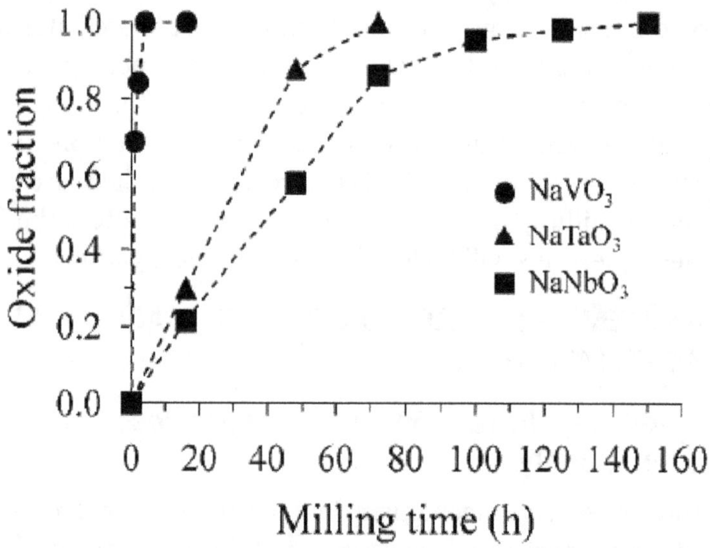

Fig. 11. Fraction of $NaMO_3$ (M = V, Nb, Ta), determined by Rietveld refinement analysis, as a function of milling time. The lines are drawn as a guide for the eye (from Rojac et al., 2011).

In order to verify whether the amorphous carbonato complex appears as a transitional phase in the three examined reactions, we performed an IR spectroscopy analysis. The results are presented in Fig. 12. In all the systems, a common trend, characteristic for the lowering of the CO_3^{2-} symmetry, is observed during milling: i) the $n_3(CO_3^{2-})$ vibration shifts gradually to higher wavenumbers and decreases in intensity until it disappears after certain milling time, ii) the n_3 vibration is replaced by new bands in the region 1650–1250 cm^{-1}, showing n_3 splitting (see 4 h, 72 h, 150 h in Fig. 12 a, b and c, respectively) and iii) the n_1 vibration is activated. Note that the n_1 activation in the case of the NA_2CO_3–V_2O_5 mixture could not be ascertained due to overlapping with the band at 1025 cm^{-1}, related to the stretching vibration of the double vanadyl V=O bonds of V_2O_5 (Fig. 12a). According to the relation between symmetry and IR vibrational spectroscopy of the CO_3^{2-} ion, described in detail in the previous section, the formation of the carbonato complex is confirmed in all the examined systems.

We note that the milling conditions for the mechanochemical synthesis of $NaNbO_3$ presented in the previous section were different from the ones that we applied for the study presented here. This is seen from the different kinetics of the formation of $NaNbO_3$, i.e., by comparing the timescale of the $NaNbO_3$ fraction-versus-time curves from Fig. 11 (closed rectangular) and Fig. 4b (open circles). Therefore, the mechanism of the mechanochemical interaction between NA_2CO_3 and NB_2O_5, in terms of the transitional carbonato complex, is qualitatively unaffected by the milling intensity.

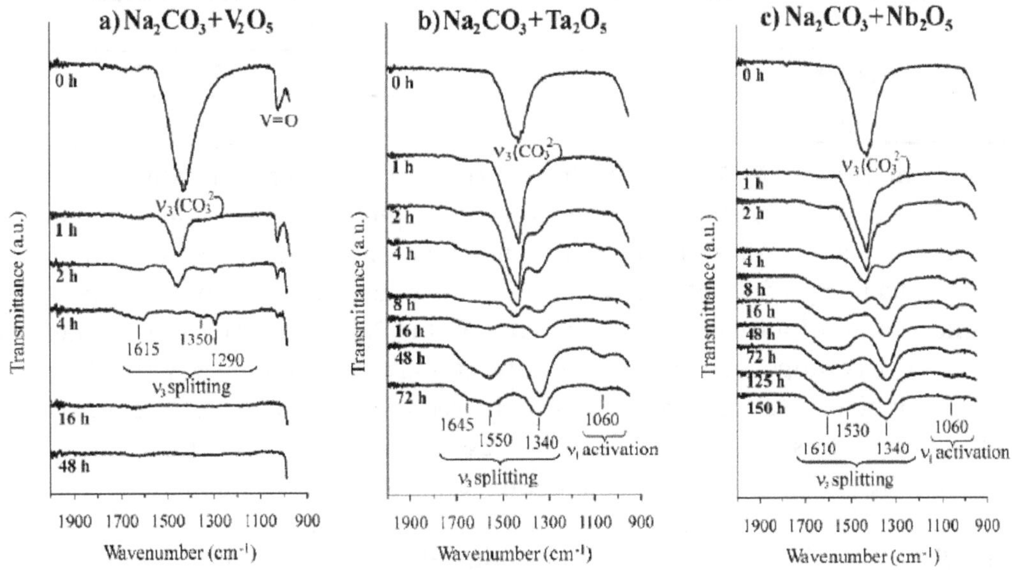

Fig. 12. FT-IR spectra of NA_2CO_3–M_2O_5 (M = V, Nb, Ta) powder mixtures after different milling times (from Rojac et al., 2011).

The results from Fig. 12 suggest a general reaction mechanism in mixtures involving CO_3^{2-} ions; in fact, in addition to the systems presented in this chapter, the carbonato complex was identified in a number of other alkaline-carbonate–transition-metal oxide mixtures, including the following:

- Li_2CO_3–NB_2O_5 (unpublished work)
- K_2CO_3–NB_2O_5 (Rojac et al., 2009)
- RB_2CO_3–NB_2O_5 (unpublished work)
- K_2CO_3–TA_2O5 (Glinsek et al., 2011)
- NA_2CO_3–K_2CO_3–NB_2O_5 (unpublished work)
- NA_2CO_3–K_2CO_3–NB_2O_5–TA_2O5 (Rojac et al., 2010)
- NA_2CO_3–K_2CO_3–Li2C2O4–NB_2O_5–TA_2O5 (Rojac et al., 2010)

A closer inspection of Fig. 12 reveals several differences between the three reaction systems. First of all, the degree of the n_3 splitting is different depending on the metal cation, i.e., V^{5+}, Nb^{5+} or Ta^{5+}, to which the CO_3^{2-} coordinate. The maximum splitting of n_3 from the spectra of the NA_2CO_3–M_2O_5 (M = V, Ta, Nb) mixtures after 4, 72 and 150 hours of milling, respectively (Fig. 12), is collected in Table 3. The maximum Dn_3 splitting is largest in the case of V_2O_5 (325 cm^{-1}), followed by TA_2O5 (305 cm^{-1}) and NB_2O_5 (270 cm^{-1}).

Table 3. Maximum splitting of n3(CO32–) vibration in NA2CO3–M2O5 (M = V, Nb, Ta) powder mixtures (from Rojac et al., 2011).

Mixture	Max Dn_3 splitting (cm^{-1})
NA_2CO_3–V_2O_5	325
NA_2CO_3–TA_2O5	305
NA_2CO_3–NB_2O_5	270

Nakamoto et al. (1957) were the first to propose the degree of n_3 splitting (Dn_3) as a criterion to distinguish between mono- and bidentate coordination in carbonato complexes. Their results showed that some bidentate cobalt carbonato complexes have Dn_3 splitting of about 300 cm^{-1}, while monodentate complexes of analogous chemical composition exhibit about 80 cm^{-1} of Dn_3. Calculations based on models of XO3 (X = C, N) groups coordinated to a metal cation confirmed the larger splitting in the case of bidentate coordination, as compared to the monodentate coordination (Britzinger & Hester, 1966; Hester & Grossman, 1966). A general relationship between the type of coordination and Dn_3 splitting, which we updated according to the critical review by Busca & Lorenzelli (1982), is shown schematically in Fig. 13a. While monodentate configurations show splitting of around 100 cm^{-1} or lower, larger Dn_3 splitting can be expected for bidentate chelate and bidentate bridged coordinations.

In addition to the type of coordination, other factors influence the degree of the n_3 splitting. As explained in the previous section, the coordination of the CO_3^{2-} ion causes a rearrangement of the C–O bonds, i.e., the C–O bond coordinated to the metal cation is typically weakened, while the others, non-coordinated, are strengthened. Calculations showed that, for a given type of coordination, this CO_3^{2-} polarization is more pronounced if the polarizing power of the central cation is high as it can attracts electrons more strongly (Britzinger & Hester, 1966). The Dn_3 splitting, which reflects the CO_3^{2-} polarization, should therefore depend on the polarizing power of the central cation. This was indeed confirmed experimentally by Jolivet et al. (1982), which identified a linear increase of Dn_3 splitting with the polarizing power of the central cation for numerous carbonato complexes having the same bidentate coordination (Fig. 13b). For those cases, the polarizing power of the cation was assumed to be proportional to e/r^2, where e and r are cation charge and radius, respectively. Therefore, the Dn$_3$ splitting criterion for distinguishing between different types of coordination (Fig. 13a) should only be applied if the polarizing power of the cation is taken into account (Fig. 13b). As pointed out by Busca & Lorenzelli (1982), low values of Dn$_3$ splitting, e.g., ~100 cm^{-1}, do not unequivocally indicate the presemce of monodentate structure, particularly in cases of metals having low polarizing power (se also Fig. 13b).

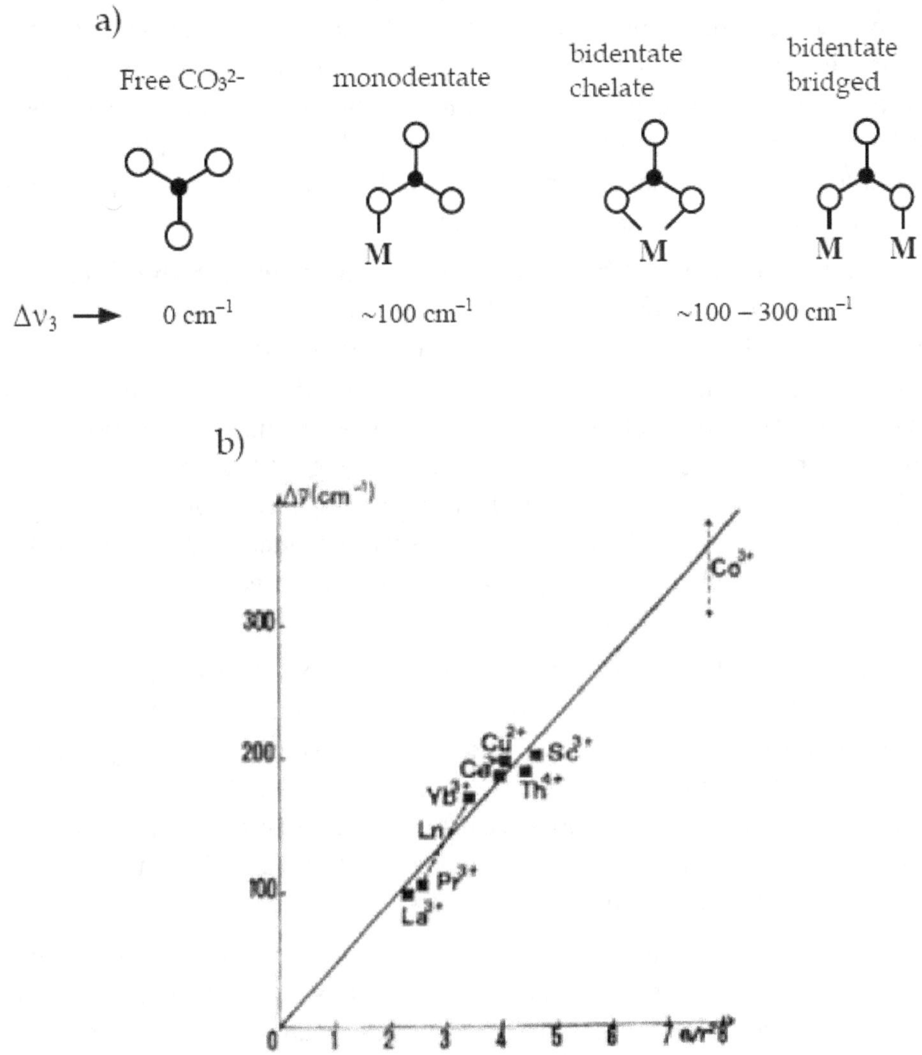

Fig. 13. a) Schematic view of the influence of the type of coordination on splitting of n3(CO_3^{2-}) vibration (from Busca & Lorenzelli, 1982) and b) correlation between splitting of n$_3$(CO_3^{2-}) vibration and polarizing power (e/r2) of the central cation for bidentate carbonato complexes. "e" and "r" denote cation charge and radius, respectively (from Jolivet et al., 1982).

By comparing with the literature data and considering the relationship shown in Fig. 13a, the maximum Dn_3 splitting in the three mixtures from Table 3, being larger than 100 cm^{-1}, might suggest bidentate chelate and/or bridged coordination. The increasing Dn_3 from the system with niobium, having the smallest Dn_3 of 270 cm^{-1}, to that with vanadium, with the largest Dn_3 of 325 cm^{-1} (Table 3), correlates with the increasing cation acidity, appearing in the order Nb^5+<Ta^{5+}<V^{5+} (see Xz/C_n values in Table 5 and refer to the next section for details). Since it is generally accepted that the acidity scales with the cation charge density, i.e., e/r (Avvakumov et al., 2001), our correlation, in principle, agrees with the one of Jolivet et al. (1982) from Fig. 13b. However, even if a correlation exists, it should be interpreted carefully Since in addition to the polarizing ability of the central cation, we should not neglect other influences on the Dn_3 splitting, such as, for example, incorporation of water molecules, as demonstrated in previous section (Fig. 10), which might differ between the three examined systems.

In addition to the n_3 splitting, another difference between the three reactions that should be noted is the much faster formation of the complex in the case of V^{5+} as central cation in comparison with Nb^5+ and Ta^{5+} (Fig. 12). This can be seen by comparing the characteristic n_3 splitting among the three systems taking into consideration the point of the transition of the original n_3 vibration into split bands as a criterion for the CO_3^{2-} coordination (Fig. 12). Whereas in the case of V_2O_5 the n_3 vibration almost completely disappeared after 4 hours of milling, giving rise to split bands (Fig. 12a), at least 16 hours were needed in the case of NB_2O_5 and TA_2O_5 (Fig. 12b and c). This is in agreement with the kinetics of NaMO3 formation (M = V, Nb, Ta), shown in Fig. 11.

In contrast to the NA_2CO_3–TA_2O5 and NA_2CO_3–NB_2O_5 systems, where the IR absorption bands of the carbonato complex are still clearly resolved after 72 and 150 hours of milling (Fig. 12b and c), these bands completely disappeared after only 16 h of milling in the case of the NA_2CO_3–V_2O_5 mixture (Fig. 12a). A reasonable explanation for the absence of the IR bands related to the complex is its decomposition. If this is the case, it suggests that the transition-metal oxide plays a role in the decomposition of the carbonato complex.

We can compare quantitatively the carbonate decomposition in the three studied systems by using thermogravimetric analysis. Similarly like explained in the previous section, by separating the mass loss related to the H_2O release from the one that is due to the CO_2 release, we can estimate the amount of the residual carbonate in the powder mixtures. The results of this analysis are shown in Fig. 14. As expected, in all the mixtures the carbonate fraction decreases with increasing milling time; this is associated with the mechanochemically driven carbonate decomposition. We can summarize the reaction as follows: after being formed (Fig.

12), the carbonato complex decomposes (Fig. 14), leading to the formation of the final NaMO3 oxides (Fig. 11). Note that, in terms of the reaction timescale, these three stages are not clearly separated, instead, they are overlapped.

The results of TG analysis (Fig. 14) reveal a substantial difference in the decomposition rate of the carbonate between the three systems: the fastest is in the case of V_2O_5, followed by TA_2O5 and NB_2O_5. Note also that the carbonate fraction reaches a plateau after prolonged milling, which we denote here as the "steady-state" milling condition. The amount of the carbonate in this "steady-state" condition depends strongly on the type of the transition- metal oxides participating in the reaction. Whereas the carbonato complex decomposed nearly completely in the case of vanadium, i.e., only 0.5% of residual carbonate was determined in the "steady-state" milling condition, 29% and 39% remained in the mixture in the case of Ta and Nb, respectively (Fig. 14 and Table 4). This is in agreement with the IR spectra from Fig 12, i.e., in contrast to the cases with Nb and Ta, no IR bands related to the carbonato complex are observed after prolonged milling of the NA_2CO_3–V_2O_5 mixture (Fig. 12a, see 16 and 48 hours). Finally, it is important to stress that the complex is amorphous and could not be analyzed using Rietveld analysis (Fig. 11); instead, we were able to follow its formation and decomposition using IR and TG analyses, respectively (Fig. 12 and 14).

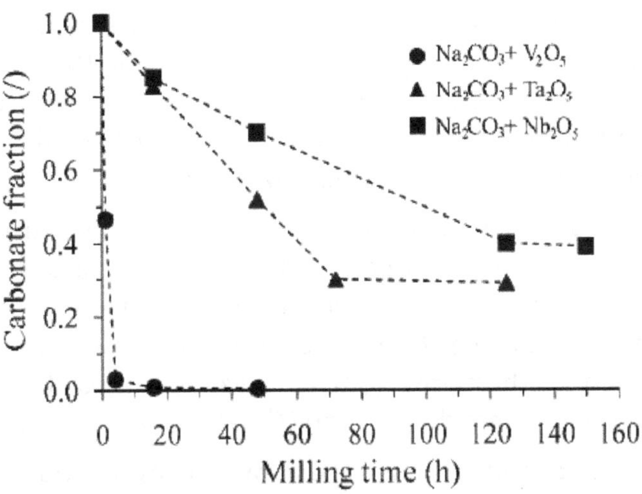

Fig. 14. Fraction of carbonate, determined by TG analysis, as a function of milling time in NA_2CO_3-M_2O_5 (M = V, Nb, Ta) powder mixtures. The lines are drawn as a guide for the eye (from Rojac et al., 2011).

From the presented results we can infer that a common mechanism, characterized by the formation of an intermediate amorphous carbonato complex, link the reactions between NA_2CO_3 and M_2O_5 (M = V, Nb, Ta); however, considerable differences exist in the rate of the formation and decomposition of this carbonato complex and, consequently, in the crystallization of the final binary compounds. The sequence of the rates of these reactions, i.e., NA_2CO_3–V_2O_5>NA_2CO_3–TA_2O5>NA_2CO_3–NB_2O_5, can be interpreted by considering the acid-base properties of the reagents involved.

Table 4. Fraction of residual carbonate in $NA_2CO^{3-}M2O5$ (M = V, Nb, Ta) powder mixtures in the "steady-state" milling conditions (see carbonate fraction after prolonged milling in Fig. 14) (from Rojac et al., 2011).

Mixture	Carbonate fraction (%)
NA_2CO_3–V_2O_5	0.5
NA_2CO_3–TA_2O5	29
NA_2CO_3–NB_2O_5	39

3.2 Acid-base mechanochemical reaction mechanism

In their extensive work on the mechanochemical reactions involving hydroxide–oxide mixtures, Senna and co-workers (Liao & Senna, 1992, 1993; Watanabe et al., 1995b, 1996, Avvakumov et al., 2001) showed that the mechanism in these mixtures is governed by an acid-base reaction between different hydroxyl groups on the solid surface. The driving force for these reactions is the acid-base potential, i.e., the difference in the acid-base properties between an acidic and basic surface –OH group, which is determined by the type of metal on which it is bound, and therefore, on the strength of the M–OH bond (M denotes the metal). For example, in the case of the M(OH)2–SiO_2 (M = Ca, Mg) mixtures, they showed experimentally that a larger acid-base potential between Ca(OH)2 and SiO_2 brought a faster mechanochemical interaction (Liao & Senna, 1993).

The acid-base reaction mechanism is not confined to the hydroxyl groups only. Thermodynamic calculations showed that a correlation exists between the Gibbs free energies of a variety of reactions between oxide compounds and the acid-base potential between the participating oxide reagents for two-component systems: the larger the potential, the more negative the value of the Gibbs energy and, thus, the faster and more complete the reaction (Avvakumov et al., 2001).

In order to fully consider the acid-base properties of oxide compounds, one should take into account that the acidity of a cation, incorporated into a certain oxide compound, depends on the oxidation state and the coordination number. For example, when the oxidation degree of the manganese ion increases by unity, the acidity increases by 2–3 times; same trend is observed when the coordination number of Si^{4+} decreases from 6 to 4. For correct comparisons, the influence of these parameters on the acid-base properties of cations should be taken into account. This can be done by introducing the electronegativity of a cation, divided by the coordination number, which defines the cation-ligand force per one bond in a coordination polyhedron; the larger the force, the larger the acidity of the cation, i.e., the stronger is the ability to attract electron pairs forming covalent bonds (Avvakumov et al., 2001).

To address the acid-base properties of the transition-metal cations, we adopted the electronegativity scale for cations derived by Zhang (1982). Table 5 shows the electronegativities Xz for V^{5+}, Ta^{5+} and Nb^5+. The ratio Xz/C_n, where C_n refers to the coordination number of the cation, taken as being indicative of the acidity of the cations in their respective oxides, is the highest for V^{5+}, followed by Ta^{5+} and Nb^5+.

Table 5. Electronegativity values Xz and Xz/Cn ratios for V5+, Nb5+ and Ta5+. Xz and Cn denote cation electronegativity defined by Zhang and coordination number, respectively. The Xz/Cn ratio is taken as a parameter proportional to cation acidity (from Rojac et al., 2011).

Cation	Xz	Xz/C_n
V^{5+}	2.02	0.40
Ta^{5+}	1.88	0.29
Nb^5+	1.77	0.27

The order of the cation acidity, i.e., V^{5+}>Ta^{5+}>Nb^5+ (Table 5) correlates with our experimental results; in fact, the reaction rate follows the same order, i.e., NA_2CO_3–V_2O_5>NA_2CO_3–TA_2O5>NA_2CO_3–NB_2O_5 (see Fig. 11). This means that the higher is the acidity of the cation involved, the faster is the reaction, including the formation and decomposition of the carbonato complex (Fig. 12 and 14), and the crystallization of the final oxides (Fig. 11). The agreement between the reaction rate sequence and the cation acidity or acid-base potential, where NA_2CO_3 is taken as basic and transition-metal oxides as acidic compound, suggests that the mechanochemical reactions studied here suit the concept of an acid-base interaction mechanism. Similar correlations between the acid-base properties and the mechanochemical reaction rate can also be found in other systems comprising CaO, as one reagent, and Al_2O_3, SiO_2, TiO_2, V_2O_5 or WO3, as the other reagent (Avvakumov et al. 1994).

We showed in the previous section that after reaching a specific milling time ("steady-state" milling condition), the fraction of the residual carbonate did not change any longer if further milling was applied (Fig. 14). These carbonate fractions correlate with the acidity of the cations as well (compare Table 4 with Table 5). Note that the small carbonate fraction in the NA_2CO_3– V_2O_5 system, i.e., 0.5 % (Table 4), is consistent with the much larger acidity of V^{5+} as compared to Ta^{5+} or Nb^5+ (see XZ/C_n values in Table 5). The results seem reasonable considering the relation that was found between the acid-base potential and the reaction Gibbs free energy; however, insufficient thermodynamic data for the systems presented here prevent us from making further steps in this direction. We note that these results carry practical consequences, i.e., they suggest that attempting to eliminate the carbonate from a mixture, characterized by a low acid-base potential, by intensifying the milling might not be successful. In fact, the residual carbonate fraction seems to be dependent on the acid-base potential, which is a characteristic of a system, rather than on the milling conditions (Rojac et al., 2008b).

Even if the rate of the three examined reactions apparently agrees with the acid-base reaction concept, we shall not neglect other parameters that could influence the course of the reaction, such as, e.g., adsorption of H_2O during milling, which might differ from one system to another. In order to directly verify this possibility, we plot in Fig. 15 the reaction rate constant versus Xz/C_n. The reaction rate constant was obtained by fitting the curves from Fig. 11 with a kinetic model proposed for mechanochemical transformations in binary mixtures (Cocco et al., 2000). A linear relationship would be expected if the reaction rate will be largely dominated by the cation acidity. While the sequence of the reaction rate constants, i.e., V>Ta>Nb, agrees with the acid-base reaction mechanism, the non-linear relationship between the rate constant and Xz/C_n from Fig. 15 suggests that, in addition to the cation acidity, probably other factors influence the reaction rate. The origin of these additional influences will be left open for further studies.

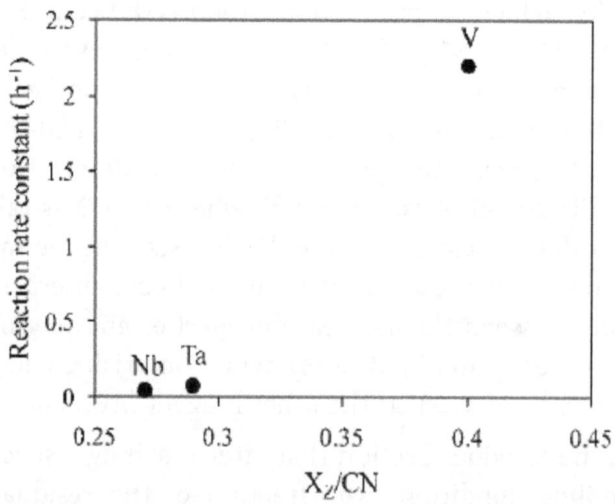

Fig. 15. Reaction rate constant versus XZ/Cn for the reactions between NA2CO3 and M2O5 (M = V, Nb, Ta) (from Rojac et al., 2011).

4. CONCLUSIONS

A systematic study of the reaction mechanism during high-energy milling of a NA_2CO_3– NB_2O_5 mixture, presented in the first part of the chapter, revealed that the synthesis of $NaNbO_3$ takes place through an intermediate amorphous stage.

Quantitative phase analysis using XRD diffraction and Rietveld refinement method showed indeed a large amount of the amorphous phase, i.e., of up of 91%, formed in the initial part of the reaction. The amount of this amorphous phase then decreased by subsequent milling, leading to the crystallization of the final $NaNbO_3$. Only limited information, including mainly the identification of the transitional nature of the amorphous phase, was obtained using quantitative XRD phase analysis.

The decomposition of the carbonate in the NA_2CO_3–NB_2O_5 mixtures was analyzed using TG analysis coupled with DTA and EGA. By following the carbonate decomposition upon annealing the mixtures milled for various periods we were able to infer about the changes occurring in the carbonate during high-energy milling. Characteristic changes in the carbonate decomposition upon increasing the milling time suggested a formation of an intermediate carbonate compound with rather defined decomposition temperature occurring in a narrow temperature range; such decomposition was found to be atypical for physical mixtures of NA_2CO_3 and NB_2O_5 powders.

A more accurate identification of the amorphous phase was made possible using IR spectroscopy analysis. Characteristic IR vibrational changes during milling, including splitting of n_3 and activation of n_1 C–O stretching vibrations of the CO_3^{2-} ion, suggested lowered CO_3^{2-} symmetry, which was interpreted as being a consequence of CO_3^{2-} coordination and formation of an amorphous carbonato complex.

Expanding the study of the NA_2CO_3–NB_2O_5 to other systems, we showed in the second part of the chapter that the mechanism involving the transitional amorphous carbonato complex is common for several alkaline-carbonate-transition-metal oxide mixtures, including NA_2CO_3–M_2O_5 (M = V, Nb, Ta). The sequence of the reaction rate in these three systems, including the formation of the complex, its decomposition and crystallization of the final NaMO3 (M = V, Nb, Ta), was interpreted by considering the acid-base reaction mechanism. The largest the acid-base potential, i.e., the difference between acidic and basic properties of the reagents involved, the faster the mechanochemical reaction.

5. REFERENCES

1. Avvakumov, E. G.; Devyatkina, E. T. & Kosova, N. V. (1994). Mechanochamical Reactions of Hydrated Oxides. Journal of Solid State Chemistry, Vol.113, No.2, pp. 379–383.
2. Avvakumov, E.; Senna, M. & Kosova, N. (2001). Soft Mechanochemical Synthesis, Kluwer Academic Publisher, Boston, USA.
3. Boldyrev, B. B. & Tkacova, K. (2000). Mechanochemistry of Solids : Past, Present, and Prospects. Journal of Materials Synthesis and Processing, Vol.8, No.3-4, pp. 121–132.

4. Brintzinger, H. & Hester, R. E. (1966). Vibrational Analysis of Some Oxyanion-Metal Complexes. Inorganic Chemistry, Vol.5, No.6, pp. 980–985.

5. Brooker, M. H. & Bates, J. B. (1971). Raman and Infrared Spectral Studies of Anhydrous Li_2CO_3 and NA_2CO_3. The Journal of Chemical Physics, Vol.54, No.11, pp. 4788–4796.

6. Buijs, K. & Schutte, C. J. H. (1961). The Infra-red Spectra and Structures of Li_2CO_3 and Anhydrous NA_2CO_3. Spectrochimica Acta B, Vol.17, No.9-10, pp. 927–932.

7. Burgio, N.; Iasonna, A; Magini, M.; Martelli, S. & Padella, F. (1991). Mechanical Alloying of the Fe-Zr System: Correlation between Input Energy and End Products. Il Nuovo Cimento, Vol.13D, No.4, pp. 459–476.

8. Busca, G. & Lorenzelli, V. (1982). Infrared Spectroscopic Identification of Species Arising from Reactive Adsorption of Carbon Oxides on Metal Oxide Surfaces. Materials Chemistry, Vol.7, No.1, pp. 89–126.

9. Cocco, G.; Delogu, F. & Schiffini, L. (2000). Toward a Quantitative Understanding of the Mechanical Alloying Process. Journal of Materials Synthesis and Processing, Vol.8, No.3-4, pp. 167–180.

10. Delogu, F. & Cocco, G. (2000). Relating Single-Impact Events to Macrokinetic Features in Mechanical Alloying Processes. Journal of Materials Synthesis and Processing, Vol.8, No.5-6, pp. 271–277.

11. Delogu, F.; Mulas, G.; Schiffini L. & Cocco, G. (2004). Mechanical Work and Conversion Degree in Mechanically Induced Processes. Materials Science and Engineering A, Vol.382, No.1-2, pp. 280–287.

12. El-Eskandarany, M. S.; Akoi, K.; Sumiyama, K. & Suzuki, K. (1997). Cyclic Crystalline- Amorphous Transformations of Mechanically Alloyed Co75Ti25. Applied Physics Letters, Vol.70, No.13, pp. 1679–1681.

13. Fujita, J.; Martell, A. E. & Nakamoto, K. (1962). Infrared Spectra of Metal Chelate Compounds. VIII. Infrared Spectra of Co(III) Carbonato Complexes. The Journal of Chemical Physics, Vol.36, No.2, pp. 339–345.

14. Gatehouse, B. M.; Livingstone, S. E. & Nyholm, R. S. (1958). The Infrared Spectra of Some Simple and Complex Carbonates. Journal of the Chemical Society, pp. 3137–3142.

15. Glinšek, S.; Malič, B.; Rojac, T.; Filipič, C.; Budič, B. & Kosec, M. (2011). KTaO3 Ceramics Prepared by the Mechanochemically Activated Solid-State Synthesis. Journal of the American Ceramic Society, Vol.94, No.5, pp. 1368–1373.

16. Goldsmith, J. A. & Ross, S. D. (1967). Factors Affecting the Infra-Red Spectra of Planar Anions with D_{3h} Symmetry – IV. The Vibrational Spectra of Some Complex Carbonates in the Region 4000–400 cm^{-1}. Spectrochimica Acta A, Vol.24, No.8, pp. 993–998.

17. Haris, M. J. & Salje, E. K. H. (1992). The Incommensurate Phase of Sodium Carbonates: An Infrared Absorption Study. Journal of Physics: Condensed Matter, Vol.4, No.18, pp. 4399–4408.

18. Healy, P. C. & White, A. H. (1972). Crystal Structure and Physical Properties of Anhydrous Sodium Copper Carbonate. Journal of the Chemical Society, Dalton Transactions, pp. 1913–1917.

19. Hester, R. E. & Grossman, W. E. L. (1966). Vibrational Analysis of Bidentate Nitrate and Carbonate Complexes. Inorganic Chemistry, Vol.5, No.8, pp. 1308–1312.

20. Iasonna, A. & Magini, M. (1996). Power Measurements during Mechanical Milling: An Experimental Way to Investigate the Energy Transfer Phenomena. Acta Materialia, Vol.44, No.3, pp. 1109–1117.

21. Iguchi, Y. & Senna, M. (1985). Mechanochemical Polymorphic Transformation and Its Stationary State between Aragonite and Calcite. I. Effects of Preliminary Annealing. Powder Technology, Vol.43, No.2, pp. 155–162.

22. Jenko, D. (2006). Synthesis of (K,Na)NbO3 Ceramics, PhD Thesis, University of Ljubljana, Ljubljana, Slovenia.

23. Jolivet, J. P.; Thomas, Y. & Taravel, B. (1980). Vibrational Study of Coordinated CO_3^{2-} Ions.

24. Journal of Molecular Structure, Vol.60, pp. 93–98.

25. Jolivet, J. P.; Thomas, Y. & Taravel, B. (1982). Infrared Spectra of Cerium and Thorium Pentacarbonate Complexes. Journal of Molecular Structure, Vol.79, pp. 403–408.

26. Kong, L. B.; Zhang, T. S.; Ma, J. & Boey, F. (2008). Progress in Synthesis of Ferroelectric Ceramic Materials Via High-Energy Mechanochemical Technique. Progress in Materials Science, Vol.53, No.2, pp. 207–322.

27. Kuscer, D.; Holc, J. & Kosec, M. (2006). Mechano-Synthesis of Lead-Magnesium-Niobate Ceramics. Journal of the American Ceramic Society, Vol.89, No.10, pp. 3081–3088.

28. Le Brun, P.; Froyen, L. & Delaey, L. (1993). The Modeling of the Mechanical Alloying Process in a Planetary Ball Mill: Comparison between Theory and In-situ Observations. Materials Science and Engineering A, Vol. 161, No.1, pp. 75–82.

29. Liao, J. & Senna, M. (1992). Enhanced Dehydration and Amorphization of Mg(OH)2 in the Presemce of Ultrafine SiO_2 Under Mechanochemical Conditions. Thermochimica Acta, Vol.210, pp. 89–102.

30. Liao, J. & Senna, M. (1993). Mechanochemical Dehydration and Amorphization of Hydroxides of Ca, Mg and Al on Grinding With and Without SiO_2. Solid State IoniCs, Vol.66, No.3-4, pp. 313–319.

31. Lin, I. J. & Nadiv, S. (1979). Review of the Phase Transformation and Synthesis of Inorganic Solids Obtained by Mechanical Treatment (Mechanochemical Reactions). Materials Science and Engineering, Vol.39, No.2, pp. 193–209.

32. Lu, L. & Lai, M. O. (1998). Mechanical Alloying, Kluwer Academic Publisher, Boston, USA. Maurice, D. R. & Courtney, T. H. (1990). The Physics of Mechanical Alloying. Metallurgical

33. Transactions A, Vol.21, No.1, pp. 289–303.

34. Nakamoto, K.; Fujita, J.; Tanaka, S. & Kobayashi, M. (1957). Infrared Spectra of Metallic Complexes. IV. Comparison of the Infrared Spectra of Unidentate and Bidentate Complexes. Journal of the American Chemical Society, Vol.79, No.18, pp. 4904–4908.

35. Nakamoto, K. (1997). Infrared and Raman Spectra of Inorganic and Coordination Compounds. Part A: Theroy and Applications in Inorganic Chemistry, Kluwer Academic Publisher, Boston, USA.

36. Rojac, T.; Kosec, M.; Šegedin, P.; Malič, B. & Holc, J. (2006). The Formation of a Carbonato Complex during the Mechanochemical Treatment of a NA_2CO_3–NB_2O_5 Mixture. Solid State IoniCs, Vol.177, No.33-34, pp. 2987–2995.

37. Rojac, T.; Benčan, A.; Uršič, H.; Malič, B. & Kosec, M. (2008a). Synthesis of a Li- and Ta- Modified (K,Na)NbO3 Solid Solution by Mechanochemical Activation. Journal of the American Ceramic Society, Vol.91, No.11, pp. 3789–3791.

38. Rojac, T.; Kosec, M.; Malič, B. & Holc, J. (2008b). The Mechanochemical Synthesis of $NaNbO_3$ Using Different Ball-Impact Energies. Journal of the American Ceramic Society, Vol.91, No.5, pp. 1559–1565.

39. Rojac, T.; Kosec, M.; Polomska, M.; Hilczer, B.; Šegedin, P. & Benčan, A. (2009). Mechanochemical Reaction in the K_2CO_3–NB_2O_5 System. Journal of the European Ceramic Society, Vol.29, No.14, pp. 2999–3006.

40. Rojac,T.;Benčan,A.&Kosec,M.(2010).MechanismandRoleofMechanochemical Activation in the Synthesis of (K,Na,Li)(Nb,Ta)O3 Ceramics. Journal of the American Ceramic Society, Vol.93, No.6, pp. 1619–1625.

41. Rojac, T.; Trtnik, Ž. & Kosec, M. (2011). Mechanochemical Reactions in NA_2CO_3–M_2O_5 (M = V, Nb, Ta) Powder Mixtures: Influence of Transition-Metal Oxide on Reaction Rate. Solid State IoniCs, Vol.190, No.1, pp. 1–7.

42. Senna, M. & Isobe, T. (1997). NMR and EPR Studies on the Charge Transfer and Formation of Complexes through Incompletely Coordinated States. Solid State IoniCs, Vol.101- 103, No.1, pp. 387–392.

43. Sopicka-Lizer, M. (2010). High-Energy Ball Milling: Mechanochemical Processing of Nanopowders, Woodhead Publishing, Boston, USA.

44. Suryanarayana, C. (2001). Mechanical Alloying and Milling. Progress in Materials Science, Vol.46, No.3-4, pp. 1–184.

45. Tacaks, L. (2004). M. Carey Lea, The First Mechanochemist. Journal of Materials Science, Vol.39, No.16-17, pp. 4987–4993.

46. Tchernychova, E.; Glinšek, S.; Malič, B. & Kosec, M. (2011). Combined Analytical Transmission Electron Microscopy Approach to Reliable Composition Evaluation of KTaO3. Journal of the American Ceramic Society, Vol.94, No.5, pp. 1611–1618.

47. Venyaminov, S. Y. & Pendergast, F. G. (1997). Water (H_2O and D2O) Molar Absorptivity in the 1000–4000 cm^{-1} Range and Quantitative Infrared Spectroscopy of Aqueous Solutions. Analytical Biochemistry, Vol.248, No.2, pp. 234–245.

48. Wang, J.; Xue, J. M.; Wan, D. M. & Gan, B. K. (2000a). Mechanically Activating Nucleation and Growth of Complex Perovskites. Journal of Solid State Chemistry, Vol.154, No.2, pp. 321–328.

49. Wang, J.; Xue, J. M. & Wan, D. (2000b). How Different is Mechanical Activation from Thermal Activation? A Case Study with PZN and PZN-Based Relaxors. Solid State IoniCs, Vol.127, No.1-2, pp. 169–175.

50. Watanabe, R.; Hashimoto, H. & Lee, G. G. (1995a). Computer Simulation of Milling Ball Motion in Mechanical Alloying (Overview). Materials Transactions JIM, Vol.36, No.2, pp. 102–109.

51. Watanabe, T.; Liao, J. & Senna, M. (1995b). Changes in the Basicity and Species on the Surface of Me(OH)2–SiO_2 (Me=Ca, Mg, Sr) Mixtures Due to Mechanical Activation. Journal of Solid State Chemistry, Vol.115, No.2, pp. 390–394.

52. Watanabe, T.; Isobe, T. & Senna, M. (1996). Mechanisms of Incipient Chemical Reaction between Ca(OH)2 and SiO_2 under Moderate Mechanical Stressing. I. A Solid State Acid-Base Reaction and Charge Transfer Due to Complex Formation. Journal of Solid State Chemistry, Vol.122, No.1, pp. 74–80.

53. Watanabe, T.; Isobe, T. & Senna, M. (1997). Mechanisms of Incipient Chemical Reaction between Ca(OH)2 and SiO_2 under Moderate Mechanical Stressing. III. Changes in the Short-Range Ordering throughout the Mechanical and Thermal Processes. Journal of Solid State Chemistry, Vol.130, No.2, pp. 248–289.

54. Watters, R. L. (2005). NIST Standard Reference Database N°69, United States, June 2005 Release (robert.watters@nist.gov).

55. Zhang, Y. (1982). Electronegativities of Elements in Valence States and Their Applications. 1. Electronegativities of Elements in Valence States. Inorganic Chemistry, Vol.21, No.11, pp. 3886–3889.

Chapter 2

UTILIZING INFRARED SPECTROSCOPY FOR THE ANALYSIS OF CHITOSAN/ CLAY NANOCOMPOSITES

1. INTRODUCTION: AN OVERVIEW

Recently, there has been significant interest in polymer/clay nanocomposites due to their ability to integrate the structure and physical and chemical characteristics of both inorganic and organic materials. The majority of research on polymer/clay nanocomposites has focused on artificial polymers, specifically thermosets like epoxy polymers, and thermoplastics such as polyethylene, polypropylene, nylon, and poly(ethylene terephthalate) (Pandey & Mishra, 2011). There has been a lack of focus on natural polymer/clay nanocomposites. Nevertheless, the chance to merge clays and natural polymers at the nanoscale, specifically chitosan, presents an appealing method to alter various characteristics of this polysaccharide. These modifications encompass mechanical and thermal behavior, solubility and swelling properties, antimicrobial activity, bioadhesion, and more (Han et al., 2010). Chitosan/clay nanocomposites are economically advantageous due to their straightforward preparation process and the utilization of affordable chemical reagents. Chitosan, derived from chitin, is a cost-effective substance due to the fact that chitin is the second most prevalent polymer in the natural world, behind cellulose (Chang & Juang, 2004). Similarly, clays are plentiful and inexpensive natural substances. Despite their attractiveness, chitosan/clay nanocomposites have not been thoroughly studied, with a very limited number of scientific papers. Furthermore, the effective formation of the nanocomposites continues to face challenges, primarily with the adequate distribution of nano-fillers inside the polymer matrix. This chapter focuses on the synthesis and characterization of chitosan/clay nanocomposites using infrared spectroscopy. Additionally, it includes the analysis of x-ray diffraction data and the evaluation of mechanical properties.

1.1 Chitosan

Chitosan is a linear polysaccharide that occurs naturally and is closely related to chitin, which is a polymer found throughout the animal kingdom. Rouget is credited with the discovery of chitosan in 1859, when he observed that heating chitin in potassium hydroxide made the polymer soluble in organic acids. In 1894, Hoppe-Seyler designated this substance as chitosan. The structure of chitosan was determined in 1950 (Dodane & Vilivalam, 1998, as quoted in Dash et al., 2011). Chitin can be derived from the exoskeletons of crustaceans, insects, fungi, and other biological sources (Wan Ngah et al., 2011). The primary commercial sources of chitin consist of the discarded shells of shrimps, lobsters, krills, and crabs. Every year, the world harvests several million tons of chitin, which makes this biopolymer a cheap and easily accessible resource (Dash et al., 2011). Chitosan is exclusively present in specific fungi (Mucoraceae) in its natural state, although it can be readily acquired through the thermochemical deacetylation of chitin in the presence of alkali (Darder et al., 2003). Various techniques have been suggested, predominantly encompassing the breakdown of the acetylated residue by the utilization of sodium or potassium hydroxide solutions, along with a combination of anhydrous hydrazine and hydrazine sulfate. The choice of conditions for deacetylation has a direct impact on both the molecular weight of the polymer and the degree of deacetylation (DD) (Dash et al., 2011; Lavorgna et al., 2010).

Chitosan is a copolymer whose chemical structure is shown in Fig. 1. The numbers on the extreme left ring are conventionally assigned to the six carbons in the glucopyranose ring, from C-1 to C-6. Substitution at C-2 may be an acetamido or amino group. Chitosan contains more than 50% (commonly 70 to 90%) of acetamido residues on the C-2 of the structural unit, while amino groups predominate in chitin. The degree of deacetylation (DD) serves as a diagnostic to classify the biopolymer as chitin or chitosan (Dash et al., 2011; Rinaudo, 2006). Notice that DD + DA =1.

The DD is the key property that affects the physical and chemical properties of chitosan, such as solubility, chemical reactivity and biodegradability and, consequently their applications. A quick test to differentiate between chitin and chitosan is based on solubility and nitrogen content. Chitin is soluble in 5% lithium chloride/N,N-dimethylacetamide solvent [LiCI/DMAc] and insoluble in aqueous acetic acid while the opposite is true of chitosan. The nitrogen content in purified samples is less than 7% for chitin and more than 7% for chitosan (Dash et al., 2011; Rinaudo, 2006).

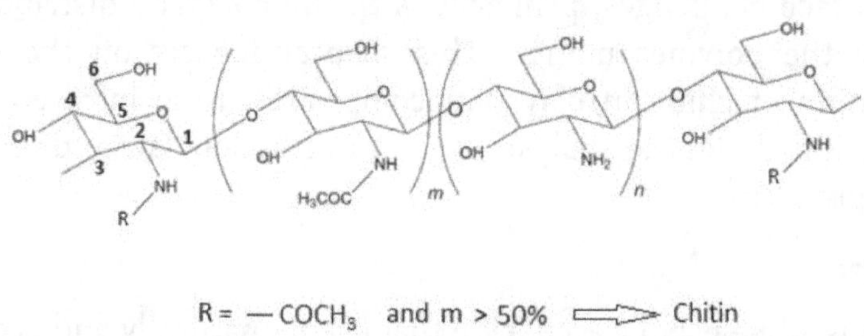

R = —$COCH_3$ and m > 50% ⟹ Chitin

R = —H and n > 50% ⟹ Chitosan

Fig. 1. Chemical structure of chitin and chitosan.

In the solid state, chitosan is a semicrystalline polymer. Its morphology has been investigated and many polymorphs are mentioned in the literature. Single crystals of chitosan were obtained using fully deacetylated chitin of low molecular weight. The dimensions the orthorhombic unit cell of the most common form were determined as a = 0,807 nm, b = 0,844 nm, c = 1,034 nm; the unit cell contains two antiparallel chitosan chains, but no water molecules (Dash et al., 2011).

The degree of acetylation (DA) and the crystallinity of chitin molecules affect the solubility in common solvents. Reducing the acetylation level in chitosan ensures the presemce of free amino groups, which can be easily protonated in an acid environment, making chitosan water soluble below pH about 6.5 (Krajewska, 2004; Lavorgna et al., 2010). In acid conditions, when the amino groups are protonated (Fig. 2), chitosan becomes a soluble polycation (Chivrac et al., 2009). The presemce of amino groups make chitosan a cationic polyelectrolyte (pKa ≈ 6.5), one of the few found in nature. Soluble chitosan is heavely charged on the NH_3+ groups, it adheres to negatively charged surfaces, aggregates with polyanionic compounds, and chelates heavy metal ions. These characteristics offer extraordinary potential in a broad spectrum of chitosan applications.

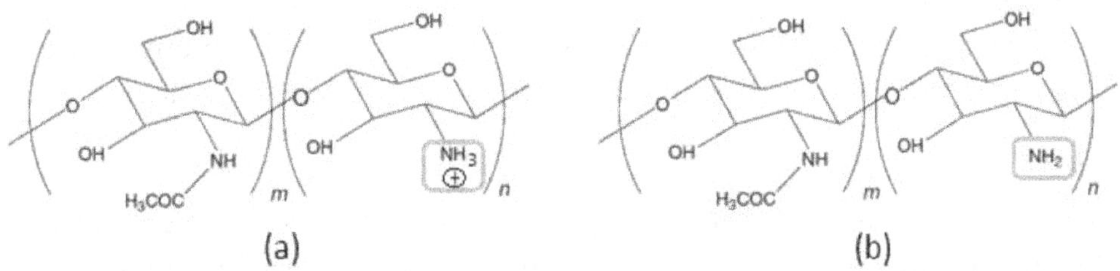

Fig. 2. Schematic illustration of chitosan: (a) at low pH (less than about 6.5), chitosan's amine groups become protonated (polycation); (b) at higher pH (above about 6.5), chitosan's aminegroups are deprontonated and reactive.

Increasingly over the last decade chitosan-based materials have been examined and a number of potential products have been developed for areas such as wastewater treatment (removal of heavy metal ions, ßocculation/coagulation of dyes and proteins, membrane puriÞcation processes), the food industry (anticholesterol and fat binding, preservative, packaging material, animal feed additive), agriculture (seed and fertilizer coating, controlled agrochemical release), pulp and paper industry (surface treatment, photographic paper), cosmetic sand toiletries (moisturizer, body creams, bath lotion) (No et al., 2000). Owing to the unparalleled biological properties, the most exciting uses of chitosan-based materials are in the area of medicine and biotechnology. Medicine takes advange of its biocompatibility, biodegradability to harmless products, nontoxicity, physiological inertness, remarkable afÞnity to proteins, hemostatic, fungistatic, antitumoral and anticholesteremic properties; it may be in drug delivery vehicles, drug controlled release systems, artiÞcial cells, wound healing ointments and dressings, haemodialysis membranes, contact lenses, artiÞcial skin, surgical sutures and for tissue engineering. In biotechnology they may Þnd application as chromatographic matrices, membranes for membrane separations, and notably as enzyme/cell immobilization supports (Felt et al., 2000; Krajewska, 2004). The current interest in medical applications of chitosan is easily understood.

Even though a number of potential products have been developed using chitosan-based materials, the tensile properties of pristine chitosan Þlms are poor (due to its crystallinity). Thermal stability, hardness, gas barrier properties and bacteriostatic activity frequently are not good enough to meet the wide ranges of demanding applications. Thus, modiÞcation (chemical modiÞcation, blending and graft copolymerization) of chitosan has gained importance as means of tailoring the material to the desired properties. In this context, synthesis of nanocomposites with layered silicate loadings was proposed as a novel approach to modify some of the properties of chitosan, including mechanical and thermal behavior (Wang et al., 2005; Wu and Wu, 2006), solubility and swelling properties in acidic media (Pongjanyakul et al., 2005), antimicrobial activity (Han et al., 2010; Wang et al., 2006) and bioadhesion (Pongjanyakul and Suksri, 2009). Chemical structure of chitosan containing multiple functional groups (hydroxyl, carbonyl, carboxyl, amine, amide) creates new possibilities for bonding chitosan to clays.

1.2 Clays

Clays are fine-grained, sedimentary rocks originated from the hydrothermal weathering volcanic volcanic ashes in akaline lakes and seas. As such, clays are classified based on their stratigraphic position, location, and mineral content. Clays contain minerals of definite crystaline structure and elementary composition, some as main components, many as impurities, which usually include organic matter in the form of humic acids. Notwithstanding the fundamental difference between clay and clay mineral, both terms are sometimes used as indistinctly, especially in the frequent occasions in which the clay has a single principal mineral component; in this sense, the clay is considered as the impure mineral and the mineral as the purified clay (Utracki, 2004).

Clays are classified on the basis of their crystal structure and the amount and locations of elelectric charge (deficit or excess) per unit cell. Crystalline clays range from kaolins, which are relatively uniform in chemical composition, to smectites, which vary in their composition, cation exchange properties, and ability to expand. The most commonly employed smectite clay for the preparation of polymeric nanocomposites is bentonite, whose main mineral component is montmorillonite (Utracki, 2004).

Montmorillonite is the name given to clay found near Montmorillonin in France, whereit was identified by Knight in 1896 (Utracki, 2004). Montmorillonite is a 2:1 layered hydrated aluminosilicate, with a triple-sheet sandwich structure consisting of a central, hydrous alumina octahedral sheet, bonded to two silica tetrahedral sheets by shared oxygen ions (Fig. 3). The unit cell of this ideal structure has a composition $[Al_2(OH)_2(Si_2O_5)_2]_2$ with a molar mass of 720 g/mol. Isomorphic substitution of Al^{3+} in the octahedral sheets by Mg^{2+} (less commonly Fe^{2+}, Mn_2^+ , and other) and, less frequently, of Si^{4+} by Al^{3+} in the tetrahedral sheet, results in a net negative charge on the crystaline layer, wich is compensated by the presemce of cations, such as Na+, K+, CA_2+, or Mg2+, sorbed between the layers and surrounding the edges. An idealised montmorillonite has 0.67 units of negative charge per unit cell, in other words, it behaves as a weak acid. These loosely held cations do not belong to the crystal structure and can be readily exchanged by other cations, organic or inorganic. The cation exchange capacity (CEC) of montmorillonite ranges from 0.8 to 1.2 meq/g of Air-dried clay, resulting in 0.6–0.9 exchangeable cations per unit cell. The layers organize themselves to form stacks with a regular gap between them, the interlayer space or gallery. The electrostatic and Van der Waals forces holding the layers together are relatively weak, and the interlayer distance varies depending on the radius of the cation present and its degree of hydration. In general, the smaller the cation and the lower its charge, the higher the clay swells in water or alcohols. For montmorillonite, the swelling capacity decreases depending on the cation chemical type according to the following trend: Li+ > Na+ > CA_2+ > Fe^{2+} > K+ (Powell et al., 1998; Tettenhorst et al., 1962, as cited in Chivrac et al., 2009). The distance between two platelets of the primary particle, called inter-layer spacing or d-spacing (d001), depends on the silicate type, on the type of the counter-cation, and on the hydration state. For instance, d001 = 0.96 nm for anhydrous sodium montmorillonite, but d001 = 1.2-1.4 nm in usual, partially hydrated conditions, as determined by x-ray diffraction techniques (Utracki, 2004).

Commercial montmorillonite is available as a powder of about 8 μm particle size, each particle containing about 3000 platelets. Montmorillonite exhibits enhanced gel strength, mucoadhesive capability to cross the gastrointestinal barrier and adsorb bacterial and metabolic toxins such as steroidal metabolites. Because of these advantages in biomedical applications, it is sometimes called a medical clay. Bentonite (named after Ford Benton, Wyoming) is rich in montmorillonite (usually more than 80%) (Utracki, 2004; Wei et al., 2009; Holzer et al., 2010; Li et al., 2010). Its color varies from white to yellow, to olive green, to brown. The names bentonite and montmorillonite are often used interchangeably. However, the terms represent materials with different degrees of purity. Bentonite is the ore that comprises montmorillonite, inessentials minerals and others impurities. Beyond quartz, kaolinite, and many other minerals often present in minute proportions (feldspar, calcite, dolomite, muscovite, chlorite, hematite, etc), organic matter is present in bentonites as intrinsic impurities composed predominantly of humic substances (Bolto et al., 2001). Since competitive reactions can take place between the organic matter present in the bentonite and the chitosan, the extent of intercalation and polymer/clay interactions can be affected. Purification capable of removing of organic matter from bentonites before intercalation is fundamental.

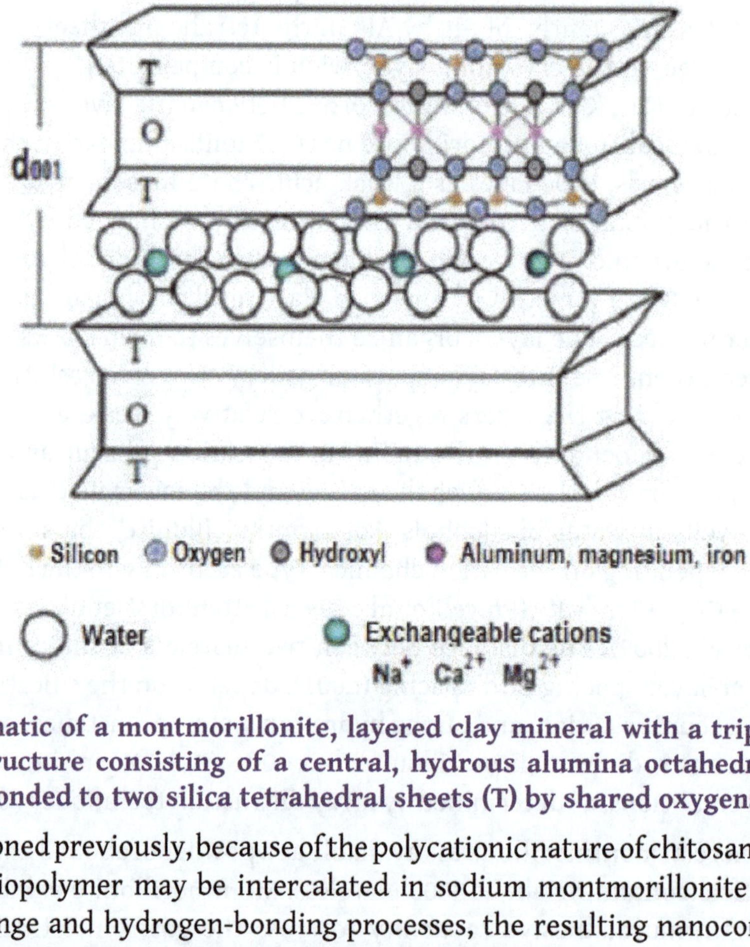

Fig. 3. Schematic of a montmorillonite, layered clay mineral with a triple-sheet sandwich structure consisting of a central, hydrous alumina octahedral sheet (O), bonded to two silica tetrahedral sheets (T) by shared oxygens.

As mentioned previously, because of the polycationic nature of chitosan in acidic media, the biopolymer may be intercalated in sodium montmorillonite through cation exchange and hydrogen-bonding processes, the resulting nanocomposites showing interesting structural and functional properties.

Chitosan/clay nanocomposites represent an innovative and promising class of materials. Potential biomedical applications of chitosan/clay nanocomposites include: the intercalation of cationic chitosan in the expandable aluminosilicate structure of the clay is expected to affect the binding of cationic drugs by anionic clay; the solubility of chitosan at the low pH of gastric ßuid may decrease the premature release of drugs in the gastric environment; cationic chitosan may result in the efÞcient transport of negatively charged drugs; the presemce of reactive amine groups on chitosan may provide ligand attachment sites for targeted drug delivery; etc. The limited solubility of a chitosan/clay nanocomposite drug carriers at gastric pH offers signiÞcant advantages for colon-speciÞc delivery of drugs that may destroyed in the acidic gastric environment or by the presemce of gastric digestive enzymes. Furthermore, the mucoadhesive properties of chitosan may enhance the bioavailability of drugs in the gastrointestional tract.

Many actual applications of chitosan/clay nanocomposites are reported in the literature. Darder et al., 2005 prepared chitosan/montmorillonite nanocomposites and used them in potentiometric sensors for anion detection. Gecol et al., 2006 investigated the removal of tungsten from water using chitosan coated montmorillonite biosorbents. Chang and Juang, 2004 studied the adsorption of tannic acid, humic acid, and dyes from water using chitosan/activated clay composites. An and Dultz, 2007 reported the adsorption of tannic acid on chitosan-montmorillonite as well Pongjanyakul et al., 2005; Wang et al., 2005; Wu and Wu, 2006; Günister et al., 2007; Khunawattanakul et al., 2008; Pongjanyakul & Suksri, 2009. Darder et al. , 2005 synthesized functional chitosan/MMT nanocomposites, successfully used in the development of bulk modified electrodes. Wang et al., 2005 reported the effect of acetic acid residue and MMT loading in the nanocomposites.

However, there are few reports on chitosan/bentonite nanocomposites (Yang & Chen, 2007; Zhang et al., 2009; Wan Ngah et al., 2010). The physical properties and biological response of chitosan strongly depend on the starting materials and nanocomposite preparation conditions. In the present study chitosan/clay nanocomposites were prepared using two kinds of clay and different chitosan/clay ratios, to evaluate how these variables affect the dispersion of clay particles into the chitosan matrix. The samples obtained were characterized by infrared spectroscopy, x-ray diffraction, and mechanical (tensile) properties.

2. EXPERIMENTAL

2.1 Materials

Chitosan was supplied by Polymar (Fortaleza, CE, Brazil) and used without purification. The chitosan was obtained by deacetylation of chitin from crab shells, with a degree of deacetylation of 86.7%. Sodium bentonite (Argel 35) was provided by Bentonit União Nordeste (Campina Grande, PB, Brazil). The clay, coded BNT, was purified according to procedure reported elsewhere (Araujo et al., 2007); the cation exchange capacity (CEC) of the purified bentonite was 0.92 meq/g (Leite et al., 2010). Sodium montmorillonite (Cloisite Na+), coded MMT, with a CEC of 0.90 meq/g was supplied by Southern Clay Products (Gonzalez, TX, USA). Both of the clays, purified sodium bentonite (BNT) and sodium montmorillonite (MMT), were screened to 200 mesh size before mixed with chitosan.

2.2 Preparation of chitosan films

Chitosan solutions were prepared by dissolving chitosan in a 1% aqueous acetic acid solution at a concentration of 1 wt% under continuous stirring at 45°C for 2 h followed by vacuum filtering to remove the insoluble residue. This solution was cast into Petri dishes (radius ~ 12 cm) and dried at 50°C for 20 h to evaporate the solvent and form the films. The dried films were soaked with an aqueous solution of 1 M NaOH for 30 min to remove residual acetic acid, followed by rinsing with distilled water to neutralize, and then dried at room temperature. The chitosan films were coded Cs.

2.3 Preparation of the chitosan/clay films

Chitosan/clay films were prepared by a casting/solvent evaporation technique. Firstly, 1% chitosan solutions were adjusted to pH = 4.9 by addition a 1M sodium hydroxide solution to form the NH_3+ groups in the chitosan structure. Given that the primary amine group in the structure of the chitosan has a pKa = 6.3, 95% of the groups amine will be protonated at the final pH = 5. of the chitosan/clay mixture (Darder et al., 2005). After, the chitosan solution was slowly added to a 1 wt% clay suspension followed by stirring at 53 ± 2°C for 4 h to obtain the films with chitosan/clay mass ratios of 1:1, 5:1 and 10:1. This chitosan/clay solution was cast into Petri dishes and dried at 50°C for 20 h to evaporate the solvent and form the films. Following the same procedure used for chitosan films, the dried films were soaked into an aqueous solution of 1 M NaOH for 30 min to remove residual acetic acid, followed by rinsing in distilled water to neutral and then dried at room temperature. The chitosan/purified sodium bentonite and chitosan/sodium montmorillonite films prepared from chitosan/clay mass ratio of the 1:1, 5:1 and 10:1 were denoted Cs1:BNT1; Cs5:BNT1; Cs10/BNT1 and Cs1:MMT1; Cs5:MMT1; Cs10:MMT1, respectively.

2.4 Characterization

Although the clay dispersion process is usually followed by x-ray diffraction and transmission electron microscopy, infrared spectroscopic techniques may shed light into the complex chemical and physical interactions involved, helping scientists and technologists to understand the mechanisms of nanocomposite formation, and leading to better products and production methods in the laboratory and the industrial plant. Furthermore, infrared spectroscopic is relatively rapid, is a common instrument found in most research laboratories, sample purity is not as critical and the method can be used with insoluble samples. This gives infrared spectroscopic methods an advantage over other methods, which require elaborate and time-consuming sample preparation.

Fourier transform infrared spectra of the chitosan films and the chitosan/clay films were collected using a Spectrum 400 Perkin Elmer operating in the range of 400-4000 cm^{-1} at a resolution of 4 cm^{-1}.

XRD patterns were obtained using a Shimadzu XRD-6000 diffractometer with $Cu_{K\alpha}$ radiation (λ = 0.154 nm, 40 kV, 30 mA) at room temperature. XRD scans were performed on sodium montmorillonite and purified sodium bentonite, chitosan films and chitosan/clay films with a 2q range between 1.5º and 12.0º, at a scanning rate of 1º/min and a scanning step of 0.02º. The basal spacing (d001) value of the layered silicates and the chitosan/layer silicate films were computed using Bragg's law.

Mechanical properties of chitosan films and chitosan/clay films were measured following ASTM D882 standard procedures. The films were cut in rectangular strips (80 ′ 10 mm) and the thickness of each sample was measured at three different locations and averaged. The tensile strength (TS), elastic modulus (EM) and elongation at break (E) of the samples were determined using a universal testing machine (EMIC, model DL1000) fitted with a load cell of 50 N, with initial gauge separation of 50 mm and a stretching speed of 5 mm/min. Reported results were the average of five independent measurements.

3. RESULTS AND DISCUSSION

3.1 Infrared spectroscopy (FTIR)

Fig. 4 shows FTIR spectra in the 4000–400 cm^{-1} wave number range for sodium montmorillonite (MMT), purified sodium bentonite (BNT), chitosan film (Cs), chitosan/MMT and chitosan/BNT films with 1:1, 5:1 and 10:1 chitosan/clay ratios, respectively (Cs1:MMT1; Cs5:MMT1; Cs10:MMT1 and Cs1:BNT1; Cs5:BNT1; Cs10:BNT1).

In the clay spectra (MMT and BNT), the characteristic absorption band at ~ 3622 cm^{-1} [nOH] is assigned to the stretching vibration of AlOH and SiOH; at ~ 3416 cm^{-1} [nOH] to the stretching vibration of H_2O; at ~ 1628 cm^{-1} [d_{HOH}] to the bending vibration of H_2O; at ~1118 cm^{-1} and at ~ 980 cm^{-1}[nSi-O] to the stretching vibration of SiO; at~913 cm^{-1} [dAl-Al-OH] to the bending vibration of AlAlOH; at~882 cm^{-1} [dAl-Fe-OH] to the bending vibration of AlFeOH; and at~841 cm^{-1} [dAl-Mg-OH] to the bending vibration of AlMgOH (Awad et al., 2004; Bora et al., 2000; Leite et al., 2010; Madejová, 2003; Xu et al., 2009).

In order to fully characterize the starting materials, a spectrum of pure chitosan was also recorded. The main bands appearing in that spectrum were due to stretching vibrations of OH groups in the range from 3750 cm^{-1} to 3000 cm^{-1}, which are overlapped to the stretching vibration of N-H; and C–H bond in $-CH_2$ (n_1 = 2920 cm^{-1}) and –CH3 (n_2 = 2875 cm^{-1})groups, respectively. Bending vibrations of methylene and methyl groups were also visible at n= 1375 cm^{-1} and n= 1426 cm^{-1}, respectively (Mano et al., 2003). Absorption in the range of 1680–1480 cm^{-1} was related to the vibrations of carbonyl bonds (C=O) of the amide group CONHR (secondary amide, n_1 = 1645 cm^{-1}) and to the vibrations of protonated amine group (d_{NH3} , n_2 = 1574 cm^{-1}) (Marchessault et al., 2006). Absorption in the range from 1160 cm^{-1} to 1000 cm^{-1} has been attributed to vibrations of CO group (Xu et al, 2005). The band located near n= 1150 cm^{-1} is related to asymmetric vibrations of CO in the oxygen bridge resulting from deacetylation of chitosan. The bands near 1080–1025 cm^{-1} are attributed to nCO of the ring COH, COC and CH_2OH. The small peak at ~890 cm^{-1} corresponds to wagging of the saccharide structure of chitosan (Darder et al., 2003; Paluszkiewicz et al., 2011; Yuan et al., 2010).The assigned characteristic FTIR absorption bands of clay (MMT and BNT) and chitosan film (Cs) derived from Fig. 4 are summarized in Table 1.

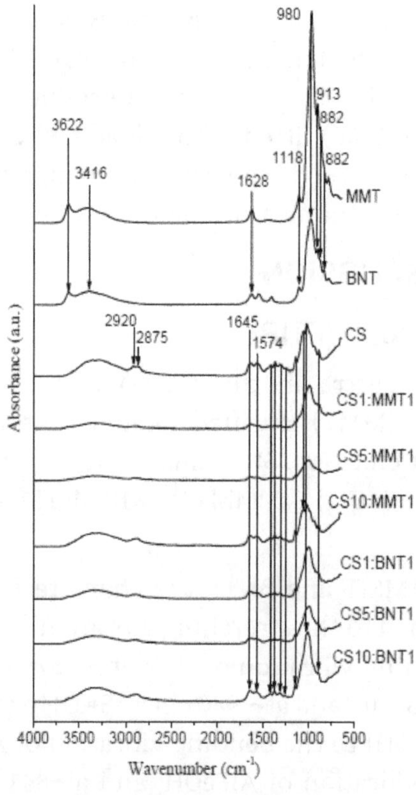

Fig. 4. FTIR spectra in the 4000–400 cm-1 wave number range for sodium montmorillonite (MMT), purified sodium bentonite (BNT), chitosan film (Cs), chitosan/MMT and chitosan/BNT films with 1:1, 5:1 and 10:1 chitosan/clay ratios, respectively (Cs1:MMT1; Cs5:MMT1; Cs10:MMT1 and Cs1:BNT1; Cs5:BNT1; Cs10:BNT1).

FTIR was also used to study the polymer/clay interaction, Since a shift in the d_{NH3} vibration may be expected when – NH_3^+ groups interact electrostatically with the negatively charged sites of the clay. In fact, a shift of the dNH band towards a lower frequency is observed in all the chitosan/clay films (Cs1:MMT1; Cs5:MMT1; Cs10:MMT1 and Cs1:BNT1; Cs5:BNT1; Cs10:BNT1) as show in Fig. 5 (spectra of Fig. 4 in the 1800–1400 cm^{-1} wavenumber range) and Table 2. Nevertheless, this shift is higher for chitosan/clay films with the lowest amounts of chitosan (Cs1:MMT1; Cs5:MMT1 and Cs1:BNT1; Cs5:BNT1), while the chitosan/clay films with the highest amounts of biopolymer (Cs10:MMT1 and Cs10:BNT1) show a frequency value that trends to that observed in the films of pure chitosan (Cs). This fact may be related to the – NH_3^+ groups that do not interact electrostatically with the clay substrate (Fig.6). Besides, the intensity of the dNH band also increases for higher amounts of intercalated chitosan (Cs10:MMT1 and Cs10:BNT1) (Fig.5). The secondary amide band (n_1) at 1645 cm^{-1} of chitosan is overlapped with the d_{HOH} bending vibration band at 1628 cm^{-1} of the water molecules associated to the chitosan/clay films, which are present as in the starting clay, as expected for a biopolymer with high water retention capability (Darder et al., 2003; Darder et al., 2005; Han et al., 2010; Paluszkiewicz et al., 2011; Tan et al., 2007; Wang & Wang, 2007). Comparing the spectra of chitosan/MMT with the spectra of chitosan/BNT we can observe that the interaction of the chitosan with both clays (MMT and BNT) is similar.

Table 1. Assignment of FTIR spectra of clays and chitosan derived from Fig. 4.

Sample	IR band (cm^{-1})	Description*
Clay (MMT and BNT)	3622	ν(O-H) for Al-OH and Si-OH
	3416	ν(O-H) for H-O-H
	1628	δ(HOH) for H-O-H
	1118 and 980	ν(Si-O) out of plane
	913	δ(AlAlOH)
	882	δ(AlFeOH)
	841	δ(AlMgOH)
Chitosan film (Cs)	3750-3000	ν(O-H) overlapped to the νS(N-H)
	2920	ν_{as}(C-H)
	2875	ν_{s}(C-H)
	1645	ν(-C=O) secondaryamide
	1574	ν(-C=O)protonated amine
	1426, 1375	δ(C-H)
	1313	ν_{s}(-CH3) tertiary amide
	1261	ν(C-O-H)
	1150, 1065, 1024	ν_{as}(C-O-C) andns(C-O-C)
	890	ω(C-H)

*ν = stretching vibration; νS = symmetric stretching vibration;
ν_{as} = asymmetric stretching vibration; ω = wagging.

Table 2. Frequency values of vibrational bands corresponding to the water molecules associated with the clay (MMT and BNT) and with the protonated amine group in the chitosan chain.

Sample	δ_{HOH} (cm^{-1})	δ_{NH3} (cm^{-1})
Clay (MMT and BNT)	1628*	-
Chitosan film (Cs)	1645	1574
Chitosan/MMT (Cs1:MMT1)	1638	1555
Chitosan/MMT (Cs5:MMT1)	1640	1558
Chitosan/MMT (Cs10:MMT1)	1645	1570
Chitosan/BNT (Cs1:BNT1)	1640	1557
Chitosan/BNT (Cs5:BNT1)	1641	1558
Chitosan/BNT (Cs10:BNT1)	1643	1570

*stretching band of secondary amide ν(-C=O).

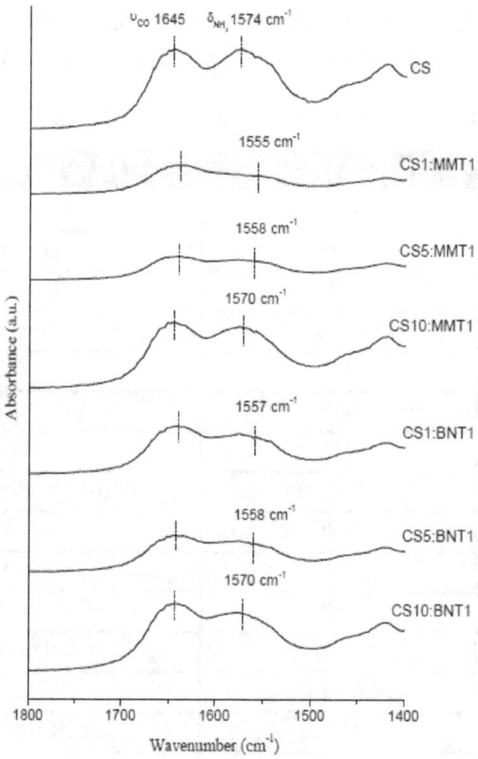

Fig. 5. IR spectra of Fig. 4 in the 1800–1400 cm^{-1} wavenumber range of chitosan film (Cs), chitosan/MMT and chitosan/BNT films prepared from 1:1, 5:1 and 10:1 chitosan–clay ratios (Cs1:MMT1; Cs5:MMT1; Cs10:MMT1 and Cs1:BNT1; Cs5:BNT1; Cs10:BNT1).

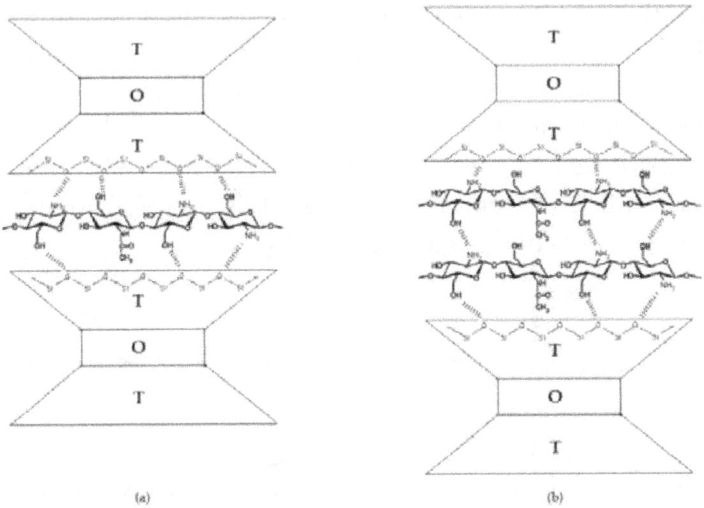

Fig. 6. Schematic illustration of the intercalation of chitosan layers into the clay inter-layer spacing for films (a) with the lowest amounts of chitosan (Cs1:MMT1 and Cs1:BNT1) and (b) with the highest amounts of biopolymer (Cs10:MMT1, Cs5:MMT1 and Cs10:BNT1, Cs5:BNT1).

3.2 X ray diffraction analysis (XRD)

XRD is the principal method that has been used to examine the distribution/ dispersion of the clay platelet stacks in the polymer matrix (Utracki, 2004). Depending on the relative distribution/dispersion of the stacks, three types of nanocomposites can be described: intercalated nanocomposites, where polymer chains are interleaved with silicate layers, resulting in a well ordered mutilayer morphology built up with alternating polymer and inorganic sheets; flocculated nanocomposites, where intercalated clay layers are sometimes bonded by hydroxylated edge-edge interactions, and exfoliated/delaminated nanocomposites, where individual clay layers are completely and homogenously dispersed in the polymer matrix (Wang et al., 2005).

FTIR data indicate that chitosan was intercalated into the MMT and BNT interlayers. However, to confirm the FTIR results, the MMT and BNT clays, as well as, chitosan/MMT and chitosan/BNT films prepared from 1:1, 5:1 and 10:1 chitosan/ clay ratios, respectively (Cs1:MMT1; Cs5:MMT1; Cs10:MMT1 and Cs1:BNT1; Cs5:BNT1; Cs10:BNT1) were analyzed by XRD and the results are shown in Figs. 7-9.

The XRD patterns of the MMT (Fig. 7) shows a reflection peak at about $2\theta = 5.9°$, corresponding to a basal spacing (d001) of 1.50 nm. After incorporating MMT within Cs, with Cs/MMT 1:1 ratio, the basal plane of MMT at $2\theta = 5.9°$ disappears, substituted by a new weakened broad peak at around $2\theta = 2.8° - 3.7°$ [Cs1:MMT1 (1), Cs1:MMT1 (2) and Cs1:MMT1 (3)]. The shift of the basal reflection of MMT to lower angle indicates the formation of an intercalated nanostructure, while the peak broadening and intensity decreases most likely indicate the disordered intercalated or exfoliated structure (Utracki, 2004; Wang et al., 2005). Similar behavior was observed for Cs/BNT (Fig. 7), i.e. the basal plane of BNT at 2θ = 6.3° disappears, substituted by a new weakened broad peak at around 2θ = 3.0° - 3.9° [Cs1:BNT1 (1), Cs1:BNT1 (2) and Cs1:BNT1 (3)]. It is suggested that the MMT and the BNT form intercalated and flocculated structures.

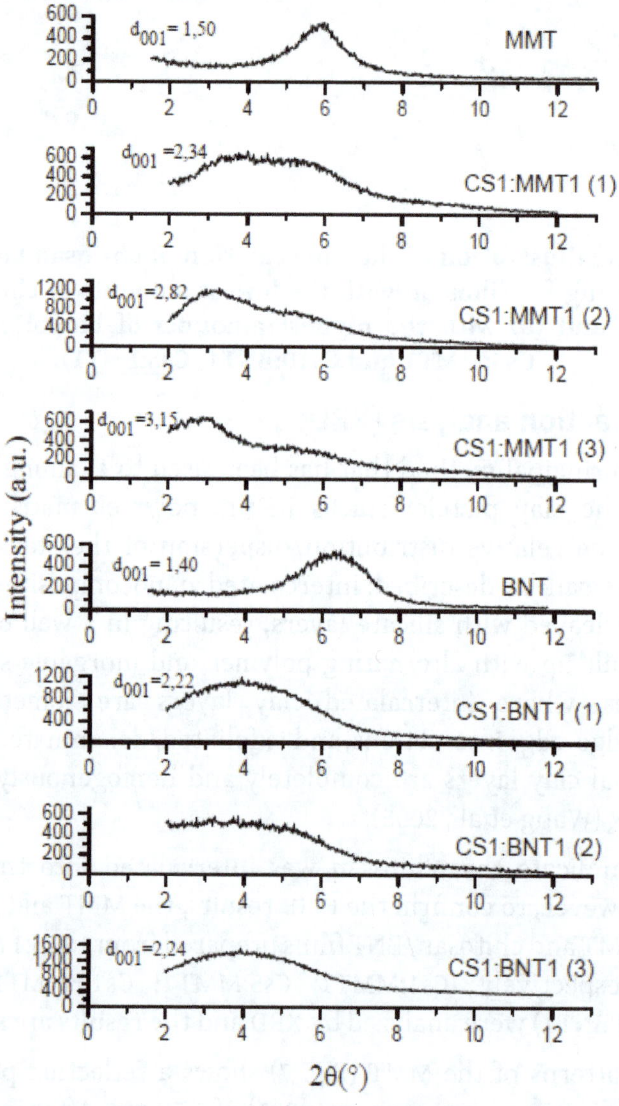

Fig. 7. XRD pattern of sodium montmorillonite (MMT), purified sodium bentonite (BNT), chitosan/MMT and chitosan/BNT films prepared from 1:1 chitosan/clay ratios in triplicate [Cs1:MMT1 (1), Cs1:MMT1 (2), Cs1:MMT1 (3) and Cs1:BNT1 (1), Cs1:BNT1 (2), Cs1:BNT1 (3)].

Fig. 8 shows he XRD patterns of the MMT, BNT, chitosan/MMT and chitosan/BNT films prepared from 5:1 chitosan/clay ratios in triplicate [Cs5:MMT1 (1), Cs5:MMT1 (2), Cs5:MMT1 (3) and Cs5:BNT1 (1), Cs5:BNT1 (2), Cs5:BNT1 (3)]. After incorporating MMT within Cs, with Cs/MMT 5:1 ratio, the basal plane of MMT at 2θ = 5.9° disappears, substituted by a new weakened broad peak at around 2θ = 2.0° - 3.8° [Cs5:MMT1 (1), Cs5:MMT1 (2) and Cs5:MMT1 (3)]. In this case the 2θ values were smaller than the values observed for chitosan/MMT prepared from 1:1 ratio (Fig. 7), indicating that exfoliated/delaminated nanocomposites were be obtained. In the same way exfoliated/delaminated nanocomposites are probably obtained for Chitosan/BNT [Cs5:BNT1 (1), Cs5:BNT1 (2) and Cs5:BNT1 (3)].

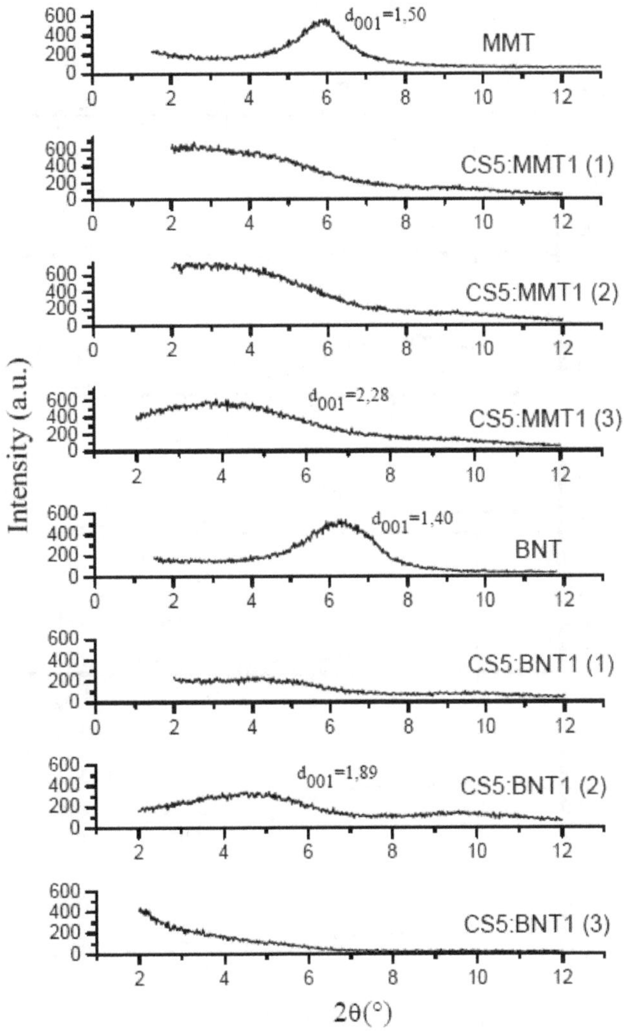

Fig. 8. XRD pattern of MMT, BNT, chitosan/MMT and chitosan/BNT films prepared from 5:1 chitosan/clay ratios in triplicate [Cs5:MMT1 (1), Cs5:MMT1 (2), Cs5:MMT1 (3) and Cs5:BNT1 (1), Cs5:BNT1 (2), Cs5:BNT1 (3)].

Fig. 9 shows he XRD patterns of the MMT, BNT, chitosan/MMT and chitosan/BNT films prepared from 10:1 chitosan/clay ratios in triplicate [Cs10:MMT1 (1), Cs10:MMT1 (2), Cs10:MMT1 (3) and Cs10:BNT1 (1), Cs10:BNT1 (2), Cs10:BNT1 (3)]. With increasing Cs content, the 2θ of (001) peak becomes lower and it is not possible to calculate the interlayer distance for each nanocomposite in the broad peaks, indicating that the MMT and the BNT forms intercalated and exfoliated structures. In all probability, exfoliated/delaminated structures were obtained in this case.

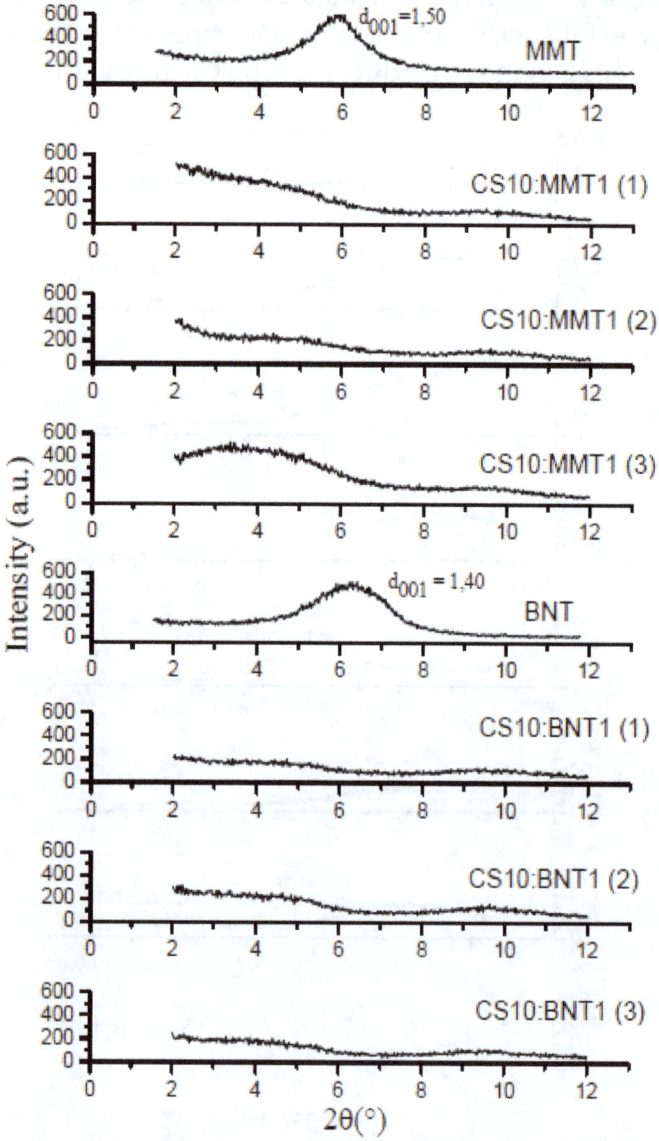

Fig. 9. XRD pattern of MMT, BNT, chitosan/MMT and chitosan/BNT films prepared from 10:1 chitosan/clay ratios in triplicate [Cs10:MMT1 (1), Cs10:MMT1 (2), Cs10:MMT1 (3) and Cs10:BNT1 (1), Cs10:BNT1 (2), Cs10:BNT1 (3)].

In summary, the morphology of the nanocomposites was affected by chitosan/clay rations. On the base of XRD patterns, it is suggested that the MMT and the BNT forms intercalated and exfoliated structures at higher Cs content (Cs10:MMT1, Cs5:MMT1 and Cs10:BNT1, Cs5:BNT1), while decreasing the Cs content (Cs1:MMT1 and Cs1:BNT1), clay layers (MMT and BNT) form intercalated and flocculated structures. According Wang et al., 2005, the formation of flocculated structure in Cs/clay nanocomposites can be due to the hydroxylated edge-edge interactions of the clay layers. Since one chitosan unit possesses one amino and two hydroxyl functional groups, these groups can form hydrogen bonds with the clay hydroxyl edge groups, which leads to the strong interactions between matrix and clay layers (Fig.6a) and corroborate FTIR results. This strong interaction is believed to be the main driving force for the assembly of MMT and BNT in the Cs matrix to form flocculated structures.

3.3 Mechanical properties

Tensile properties of the chitosan film (Cs) and chitosan/clay films prepared from 1:1, 5:1 and 10:1 chitosan/clay ratios, respectively (Cs1:MMT1; Cs5:MMT1; Cs10:MMT1 and Cs1:BNT1; Cs5:BNT1; Cs10:BNT1) are collected in Table 3. The tensile strength (TS) and elastic modulus (EM) of chitosan films increase by the formation of nanocomposites, particularly for chitosan/clay prepared from 5:1 rations. The increase in the TS and EM of such nanocomposite films can be attributed to the high rigidity and aspect ratio of the nano- clay as well as the high affinity between the biopolymer and the clay. On the other hand, the chitosan/clay nanocomposites have shown significant decrease in elongation at break (E_b). This reduction can be attributed to the restricted mobility of macromolecular chains.

Table 3. Tensile properties of chitosan and chitosan/clay films.

Sample	TS (MPa)	EM (MPa)	E_b (%)
Chitosan film (Cs)	44.5 + 4.5	1774 + 63	7.7 + 0.5
Chitosan/MMT (Cs1:MMT1)	84.9 + 3.7	5214 + 112	3.3 + 0.5
Chitosan/MMT (Cs5:MMT1)	79.1 + 1.1	4449+ 329	4.6 + 0.7
Chitosan/MMT (Cs10:MMT1)	68.5 + 1.4	3536 + 180	4.6 + 0.8
Chitosan/BNT (Cs1:BNT1)	49.6 + 4.9	4075 + 73	2.4 + 0.9
Chitosan/BNT (Cs5:BNT1)	62.1 + 4.5	3106 + 50	6.8 + 0.8
Chitosan/BNT (Cs10:BNT1)	40.4 + 1.8	2421 + 87	5.9 + 0.8
(TS = tensile strength, EM = elastic modulus (EM), E_b = elongation at break)			

4. CONCLUSIONS

In this study chitosan/clay nanocomposites were successfully prepared by the solution intercalation process. It was found that clay dispersion is affected by the kind of clay and the chitosan/clay ratio. Since the nanocomposites prepared with purified bentonite (BNT) showed similar behavior to that prepared with montmorillonite, less expensive bentonite may be employed in the preparation of chitosan/clay nanocomposites.

The intercalation of the cationic biopolymer chitosan into layered silicate clays (montmorillonite and bentonite) through a cation exchange process results in nanocomposites with interesting structural and functional properties. The clay reduces the film-forming capability of chitosan leading to compact, robust, and handy three- dimensional nanocomposites. The techniques employed in the characterization of the nanocomposites, infrared spectroscopy, x-ray diffraction, and mechanical properties in tension, confirm the high affinity between the clay substrate and the biopolymer, as well as the special arrangement of chitosan as a bilayer when the biopolymer amount exceeds the cation exchange capacity of the clay. The intercalation of the first layer of chitosan takes place mainly by electrostatic interactions between positive ammonium groups in the chitosan chain and negative sites in the clay. In contrast, hydrogen bonds between amino and hydroxyl groups of chitosan and the clay substrate are established in the adsorption of the second layer.

5. REFERENCES

1. An JH., Dultz S. (2007). Adsorption of tannic acid on chitosan–montmorillonite as a function of pH and surface charge properties. Applied Clay Science. Vol. 36, pp. 256– 64.
2. Araujo, P. E. R.; Araújo, S. S.; Raposo, C. M. O.; Silva, S. M. L. (2007). 23th Polymer Processing. Society Annual Meeting. Vol. 23, pp.1.
3. Awad, W. H.; Gilman, J. W.; Nyden, M.; Harris, R. H.; Sutto, T. E.; Callahan, J.; Trulove, P. C.; Delong, H. C.; Fox, D. M. (2004). Thermal degradation studies of alkyl- imidazolium salts and their application in nanocompósitos. Thermochimica Acta. Vol. 409, p. 3-11.
4. Bolto, B.; Dixon, D.; Eldridge, R.; King, S. (2001). Cationic polymer and clay or metal oxide combinations for natural organic matter removal. Water Research.Vol. 35, p. 2669- 2676.
5. Bora, M.; Ganguli, J. N.; Dutta, D. K. (2000). Thermal and spectroscopic studies on the decomposition of [Ni{di(2-aminoethyl)amine}2]- and [Ni(2,2":6",2"-terpyridine)2]- Montmorillonite intercalated composites Thermochimica Acta. Vol. 346, p.169-175.

6. Chang, M. Y.; Juang, R. S. (2004). Adsorption of tannic acid, humic acid and dyes from water using the composite of chitosan and activated clay. Journal of Colloid and Interface Science. Vol. 278, pp.18–25.

7. Chivrac, F.; Pollet, E.; Avérous, L. (2009). Progress in nano biocomposites based on polysaccharides and nanoclays. Materials Science and Engineering R. Vol. 67, pp. 1–17.

8. Darder, M.; Colilla, M.; Ruiz-hitzky, E. (2003). Biopolymer-Clay Nanocomposites Based on Chitosan Intercalated in Montmorillonite. ChemicalMaterials. vol. 15, pp. 3774–3780.

9. Darder, M.; Colilla, M.; Ruiz-Hitzky, E. (2005). Chitosan clay nanocomposites: application as Electrochemical sensors. Applied Clay Science. Vol.28, pp.199-208.

10. Dash, M.; Chiellini, F.; Ottenbrite, R.M.; Chiellini E. (2011). Chitosan—A versatile semi- synthetic polymer in biomedical applications. Progress in Polymer Science, Vol. 36, pp. 981–1014.

11. E. S. Costa; H. S. Mansur. (2008). Preparação e caracterização de blendas de quitosana/poli (álcool vinílico) reticuladas quimicamente com glutaraldeído para aplicação em engenharia de tecido. Química Nova. Vol.31, pp.1460-1466.

12. Felt, O.; Carrel, A.; Baehni, P.; Buri, P.; Gurny, R. (2000) Chitosan as tear substitute: a wetting agent endowed with antimicrobial efficacy. Journal of Ocular Pharmacology and TherapeutiCs. Vol. 16, pp. 261–270.

13. Gecol, H.; Miakatsindila, P.; Ergican, E.; Hiibel, S. R. (2006). Biopolymer coated clay particles for the adsorption of tungsten from water. Desalination. Vol. 197, pp. 165–78.

14. Günister, E., Pestreli, D., Ünlü, C.H.; Güngör, N. (2007). Synthesis and characterization of chitosan-MMT biocomposite systems. Carbohydrate Polymers. Vol. 67, pp.358-365.

15. Han, Y.; Lee, S.; Choi, K. H. (2010). Preparation andcharacterizationofchitosan–claynanocompositeswith antimicrobial activity. Journal of Physics and Chemistry of Solids. Vol. 71, pp. 464–467.

16. Holzer, L.; Münch, B.; Rizzi, M.; Wepf, R.; Marschall, P.; Graule, T. (2010). 3D-microstructure analysis of hydrated bentonite with cryo-stabilized pore water. Applied Clay Science. Vol. 47, pp. 330–342.

17. Khunawattanakul , W.; Puttipipatkhachorn , S.; Rades , T.; Pongjanyakul, T. (2008). Chitosan-magnesium aluminum silicate composite dispersions: characterization of rheology, ßocculate size and zeta potential. International Journal of Pharmaceutics. Vol.351, pp. 227–235.

18. Krajewska, B. (2004). Application of chitin and chitosan-based materials for enzyme immobilizations: a review. Enzyme and Microbial Technology. Vol. 35, pp. 126-134.

19. Lavorgna, M.; Piscitelli, F.; Mangiacapra, P.; Buonocore, G. (2010). Study of the combined effect of both clay and glycerol plasticizer on the properties of chitosan Þlms. Carbohydrate Polymers. Vol. 82, pp. 291–298.

20. Leite, I. F.; Soares, A. P. S.; Carvalho, L.H.; Malta, O. M. L.; Raposo, C. M. O.; Silva, S. M. S. (2010). Characterization of pristine and purified organobentonites. Journal of Thermal Analysis and Calorimetry. Vol. 100, pp.563.

21. Li, Q.; Yue, Q. Y.; Sun, H. J.; Su, Y.; Gao, B. Y. (2010). A comparative study on the properties, mechanism and process designs for the adsorption of non-ionic or anionicdyes onto cationic-polymer/bentonite. Journal of Environmental Management. Vol. 91, pp.1601– 1611.

22. Madejová, J. (2003). FTIR techniques in clay mineral studies. Vibrational Spectroscopy. Vol.31, pp.1-10.

23. Mano, J.F.; Koniarova, D.; Reis, R.L. (2003).Journal of Materials Science: Materials in Medicine. Vol. 14, pp.127–135.

24. Marchessault, R.H.; Ravenelle, F.; Zhu, X.X. (2006). Polysaccharides for drug delivery and pharmaceutical applications, American Chemical Society.

25. No, H. K.; Lee, K. S.; Meyers, S. P. (2000). Correlation between physicochemical characteristics and binding capacities of chitosan products. Journal Food Science. Vol. 65, pp. 1134–1137.

26. Pandey, S. Mishra S. B., (2011). Organic–inorganic hybrid of chitosan/organoclay bionanocomposites for hexavalent chromium uptake, Journal of Colloid and Interface Science. Vol. 361, pp. 509–520.

27. Paluszkiewicz, C.; Stodolak, E.; Hasik, M.; Blazewicz, M. (2011). FT-IR study of montmorillonite–chitosan nanocomposite materials, Spectrochimica Acta Part A. Vol. 79, pp. 784–788.

28. Pongjanyakul, T.; Priprem, A.; Puttipipatkhachorn, S. (2005). Investigation of novelalginate- magnesium aluminum silicate microcomposite Þlms for modi Þed-releasetablets. Journal Control. Release. Vol. 107, pp. 343– 356.

29. Pongjanyakul, T.; Suksri, H. (2009). Alginate –magnesium aluminum silicate Þlms forbuccal delivery of nicotine. Colloids and Surfaces B: Biointerfaces. Vol. 74, pp. 103–113.

30. Rinaudo, M. (2006). Chitin and chitosan: Properties and applications. Progress in Polymer Science.Vol. 31, pp. 603-632.

31. Tan, W.; Zhang, Y.; Szeto, Y.; Liao, L. (2007). A novel method to prepare chitosan/montmorillonite nanocompósitos in the presemce of hydroxyl-aluminum oligomeric cations. Composites Science and Technology.

32. Utracki, L. A. (2004). Basic Elements of Polymeric Nanocomposite Technology, In: Clay- containing Polymeric Nanocomposites, pp. 73-96, Rapra Technology Limited, England.

33. Wan Ngah, W. S.; Ariff, N. F. M.; HanaÞah, M. A. K. M. (2010).Preparation, characterization, and environmental application of crosslinked chitosan-coated bentonite for tartrazine adsorption from aqeous solutions. Water, Air and Soil Pollution. Vol. 206, pp.225–236.

34. Wan Ngah, W. S.; Teong, L. C.; HanaÞah, M. A. K. M. (2011). Adsorption of dyes and heavy metal ions by chitosan composites: A review. Carbohydrate Polymers. Vol. 83, pp. 1446–1456.

35. Wang, L.; Wang, A. (2007). Adsorption characteristics of Congo Red onto the chitosan/montmorillonite nanocomposite. Journal of Hazardous Materials. Vol. 147, pp. 979–985.

36. Wang, S. F.; Shen, L.; Tong, Y. J.; Chen, L.; Phang, I. Y.; Lim, P. Q. (2005). Biopolymer chitosan/montmorillonite nanocomposites: Preparation and characterization. Polymer Degradation and Stability. Vol. 90, pp.123-131.

37. Wang, X.; Du, Y.; Yang, J.; Wang, X.; Shi, X.; Hu Y. (2006). Preparation, characterization and antimicrobial activity of chitosan/layered silicate nanocompósitos. Polymer. Vol. 47, pp. 6738-6744.

38. Wei, J. M.; Zhu, R. L.; Zhu, J. X.; Ge, F.; Yuan, P.; He, H. P. (2009). Simultaneous sorption of crystal violet and 2-naphthol to bentonite with different CECs. Journal of Hazardous Materials. Vol. 166, pp.195–199.

39. Wu, T. M.; Wu, C. Y. (2006). Biodegradable poly (lactic acid)/chitosan-modiÞed montmorillonite nanocomposites: preparation and characterization. Polymer Degradation and Stability. Vol. 91, pp. 2198–2204.

40. Xu, X.; Ding, Y.; Qian, Z.; Wang, F.; Wen, B.; Zhou, H.; Zhang, S.; Yang, M. (2009). Degradation of poly(ethylene terephthalate)/clay nanocomposites during melt extrusion: Effect of clay catalysis and chain extension. Polymer Degradation and Stability. Vol.94, pp.113-123.

41. Xu, Y.; Kim, K.; Hanna, M.; Nag, D. (2005). Industrial Crops and Products. Vol. 21, pp. 185- 192.

42. Yang, Y. Q.; Chen, H. J. (2007). Study on the intercalation organic bentonite and its adsorption. Journal of Xinyang Normal University. Vol. 20, pp. 338–340.

43. Yuan, Q.; Shah, J.; Hein, S.; Misra, R.D.K. (2010). Controlled and extended drug release behavior of chitosan-based nanoparticle carrier. Acta Biomaterialia. Vol. 6, pp. 1140– 1148.

44. Zhang, A. C.; Sun, L. S.; Xiang, J.; Hu, S.; Fu, P; Su, S. (2009). Removal of elemental mercury from coal combustion flue gas by bentonite-chitosan and their modifier. Journal of Fuel Chemistry and Technology. Vol. 37, pp. 489–495.

3 Chapter

STRUCTURAL AND OPTICAL BEHAVIOR OF VANADATE-TELLURATE GLASSES CONTAINING PBO OR SM2O3

1. INTRODUCTION

Tellurium oxide based glasses are of scientific and technological interest due to their unique properties such as chemical durability, electrical conductivity, transmission capability, high refractive indices, high dielectric constant and low melting points [1-3].

Tellurate glasses have recently gained wide attention because of their potential as hosts of rare earth elements for the development of fibres and lasers covering all the main telecommunication bands and promising materials for optical switching devices [4, 5]. Recently, tellurate glasses doped with heavy metal oxides or rare earth oxides have received great scientific interest because these oxides can change the optical and physical properties of the tellurate glasses [5].

Vanadium tellurate glasses showed better mechanical and electrical properties due to the V_2O_5 incorporated into the tellurate glass matrix. In the case of V_2O_5 contents below 20mol%, the three-dimensional tellurate network is partially broken by the formation of [TeO3] trigonal pyramidal units, which in turn reduce the glass rigidity and. When the V_2O_5 concentration is above 20mol%, the glass structure changes from the continuous tellurate network to the continuous vanadate network [6].

Due to the large atomic mass and high polarizability of the Pb+2 ions, heavy metal oxide glasses with PbO possess high refractive index, wide infrared Transmittance, and hence they are considered to be promising glass hosts for photonic devices [7, 8]. The special significance of PbO is that it contributes to form stable glasses over a wide range of concentrations due to its dual role as glass modifier and glass former.

Rare-earth ions doped glasses have been prepared and characterized to understand their commercial applications as glass lasers and also in the production of wide variety of other types of optical components [9].

The luminescence spectral properties of rare earth ions such as Eu+3 (4f6) and Tb+3 (4f 8), Sm+3 (4f5) and Dy+3 (4f9) in the heavy-metal borate glasses have shown interesting and encouraging results [10, 11]. Eu+3 and Tb+3 ions have shown prominent emissions (red and green) in the visible wavelength region, while Sm+3 and Dy+3 show strong absorption bands in the NIR range (800–2200 nm) and intense emission bands in the visible region (550–730 nm) [11].

The present work deals with the role of lead and samarium ions in the short-range structural order of the vanadate-tellurate glass network. Lead and samarium-activated vanadate- tellurate glasses have been investigated using infrared spectroscopy and ultraviolet-visible spectroscopy. The main goal is to obtain information about the influence of the radii and concentration of lead and samarium ions on the TeO4/TeO3 and VO_5/VO_4 conversion in vanadate-tellurate glasses and especially to illuminate aspects of the vanadate glass network using DFT calculations.

2. EXPERIMENTAL PROCEDURE

Glasses were prepared by mixing and melting of appropriate amounts of lead (IV) oxide, tellurium oxide (IV), lead (II) oxide or samarium (III) oxide of high purity (99,99%, Aldrich Chemical Co.). Reagents were melted at 8750C for 10minutes and quenched by pouring the melts on stainless steel plates.

The samples were analyzed by means of X-ray diffraction using a XRD-6000 Shimadzu diffractometer, with a monochromator of graphite for the Cu-Kα radiation (λ=1.54Å) at room temperature.

The FT-IR absorption spectra of the glasses in the 370-1100cm^{-1} spectral range were obtained with a JASCO FTIR 6200 spectrometer using the standard KBr pellet disc technique. The spectra were carried out with a standard resolution of 2cm^{-1}.

UV-Visible absorption spectra measurements in the wavelength range of 250-1050nm were performed at room temperature using a Perkin-Elmer Lambda 45 UV/VIS spectrometer equipped with an integrating sphere. These measurements were made on glass powder dispersed in KBr pellets. The optical absorption coefficient, α, was calculated from the absorbance, A, using the equation:

$$\alpha = 2.303\ A/d$$

where d is the thickness of the sample.

The starting structures have been built using the graphical interface of Spartan'04 [12] and preoptimized by molecular mechaniCs. Optimizations were continued at DFT level (B3LYP/CEP-4G/ECP) using the Gaussian'03 package of programs [13].

It should be noticed that only the broken bonds at the model boundary were terminated by hydrogen atoms. The positions of boundary atoms were frozen during the calculation and the coordinates of internal atoms were optimized in order to model the active fragment flexibility and its incorporation into the bulk.

3. RESULTS AND DISCUSSION

The vitreous or/and crystalline nature of the xPbO·(100-x)(3TeO2·2V_2O_5) and xSm_2O_3·(100- x)(3TeO_2·2V_2O_5) samples with various contents of lead or samarium oxide (0≤x≤50mol%) was tested by X-ray diffraction. The X-ray diffraction patterns of the studied samples are shown in Fig. 1. The X-ray diffraction patterns did not reveal the crystalline phases in the samples with 0≤x≤50mol% PbO while in the samples with x≥40mol% Sm_2O_3 the presemce of the $SmVO_4$ crystalline phase was detected.

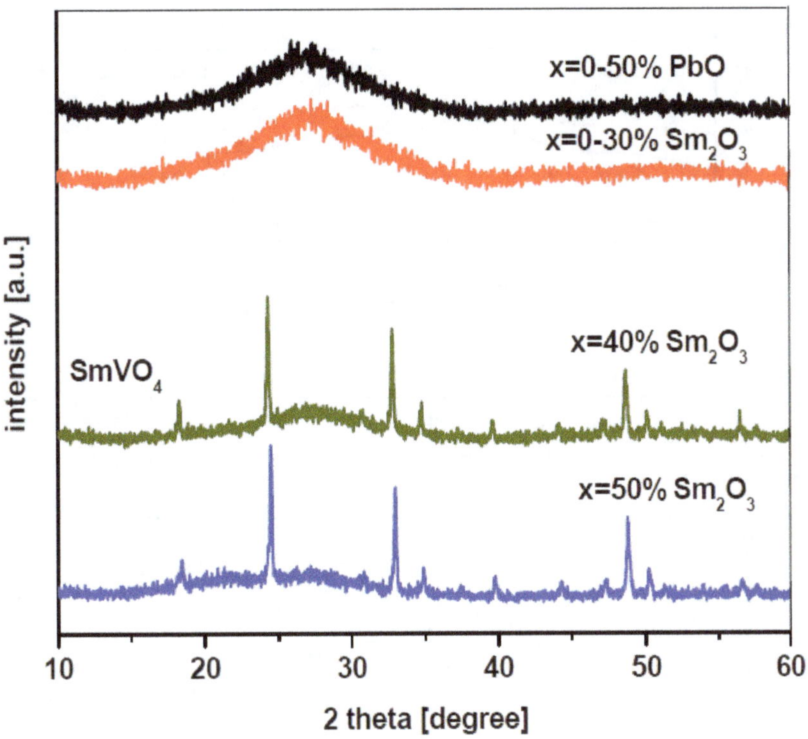

Fig. 1. X-ray diffraction patterns for xPbO (or xSm2O3)·(100-x)[3TeO2·2V2O5] samples where 0≤x≤50mol%.

3.1 FTIR spectroscopy

The absorption bands located around 460cm^{-1}, 610-680 and 720 to 780cm^{-1} are assigned to the bending mode of Te-O-Te or O-Te-O linkages, the stretching mode of [TeO4] trigonal pyramids with bridging oxygen and the stretching mode of [TeO3] trigonal pyramids with non-bridging oxygen, respectively [14-18].

The IR spectrum of the pure crystalline and amorphous V_2O_5 is characterized by the intense band in the 1000-1020cm^{-1} range which is related to vibrations of isolated V=O vanadyl groups in [VO_5] trigonal bipyramids. The band located at 950-970cm^{-1} was attributed to the [VO_4] units [19-21].

The examination of the FTIR spectra of the xM_aO_bá(100-x)[3TeO2á2V_2O_5] where M_aO_b = PbO or Sm_2O_3 glasses and glass ceramics shows some changes in the characteristic bands corresponding to the structural units of the glass network (Fig. 2). These modifications can be summarized as follows:

1. x =10mol% M_aO_b (1)

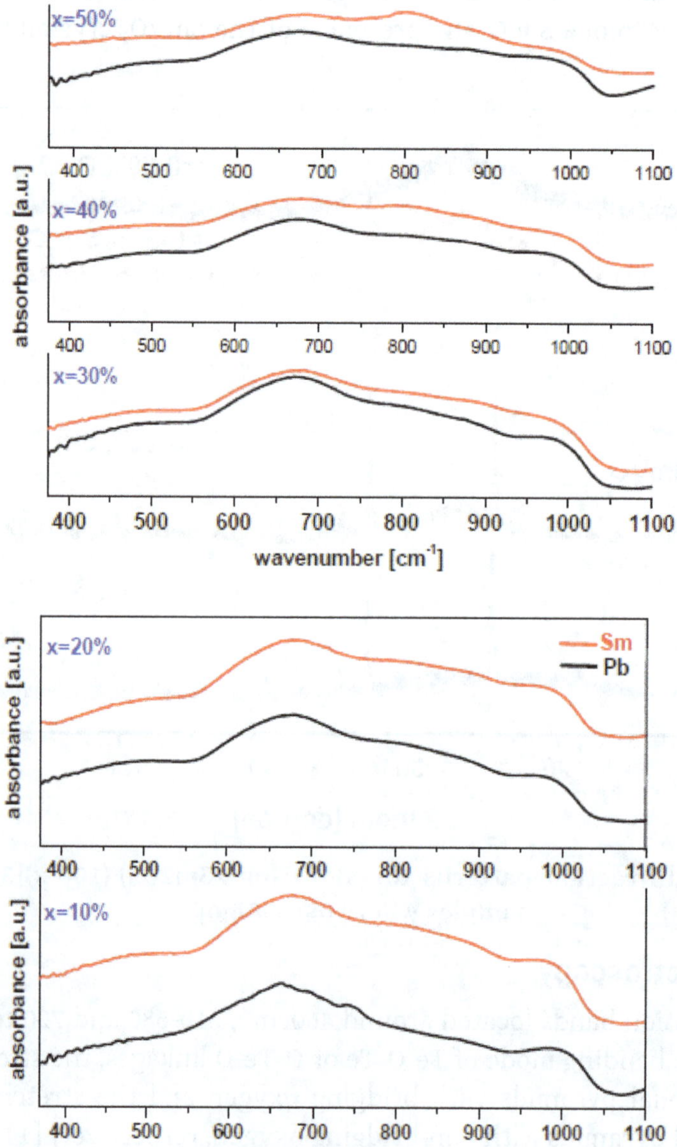

Fig. 2. FTIR spectra of xPbO (or xSm2O3)·(100-x)[3TeO2·2V2O5] samples where 0≤x≤50mol%.

The incorporation of network modifier lead ions into the vanadate-tellurate glasses enhances the breaking of axial Te-O-Te linkages in the [TeO4] trigonal bypiramidal structural units. As a consequence, three-coordination tellurium is formed and accumulated. The band centered at about 750cm^{-1} indicates the presemce of the [TeO3] structural units [17,18]. The [TeO3] structural units are expected to participate in the depolymerization of the glass network because they create more bonding defects and non-bonding oxygens.

The intensity of the band corresponding to the [TeO3] units (located at about ~750cm^{-1}) decreases and a new band located at about 800cm^{-1} appears with the adding of the Sm_2O_3 content. In detail, the band situated at about 800cm^{-1} is due to the asymmetric stretching of VO_4-3 entity from orthovanadate species [22, 23].

2. 10 ≤x ≤30 mol% M_aO_b (2)

An increasing trend was observed in the strength of the bands centered at ~1020cm^{-1}. The feature of the band located at about 875cm^{-1} comes up in intensity. This effect is more pronounced when adding of samarium ions in the matrix network. This band is attributed to the vibrations of the V-O bonds from the pyrovanadate structural units.

The gradual addition of the samarium (III) oxide leads not only to a simple incorporation of these ions in the host glass matrix but also generates changes of the basic structural units of the glass matrix. Structural changes reveal that the samarium ions causes a change from the continuous vanadate-tellurate network to a continuous samarium-vanadate-tellurate network interconnected through Sm-O-V and Te-O-Sm bridges. Then, the surplus of non- bridging oxygens is be converted to bridging ones leading to the decrease of the connectivity of the network.

3. x ≥ 40mol% M_aO_b (3)

By increasing the Sm_2O_3 content up to 40mol%, the evolution of the structure can be explained considering the higher capacity of migration of the samarium ions inside the glass network and the formation of the $SmVO_4$ crystalline phase, in agreement with XRD data. The accumulation of oxygen atoms in the glass network can be supported by the formation of ortho- and pyro-vanadate structural units.

On the other hand, the lead oxide generates the rapid deformation of the Te-O-Te linkages yielding the formation of [TeO3] structural units. Further, the excess of oxygen can be accommodated in the host matrix by conversion of some [VO_4] structural units into [VO_5] structural units.

The broader band centered at ~ 670-850cm^{-1} can be attributed to the Pb-O bonds vibrations from the [PbO4] and [PbO3] structural units. The absorption band centered at about 470cm^{-1} may be correlated with the Pb-O stretching vibration in [PbO4] structural units [24-26]. The increase in the intensity of the bands situated between 650 and 850cm^{-1} show that the excess of oxygen in the glass network can be supported by the increase of [PbOn] structural units (with n=3 and 4).

In brief, the variations observed in the FTIR spectra suggest a gradual inclusion of the lead ions in the host vitreous matrix with increasing of the PbO content up to 50mol%, while progressive adding of samarium oxide determines the increase in the intensity of the bands due to the ortho- and pyrovanadate structural units. The mechanisms of incorporation of the lead and samarium ions in the host matrices can be summarized as following:

1. For the samples with lead oxide, the Pb+2 ions can occupy a position in the chain itself and their influence on the V=O bonds is limited. Since the V=O mode position from about 1020cm^{-1} is preserved it can be concluded that the V=O bond is not directly influenced and the coordination number and symmetry of the [VO_4] and [VO_5] structural units do not change significantly. The increase in the intensity of the bands situated at 650 and 850cm^{-1} show that the PbO acts as a network former with a moderate effect on the vanadate-tellurate network.
2. By increasing the samarium oxide content up to 30mol%, Sm+3 ions located between the vanadate chains may affect the isolated V=O bonds yielding to the depolimerization of the vanadate network in shorter and isolated chains formed of ortho- and pyro- vanadate structural units. As a result they are markedly elongated and the vibrations frequency shifts toward lower wavenumbers. The increase of the content of samarium ions produces a strong depolymerization of the network leading to formation of $SmVO_4$ crystalline phase, in agreement with the XRD data. The combined XRD and IR spectroscopy data show that the Sm_2O_3 acts as a network modifier with a strong effect about vanadate network.

3.2 UV-VIS spectroscopy

Optical absorption in solids occurs by various mechanisms, in all of which the phonon energy will be absorbed by either the lattice or by electrons where the transferred energy is covered. The lattice (or phonon) absorption will give information about atomic vibrations involved and this absorption of radiation normally occurs in the infrared region of the spectrum. Optical absorption is a useful method for investigating optically induced transitions and getting information about the energy gap of non-crystalline materials and the band structure. The principle of this technique is that a photon with energy greater than the band gap energy will be absorbed [27].

One of the most important concerns in rare earth doped glasses is to define the dopant environment. Hypersensitive transitions are observed in the spectra of all rare earth ions having more than one f electrons. Hypersensitive transitions of rare earth ions manifest an anomalous sensitivity of line strength to the character of the dopant environment [28, 29].

The measured UV-VIS absorption spectra of the lead and samarium-vanadate-tellurate glasses are shown in Fig. 3. The spectra show that the maxima of the absorption are located in the UV region for all investigated glasses containing PbO or Sm_2O_3.

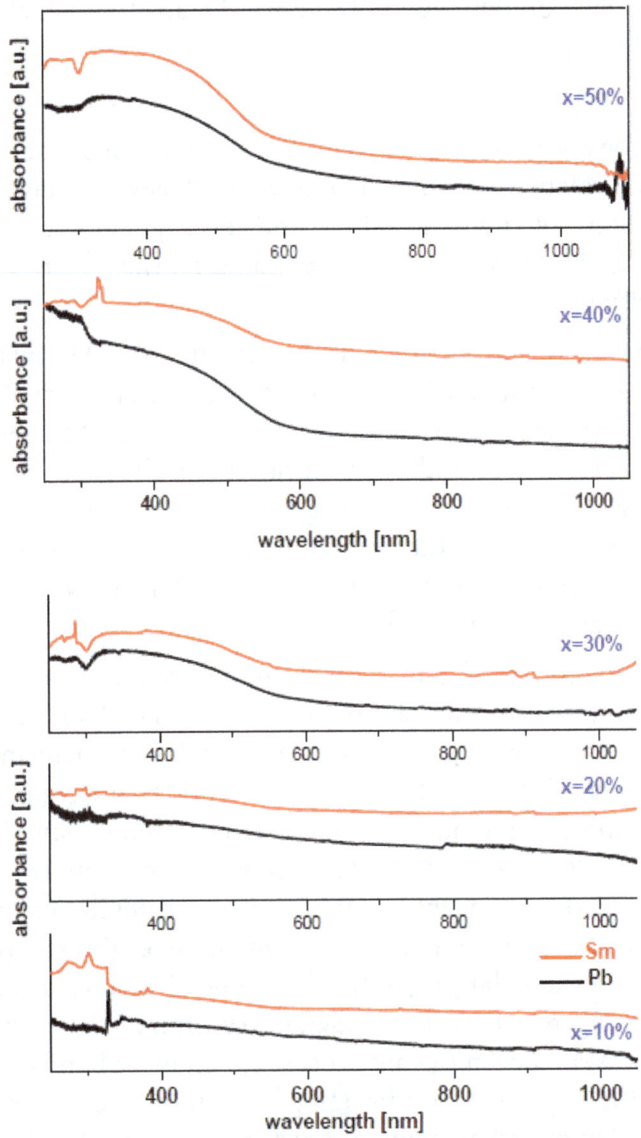

Fig. 3. UV-VIS absorption spectra of xPbO (or xSm2O3)·(100-x)[3TeO2·2V2O5] samples where 0≤x≤50mol% in function of lead (II) or samarium (III) oxide content.

The measurements of optical absorption and the absorption edge are important especially in connection with the theory of electronic structure of amorphous materials. The energy gap, Eg, is an important feature of semiconductors which determines their applications in optoelectronics [32]. Observations of the variation of Eg with increase in the modifier content can be attributed to the changes in the bonding that takes place in the glass

The nature of the optical transition involved in the network can be determined on the basis of the dependence of absorption coefficient (α) on phonon energy (hυ). The total absorption could be due to the optical transition which is fitted to the relation:

$$\alpha h\upsilon = \alpha_0 (h\upsilon\text{-}Eg)^n$$

where Eg is the optical energy gap between the bottom of the conduction band and the top of the valence band at the same value of wavenumber, $\alpha 0$ is a constant related to the extent of the band tailing and the exponent n is an index which can have any values between ½ and 2 depending on the nature of the interband electronic transitions.

Extrapolating the linear portion of the graph $(\alpha h\nu)2 \rightarrow 0$ to hν axis, the optical band gaps, Eg are determined with increasing PbO and Sm_2O_3 content (Figs. 4 and 5). The optical band gap increases gradually from 1.84eV to 2.09eV and 2.21eV, respectively, by adding of PbO and Sm_2O_3. In either case the values are systematically increasing with the increase of x. It is to be noted that the curves are characterized by the presemce of an exponential decay tail at low energy. These results indicate the presemce of a well defined $\pi \rightarrow \pi^*$ transition associated with the formation of conjugated electronic structure [33].

The increase of the band gap may occur due to variation in non-bridging oxygen ion concentrations. In metal oxides, the valence band maximum mainly consists of 2p orbital of the oxygen atom and the conduction band minimum mainly consists of ns orbital of the metal atom. The non-bridging oxygen ions contribute to the valence band maximum. The non-bridging orbitals have higher energies than bonding orbitals. When a metal-oxygen bond is broken, the bond energy is released. The increase in concentration of the non- bridging oxygen ions results in the shift of the valence band maximum to higher energies and the reduction of the band gap. Thus, the enlarging of band gap energy due to increase in the PbO or Sm_2O_3 content suggests that non-bridging oxygen ion concentration decreases with increasing the PbO or Sm_2O_3 content that expands the band gap energy. In the glasses doped with Sm_2O_3, the non-bridging oxygen ions concentration decreases due to the formation of orthovanadate structural units. In the glasses doped with PbO, the non- bridging oxygen ions concentration decreases also because the lead atoms act as network formers and the accommodation with the excess of oxygen ions is possible by the increase of the polymerization degree of the network by Pb-O-Te and Pb-O-V linkages.

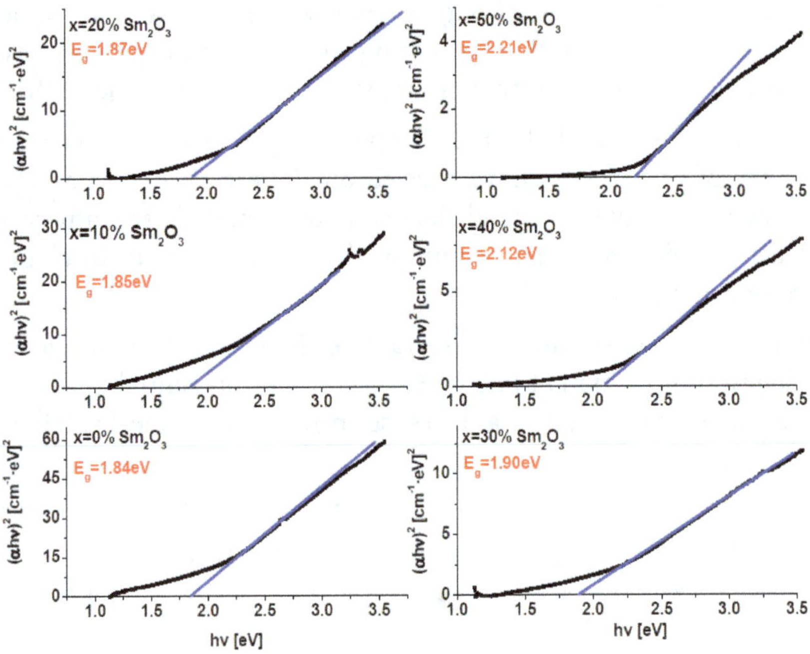

Fig. 4. Plots of (αhν)2 versus hν for xPbO·(100-x)[3TeO2·2V2O5] glasses where 0≤x≤50mol%

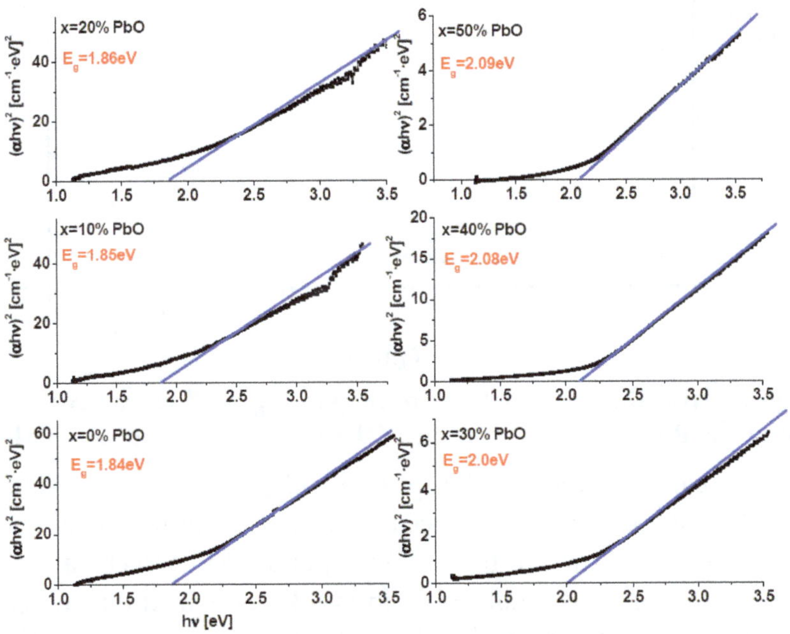

Fig. 5. Plots of (αhν)2 versus hν for xSm2O3·(100-x)[3TeO2·2V2O5] samples where 0≤x≤50mol%.

The existence and variation of optical energy gap may be also explained by invoking the occurrence of local cross linking within the amorphous phase of the matrix network, in such a way as to increase the degree of ordering in these parts.

Refractive index is one of the most important properties in optical glasses. A large number of researchers have carried out investigations to ascertain the relation between refractive index and glass composition. It is generally recognized that the refractive index, n, of many common glasses can be varied by changing the base glass composition [34].

The observed decrease in the refractive index of the studied glasses accompanying to the addition of PbO or Sm_2O_3 content presented in the Fig. 6 can be considered as an indication of a decrease in number of non-bridging oxygen ions.

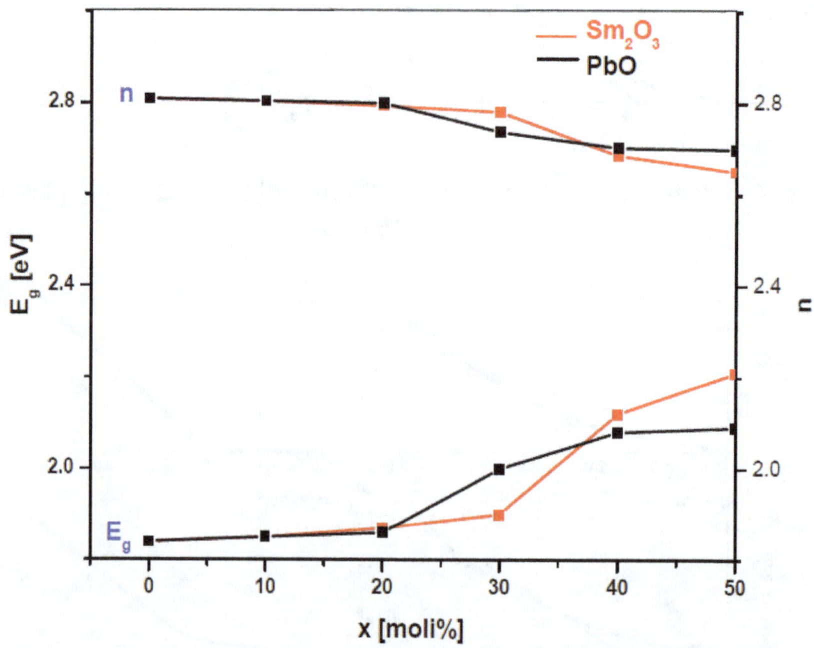

Fig. 6. The relationship between the optic gap, Eg, and refractive index, n, and the and the PbO or Sm2O3 contents (the line is only a guide for the eye).

In brief, we can conclude that the optical band gap increases with increasing the PbO or Sm_2O_3 contentof the glass. Since the basic structural units of the vanadate-tellurate glasses are known to be the [TeO3] and $[VO_4]$ structural units and the internal vibrations of these molecular units take part in the transitions. In this work, the increase of the optical band gap, Eg, to larger energies with increasing the PbO or Sm_2O_3 content is probably related to the progressive decrease in the concentration of non-bridging oxygen. This decrease in turn gives rise to a possible decrease in the bridging Te-O-Te and V-O-V linkages. The shift is attributed to structural changes which are the the result of the different (interstitial or substitutional) site occupations of the Pb+2 or Sm+3 ions which are added to the vanadate- tellurate matrix and modify the network.

We assume that as the cation concentration increases, the Te-O-Te and V-O-V linkages develop bonds with Pb+2 or Sm+3 ions, which in turn leads to the gradual breakdown of the glass network. This effect seems more pronounced in doping of the network with the Sm+3 ions. These results are in agreement with XRD data which indicate the higher affinity of the samarium ions to attract structural units with negative charge yielding the formation of the $SmVO_4$ crystalline phase for samples with x>30mol%.

3.3 DFT calculations

In this section, the purpose of the present paper was to continue the investigation of the structure of the vanadate-tellurate network and especially to illuminate structural aspects of the vanadate network using quantum-chemical calculations because the coordination state of vanadium atoms is not well understood. Figure 7 shows the optimized structure proposed to the 3TeO2á2V_2O_5 glass network.

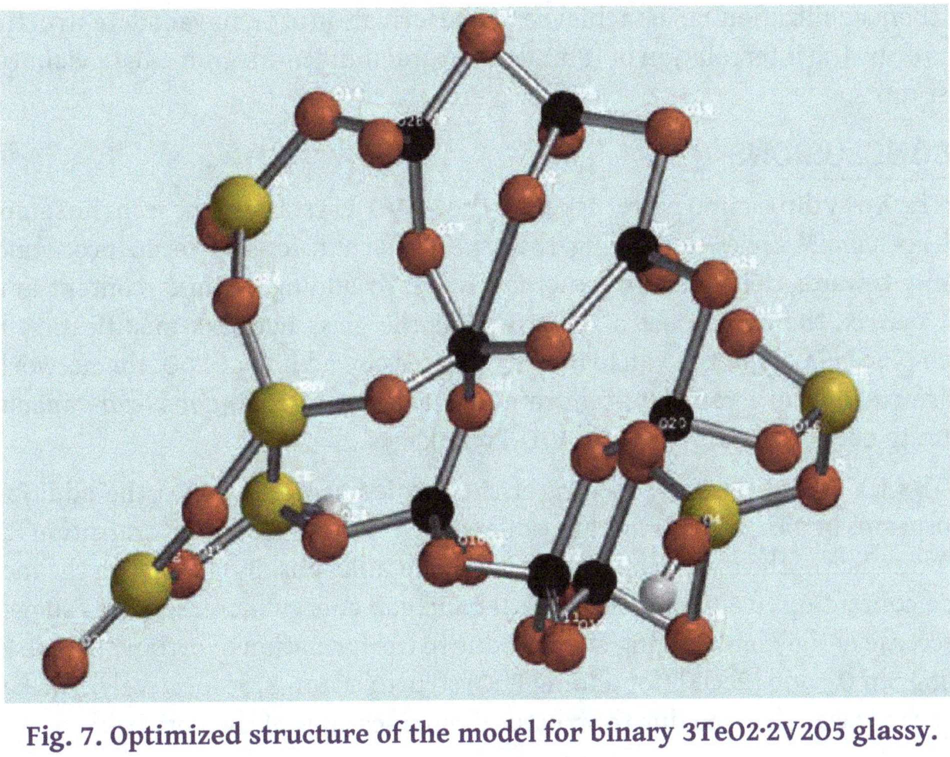

Fig. 7. Optimized structure of the model for binary 3TeO2·2V2O5 glassy.

Analyzing the structural changes resulted from the geometry optimization of our model, we found that the vanadium ions are distributed into two crystallographic sites: the [VO_4] tetrahedral and [VO_5] square pyramidal units. The vanadium tetrahedrons are very regular with vanadium-oxygen distances ranging from 1.57 to 1.82Å and O-V-O angles ranging from 1040 to 1100 (with a mean value (109.50) very close to the ideal value (109.280) corresponding to the tetrahedral geometry). In our model, the V-O interatomic bond distances are ranging from 1.60 to 1.65Å, 1.75 to 1.80Å (the average V-O distance is 1.72Å) and O-V-O angles values are ranging from 101 to 1120. This result show that the [VO_4] tetrahedrons are easy distorted around the vanadium center.

The [VO_5] square pyramidal units are considerably distorted around the vanadium center and the V-O bond distances are ranging from 1.65 to 2.30Å. Such a behavior was reported for the two-dimensional layered vanadate compounds [35, 36]. This shows that there is instability in the nonequivalent V-O bonds in the polyhedron. In essence, this is due to the displacement of the vanadium atom from the centre of the polyhedron, whose asymmetry strongly depends on the manner of connection with the surrounding polyhedron. This deformation will be expressed more clearly in the formation of the vitreous matrix.

This structural model shows a very complex behavior of the vanadium atoms and their stabilization can be achieved by the formation of orthovanadate structural units or by the intercalation of [PbO3] structural units in the immediate vicinity of these units.

4. CONCLUSIONS

The X-ray diffraction patterns reveal the $SmVO_4$ crystalline phase in the samples with x>30mol% Sm_2O_3 indicating that the samarium ions have an pronounced affinity towards the vanadate structural units. By adding of Sm_2O_3 content in the host matrix, the FTIR spectra suggests that the glass network modification has taken place mainly in the vanadate part whereas by adding of PbO, the network is transformed from a vanadate-tellurate network into a continuous lead-vanadate-tellurate network by Te-O-Pb and V-O-Pb linkages.

The UV-VIS absorption spectra of the studied samples reveal the additional absorptions in the 250-1050nm range due to the generation of $n \rightarrow \pi^*$ transitions and the presemce of the transition or rare earth metallic ions. By increasing the metal oxide content up to 50mol%, the optical band gap energy increases. This suggests a decrease of the non-bridging oxygens due to the formation of orthovanadate (for adding Sm_2O_3) and [PbO3] (for adding PbO) structural units, respectively. The band gap energy was changed due to structural modifications of the network.

Our DFT investigations show that the penta-coordinated vanadium atoms show a unique influence on the structural properties of the glasses.

5. REFERENCES

1. S. Tanaba, K. Hirao, N. Soga, J. Non-Cryst. Solids 122 (1990) 79.
2. H. Nasu, O. Matsusita, K. Kamiya, H. Kobayashi, K. Kubodera, J. Non-Cryst. Solids 124 (1990) 275.
3. B. Eraiah, Bull. Mater. Sci. 29(4) (2006) 375.
4. G. Nunziconti, S. Bemeschi, M. Bettinelli, M. Brei, B. Chen, S. Pelli, A. Speghini, G. C. Righini, J. Non-Cryst. Solids 345&346 (2004) 343.
5. M. Ganguli, M. Bhat Harish, K. J. Rao, Phys. Chem. Glasses 40 (1999) 297.
6. S. Jayaseelan, P. Muralidharan, M. Venkateswarlu, N. Satyanarayana, Mater. Chem. Phys. 87 (2004) 370.
7. G. A. Kumar, A. Martinez, A. Mejia, C. G. Eden, J. Alloys Compd. 365 (2004) 117.
8. J. Yang, S. Dai, Y. Zhou, L. Wen, L. Hu, Z. H. Jiang, J. Appl. Phys. 93 (2003) 977.
9. W. A. Pisarski, T. Goryczka, B. Wodecka-Dus, M. Plonska, J. Pisarska, Mater. Sci. Eng. 122 (2005) 94.
10. G. Lakshminarayana, S. Buddhudu, Mater. Chem. Phys. 102 (2007) 181.
11. Thulasiramudu, S. Buddhudu, Spectrochim Acta A 67 (2007) 802.
12. Spartan'04, Wavefunction Inc., 18401 Von Karman Avenue, Suite 370 Irvine, CA 92612.
13. M. J. Frisch, G. W. Trucks, H. B. Schlegel, G. E. Scuseria, M. A. Robb, J. R. Cheeseman, J.
14. Montgomery, Jr., T. Vreven, K. N. Kudin, J. C. Burant, J. M. Millam, S. S. Iyengar, J. Tomasi, V. Barone, B. Mennucci, M. Cossi, G. Scalmani, N. Rega, G. A. Petersson, H. Nakatsuji, M. Hada, M. Ehara, K. Toyota, R. Fukuda, J. Hasegawa, M. Ishida, T. Nakajima, Y. Honda, O. Kitao, H. Nakai, M. Klene, X. Li, J. E. Knox, H. P. Hratchian, J. B. Cross, C. Adamo, J. Jaramillo, R. Gomperts, R. E. Stratmann, O. Yazyev, A. J. Austin, R. Cammi, C. Pomelli, J. W. Ochterski, P. Y. Ayala, K. Morokuma, G. A. Voth, P. Salvador, J. J. Dannenberg, V. G. Zakrzewski, S. Dapprich, A. D. Daniels, M. C. Strain, O. Farkas, D. K. Malick, A. D. Rabuck, K. Raghavachari, J. B. Foresman, J. V. Ortiz, Q. Cui, A. G. Baboul, S. Clifford, J. Cioslowski, B. B. Stefanov, G. Liu, A. Liashenko, P. Piskorz, I. Komaromi, R. L. Martin, D. J. Fox, T. Keith, M. A. Al-Laham, C. Y. Peng, A. Nanayakkara, M. Challacombe, P. M. W. Gill, B. Johnson, W. Chen, M. W. Wong, C. Gonzalez, and J.
15. Pople, Gaussian 03, Revision A.1, Gaussian, Inc., Pittsburgh PA, 2003.
16. S. Rada, M. Culea, E. Culea, J. Phys. Chem. A 112(44) (2008) 11251.

17. S. Rada, M. Neumann, E. Culea, Solid State IoniCs 181 (2010) 1164.
18. S. Rada, E. Culea, M. Rada, Mater. Chem. Phys. 128(3) (2011) 464.
19. S. Rada, E. Culea, M. Culea, Borate-Tellurate Glasses: An Alternative of Immobilization of the Hazardous Wastes, Nova Science Publishers Inc., New York, 2010.
20. S. Rada, E. Culea, J. Molec. Struct. 929 (2009) 141.
21. M. Rada, V. Maties, S. Rada, E. Culea, J. Non-Cryst. Solids 356 (2010) 1267.
22. S. Rada, R. Chelcea, E. Culea, J. Molec. Model. 17 (2011) 165.
23. S. Rada, E. Culea, M. Culea, J. Mater. Sci. 43(19) (2008) 6480.
24. K. V. Ramesh, D. L. Sastry, J. Non-Cryst. Solids 352 (2006) 5421.
25. K. Gatterer, G. Pucker, H. P. Fritzer, Phys. Chem. Glasses 38 (1997) 293.
26. S. Rada, M. Culea, E. Culea, J. Non-Cryst. Solids 354(52-54) (2008) 5491.
27. S. Rada, M. Culea, M. Neumann, E. Culea, Chem. Phys. Letters 460 (2008) 196.
28. M. Rada, V. Maties, M. Culea, S. Rada, E. Culea, Spectrochim. Acta A 75 (2010) 507.
29. M. T. Abd El-Ati, A. A. Higazy, J. Mater. Sci. 35 (2000) 6175.
30. S. N. Misra, S. O. Sommerer, Appl. Spectrosc. Rev. 26 (1991) 151.
31. V. K. Tikhomirov, M. Naftaly, A. Jha, J. Appl. Phys. 86 (1999) 351.
32. D. Ehrt, J. Non-Cryst. Solids 348 (2004) 22.
33. M. Nassar, N. A. Ghoneim, J. Non-Cryst. Solids 46 (1981) 181.
34. T. Aoki, Y. Hatanaka, D.C. Look, Appl. Phys. Lett. 76 (2000) 3257.
35. W. R. Salaneck, C. R. Wu, J. L. Bredas, J. J. Ritsko, Chem. Phys. Lett. 127 (1986) 88.
36. R. El-Mallawany, J. Appl. Phys. 72 (1992) 1774.
37. Sun, E. Wang, D. Xiao, H. An, L. Xu, J. Molec. Struct. 840 (2007) 53.
38. Y. Zhou, H. Qiao, Inorg. Chem. Comm. 10 (2007) 1318.

4 Chapter

WATER IN ROCKS AND MINERALS – SPECIES, DISTRIBUTIONS, AND TEMPERATURE DEPENDENCES

1. INTRODUCTION

Water is ubiquitously distributed in the interior of the earth, as various forms in rocks and minerals: In rocks, aggregates of minerals, fluid water in which molecular H_2O is clustered, is trapped at intergranular regions and as fluid inclusions (e.g., Hiraga et al., 2001). In minerals, water is located as the form of –OH in their crystal structures as impurities. Surprisingly, such water species are distributed over five times in the earth's interior than ocean water (e.g., Jacobsen & Van der Lee, 2006), and play very important roles on earth dynamiCs such as deformation and reactions of minerals and rocks (e.g., Thompson & Rubie, 1985; Dysthe & Wogelius, 2006). In the filed of earth Sciences therefore, people are trying to measure properties of water in rocks and minerals such as species, contents, distribution, thermal behaviour, migration rate, etc.

Infrared (IR) spectroscopy is a powerful tool to quantitatively measure these properties of water in rocks and minerals (See Aines & Rossman, 1984; Keppler & Smyth, 2006 for IR spectra of various rocks and minerals). In this chapter, for an advanced IR spectroscopic measurement, I introduce in-situ high temperature IR spectroscopy to investigate above matters. First, I use chalcedonic quartz, which contains fluid water at intergranular regions and –OH in quartz crystal structures. Next, I use beryl, a typical cyclosilicate which contains isolated (not clustered) H_2O molecules in open cavities of the crystal structure. Changes of the states of water in chalcedonic quartz and beryl by temperature changes and dehydration will be discussed. Finally, I perform two-dimensional IR mappings for naturally deformed rocks to investigate water distribution in polymineralic mixtures, and discuss possible water transportation during rock deformation.

2. METHODS

Transmitted IR spectra for rocks and minerals are generally measured by making thin sections of samples with thicknesses of from 20 to 200 μm, which depend on concentrations and absorption coefficients based on Beer-Lambert law. A Fourier Transform IR microspectrometer totally used in this study is equipped with a silicon carbide (globar) IR source and a Ge-coated KBr beamsplitter. IR light through a sample is measured using a mercury-cadmium-telluride detector.

In-situ high temperature IR spectra were measured for a sample on a heating stage which was inserted into the IR path. The sample was heated at 100 °C/minute to desired temperature, and spectra were collected at about 1 minute. Mapping measurements were carried out using an auto XY-stage under atomospheric condition.

3. WATER IN ROCKS: AS AN EXAMPLE OF CHALCEDONIC QUARTZ

Fluid water and -OH are trapped in chalcedonic quartz, an aggregate of microcrystalline quartz (SiO_2) grains (sometimes called as chalcedony or agate which has different transparency and is treated as a gem). Fluid water in chalcedonic quartz is dominantly trapped at intergranular regions such as grain boundaries and triple junctions of grains. IR spectra of chalcedonic quartz have been measured at room temperature (RT) (Frondel, 1982; Graetsch et al., 1985). In this section, I measure in-situ high temperature IR spectra for chalcedonic quartz as a representative material that abundantly contains fluid water and structurally-trapped -OH.

3.1 Typical IR spectrum at RT and water species

Figure 1 is a typical IR spectrum at RT for a thin section (ca. 100 μm thickness) of chalcedonic quartz. The band due to fluid water shows an asymmetric broad band ranging from 2750 to 3800 cm^{-1} with a shoulder around 3260 cm^{-1}: this band feature is the same with that of simple fluid water which exists everywhere around us (e.g., Eisenberg & Kauzman, 1969). -OH in chalcedonic quartz is mainly trapped as Si-OH by breaking the network of SiO_2 bonds (Kronenberg & Wolf, 1990), and the OH stretching band is sharp (3585 cm^{-1} at RT). The bending mode of fluid water, which should be seen at around 1600 cm^{-1}, is hindered by many sharp Si-O stretching bands. Combination modes of the stretching and bending modes of fluid water and Si-OH are clearly seen for a thicker sample (1 mm in Fig. 1), and they are detected at 5200 cm^{-1} and 4500 cm^{-1} respectively. Contents of fluid water and Si-OH can be calculated from these bands' heights using their molar absorption coefficients of 0.761 and 1.141 L mol–1 cm^{-1}, respectively (Scholze, 1960; Graetsch et al., 1985); in this case 0.32 wt % H_2O and 0.28 wt % Si-OH, respectively (Fukuda et al., 2009a).

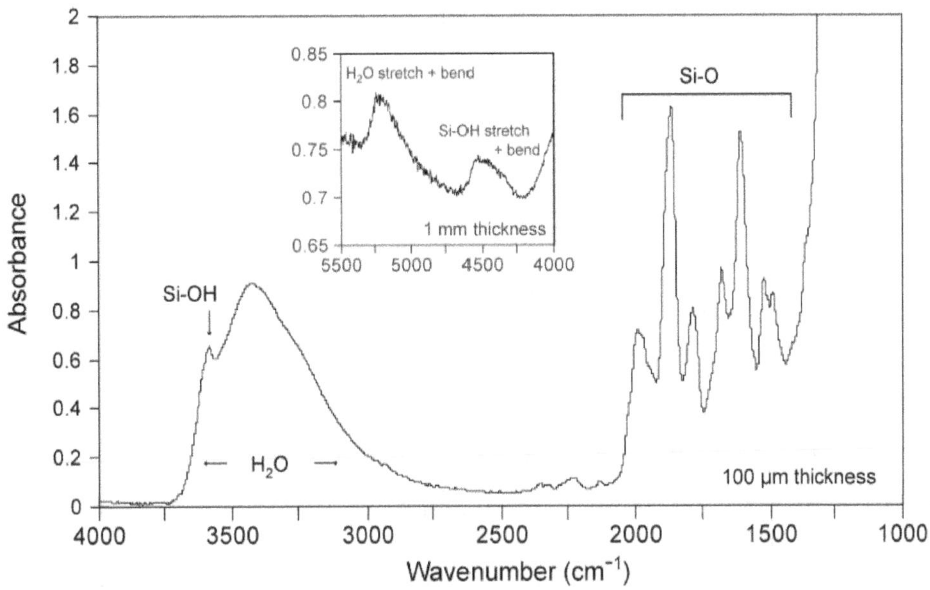

Fig. 1. Typical IR spectrum of chalcedonic quartz at RT. The sample thickness of 100 μm for 4000–1000 cm-1 and 1 mm for 5500–4000 cm-1 (combination modes of the stretching and bending modes of fluid water and Si-OH).

3.2 Water vibrations at high temperatures

High temperature IR spectra were measured for the sample set on the heating stage (Fig. 2). With increasing temperature up to 400 °C, the broad band due to fluid water dominantly and asymmetrically shifts to high wavenumbers. Contrary to this, the band due to Si-OH slightly shifts to high wavenumbers. After quenching to RT from high temperatures, these water bands do not change from those before heating, indicating that vibrational states of fluid water and Si-OH are changed at high temperatures without dehydration. The following is discussion for changes in the vibrational states of water.

Since vibrational energy of OH stretching of Si-OH is structurally limited within quartz crystal structures, the band is sharp even at high temperatures. The deviation of the wavenumber from free -OH stretching (around 3650 cm^{-1}; summarized in Libowitzky, 1999) can be explained by the work of hydrogen bond in Si-OH...O-Si in quartz crystal structures, which weaken OH vibrational energy. With increasing temperature, the hydrogen bond distance is extended due to thermal expansion of quartz crystal structure (e.g., Kihara, 2001). Resultantly, the band due to Si-OH slightly shifts to high wavenumber and the band height is not so decreased (3599 cm^{-1} at 400 °C; Fukuda & Nakashima, 2008).

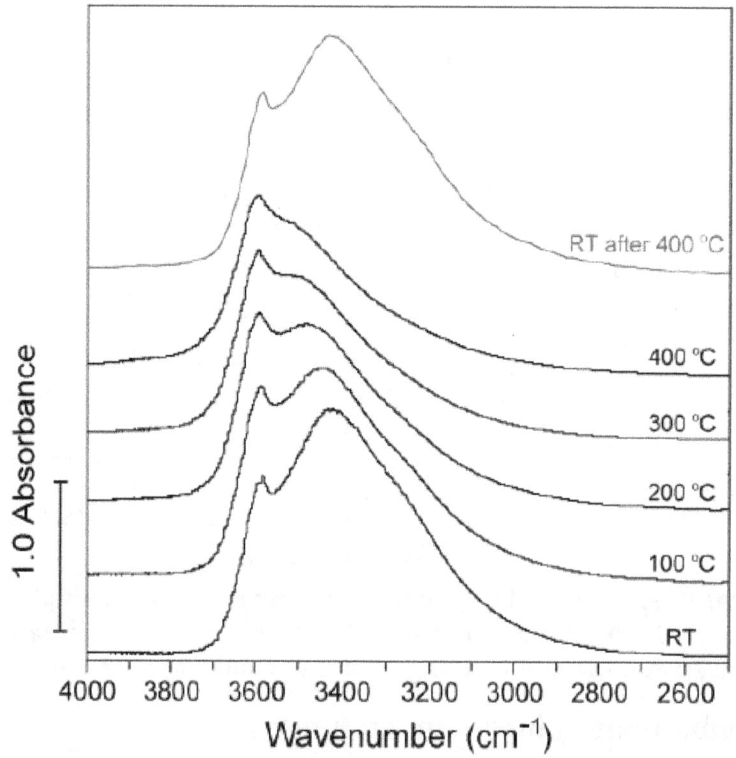

Fig. 2. In-situ high temperature IR spectra of a chalcedonic quartz (black lines) in the water stretching region (Replotted from Fukuda et al., 2009c). The spectrum at RT after heating at 400 °C is shown as gray line on the top, showing no significant change from the spectrum before heating.

In fluid water, H_2O is clustered and networked by various hydrogen bond strengths (e.g., Brubach et al., 2005). Therefore, fluid water shows the broad band. With increasing temperatures, the average coordination numbers of a H_2O molecule to adjacent H_2O molecules at confined intergranular regions of chalcedonic quartz are reduced due to increases of vibrational energies without dehydration. The average coordination number of a single H_2O molecule in fluid water is 2–3 molecules at RT (Brubach et al., 2005), and 1–2 above supercritical temperature (Nakahara et al., 2001). This leads to significant shifts of wavenumbers to higher. Also, band heights of fluid water are decreased with increasing temperatures. For example, the maximum band height at 400 °C is approximately 50 % of that at RT. This is also because of decreases of average numbers of H_2O in areas that IR light captures (i.e., density; Schwarzer et al., 2005).

3.3 Dehydration behaviour

When the sample is kept at high temperatures, dehydration occurs. High temperature IR spectra were continuously measured to monitor dehydration.

Figure 3 shows dehydration behaviour measured at 500 °C (Fig. 3a) and 400 °C (Fig. 3b). Both of the experiments were performed by heating during 500 minutes in total. Broad bands around 3800-3000 cm^{-1} in both spectra decrease with keeping at high temperatures, and the wavenumbers of the broad bands are not changed during heating. IR spectra at RT after heating (gray spectra in Fig. 3) also show decreases of fluid water. This indicates that fluid water was dehydrated through intergranular regions which are fast paths for mass transfers (See Ingrin et al., 1995; Okumura and Nakashima, 2004; Fukuda et al., 2009c for estimation of water diffusivity). Over 50 % of the band areas are decreased during the heating in 84 minutes at 500 °C. Band areas of 80 % are decreased in 250 minutes, and the features of the spectra are not changed after that. The RT temperature spectrum after 500 minutes heating at 500 °C also shows significant reduction of board band due to fluid water. This remained band due to fluid water may reflect fluid inclusions, which is tightly trapped at open spaces in crystal structures. On the other hand, 60 % of the band areas are preserved after heating in 500 minutes at 400 °C, and the RT spectrum after 500 minutes heating at 400 °C still shows a strong signal of fluid water.

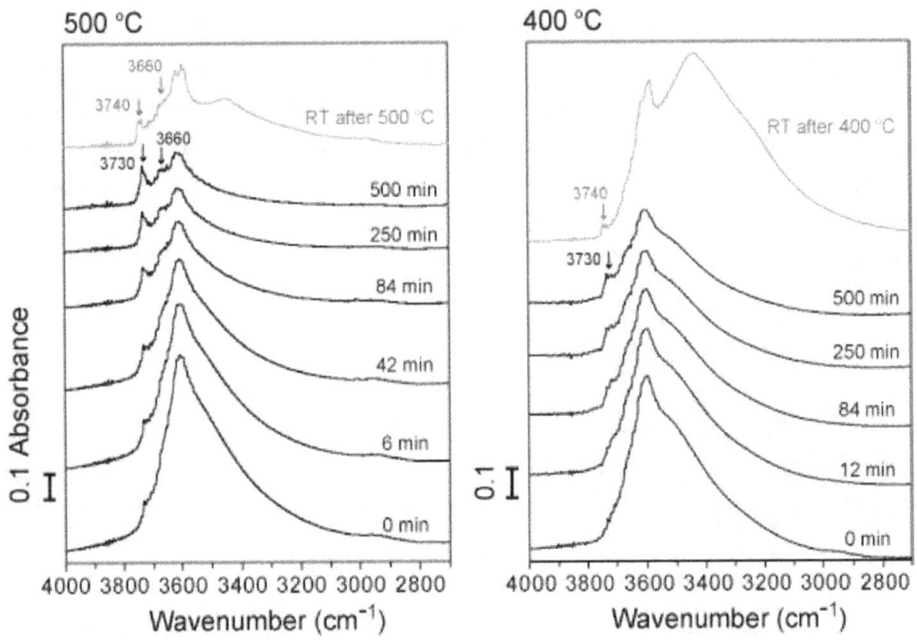

Fig. 3. Dehydration behaviour at 500 °C (left) and 400 °C (right) in the water stretching region (after Fukuda et al., 2009c). The spectra at RT after heatings are also shown as gray lines. The integrated heating times are shown at the right of each spectrum.

Since contents of fluid water are reduced by dehydration, different degrees of hydrogen bonds of Si-OH at intergranular regions (surface silanol) to fluid water are formed. This leads to the appearances of several -OH bands (3660 and 3730 cm^{-1} at 500 °C, and possibly the split of the band at 3585 cm^{-1} at RT) (Yamagishi et al., 1997). The appearance of the band at 3730 cm^{-1} at 0 minute heating at 500 °C is due to slight dehydration during heating from RT to 500 °C at 100 °C/minute. The wavenumbers of these new -OH bands at high temperature are slightly different from those at RT, presumably due to thermal expansions of crystals and changes of hydrogen bond distances at high temperature.

4. H_2O MOLECULES IN MINERALS: AS AN EXAMPLE OF BERYL

In addition to fluid water in rocks and -OH in mineral crystal structures as described above, isolated (not clustered) H_2O molecules are incorporated in open cavities of crystal structures, and they are sometimes coupled with cations. H_2O in open cavities has well been studied for beryl, a typical cyclosilicate (after Wood & Nassau, 1967). Ideal chemical formula of beryl is $Be_3Al_2Si_6O_{18}$, and six-membered SiO4 rings are stacked along the crystallographic c-axis and make a pipe-like cavity called a channel (Fig. 4) (Gibbs et al., 1968). Isolated H_2O is trapped in the channel, and forms two kinds of orientations, depending on whether it coordinates to a cation (called type II) or not (called type I) (after Wood & Nassau, 1967). Such cations are trapped in the channels to compensate the electrical charge balances caused by Be2+-Li+ and Si^{4+}-Al^{3+} substitutions and lacks of Be^{2+} in the crystal structure of beryl. The cations in the channels are assumed to be mainly Na+, and some other alkali cations may be incorporated (Hawthorne and Černý, 1977; Aurisicchio et al., 1988; Artioli et al., 1993; Andersson, 2006). In this section, I introduce polarized IR spectra of beryl, and discuss changes of the states of type I/II H_2O in the channels by temperature changes and dehydration.

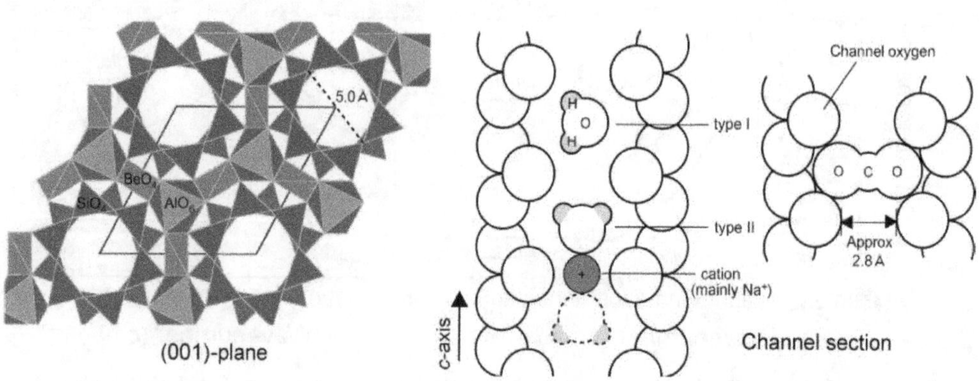

Fig. 4. Crystal structure of beryl. The (001)-plane (i.e., viewed down from the c-axis) and the channel section. Positions of type I/II H_2O, a cation, and CO2 are also shown. Modified after Fukuda & Shinoda (2011).

4.1 Chemical composition of the sample

The chemical composition of the natural beryl sample used in this study was analyzed by X- ray wavelength dispersive spectroscopy for major atomic contents, inductivity coupled plasma-atomic emission spectroscopy for Be content, and atomic absorption spectroscopy for Li and Rb contents (Table 1). The type I/II H_2O contents were determined from intensities of IR bands due to the asymmetric stretching of type I and the symmetric stretching of type II in a polarized IR spectrum at RT (See the spectrum in the next section), using their molar absorption coefficients of 206 L mol–1 cm^{-1} and 256 L mol–1 cm^{-1}, respectively (Goldman et al., 1977). The CO_2 content was also calculated for the band at 2360 cm^{-1} from 800 L mol–1 cm^{-1} in Della Ventura et al. (2009). In the chemical composition, the Li and Na contents are relatively high in addition to the Si, Al and Be contents of major atoms. Be content (2.893 in 18 oxygen), which is lower than the ideal composition of beryl (Be3Al2Si6O18), must be replaced by Li, and Na must be incorporated as Na+ in the channels to compensate the electrical charge balance. However, Li+ may be also incorporated in the channels, Since the Li content is not completely explained by Be2+-Li+ substitution (e.g., Hawthorne and Černý, 1977).

Table 1. Chemical composition of the beryl sample used in this study.

wt%		180 base	
$Si02_2$	64.45	Si	5.961
Al_2O_3	18.01	Al	2.006
BeO	13.08	Be	2.893
MgO	0.01	Mg	0.002
FeO	0.16	Fe	0.013
TiO_2		Ti	—
MnO	0.02	Mn	0.002
CaO	0.01	Ca	0.001
NiO	0.01	Ni	0.001
Cr_2O_3		Cr	—
ZnO	0.17	Zn	0.012
Li_2O	0.50	Li	0.185
Na_2O	0.54	Na	0.097
K_2O	0.04	K	0.004
Rb_2O	0.02	Rb	0.001
H_2O	1.80		
type II	1.80		
type II	1.24		
CO_2	0.01		
total	100.08		

4.2 Typical polarized IR spectra of beryl at RT and types of H_2O

Polarized IR spectra were measured by inserting a wire grid IR polarizer to IR light through the sample. Electric vector of IR light, E to the c-axis (i.e., the direction of the arraignments of the channels) were gradually changed and spectra were obtained (Fig. 5). Fundamental vibrations of type I/II H_2O (asymmetric stretching; v_3, symmetric stretching; v_1, and bending modes; ν2) can be detected under different polarized conditions, which correspond to the orientations of type I/II H_2O in the channels (Fig. 4) and IR active orientations of their vibrational modes: In a sample section of the (100)-plane (i.e., the section including the alignment of channels), the bands due to the ν_3 mode of type I (referred to as v_3-I hereafter), ν_1-II, and ν_2-II are dominantly detected under E//c-axis (Fig. 5a). Their wavenumbers are 3698, 3597, and 1628 cm^{-1}, respectively.

Under E⊥c-axis in the (100)-section (bottom spectra in Fig. 5a) or in the (001)-section (Fig. 5b), the bands due to ν_3-II (3661 cm^{-1}), ν_1-I (3605 cm^{-1}), and v_2-I are dominant. The v_2-I somewhat shows three bands at 1640, 1600, and 1546 cm^{-1} (e.g., Wood & Nassau 1967; Charoy et al., 1996; Łodziński et al., 2005), and I refer these three bands to the ν_2-I related bands. The asymmetric stretching mode of CO_2 molecules is detected at 2360 cm^{-1} under E⊥c-axis. These H_2O and CO_2 bands are not changed at any angles of E to the sample in the (001)-plane, corresponding to that these molecules are isotropically distributed due to the hexagonal symmetry of beryl. There are other unassigned bands; for example the sharp band at 3594 cm^{-1} can be seen. This band has been argued and might be due to Na+-OH in the channels.

Ideally, the v_3, v_1, and ν2 modes of isolated free H_2O are detected at 3756, 3657, and 1595 cm^{-1}, respectively at RT (Eisenberg & Kauzman, 1969). That the fundamental vibrations of type I/II H_2O in beryl are deviated from those for ideal value, is due to interaction of type I/II H_2O with channel oxygens and cations (Fig. 5). The wavenumbers of the stretching modes of type I/II H_2O at RT are lower than those of free H_2O, while those of the bending modes are higher. Falk (1984) experimentally demonstrated reverse correlations of band shifts between stretching and bending modes. The lower wavenumbers of the stretching modes than those of isolated water molecules are due to weak hydrogen bonds between type I/II and channel oxygens, similarly to the case for chalcedonic quartz. The reverse wavenumber shifts from ideal H_2O between stretching and bending modes is mainly explained by changes in H-H repulsion constants in a simple spring model (Fukuda & Shinoda, 2008).

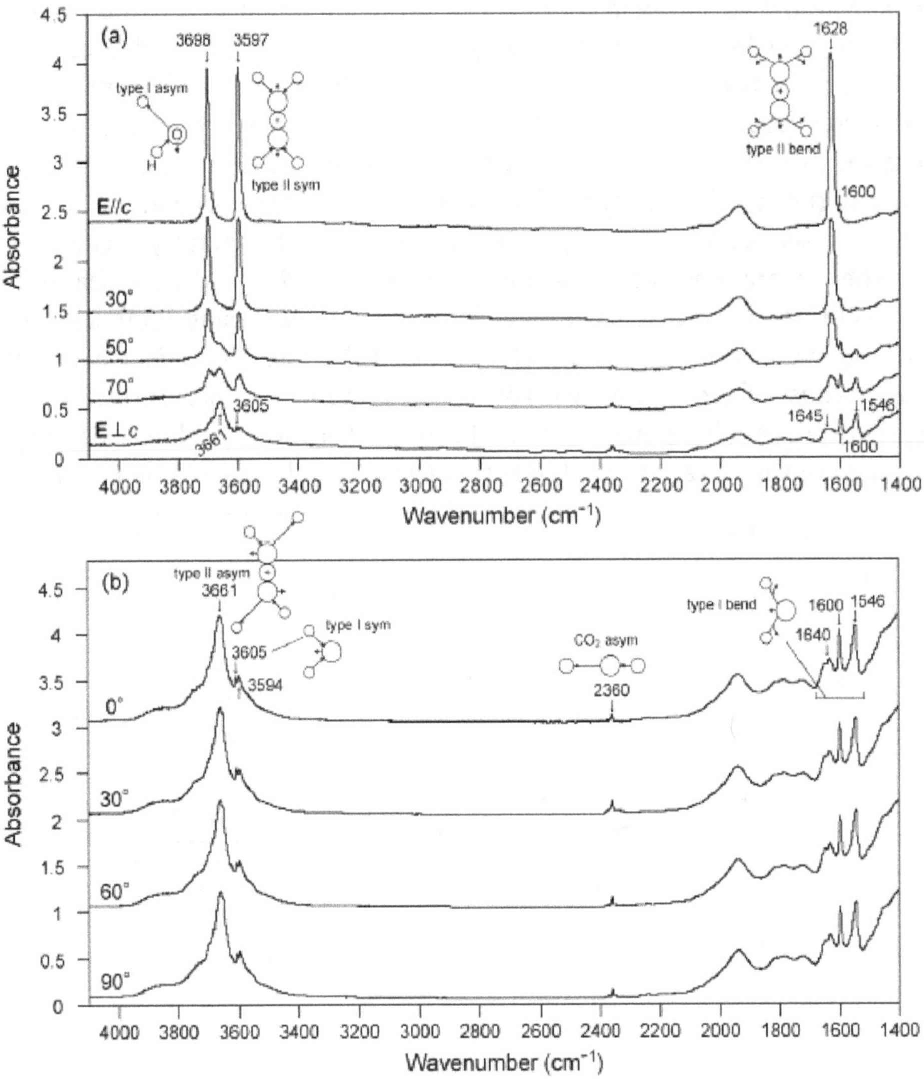

Fig. 5. Polarized IR spectra for natural beryl under different polarized conditions at RT. (a) From E//c-axis to E⊥c-axis in the (100)-section (the sample thickness of 20 μm). The angle of the c-axis (i.e., the direction of the channels) respective to E is shown on the left of each spectrum. (b) Under E⊥c-axis in the (001)-section (the sample thickness of 120 μm).

4.3 Water vibrations at high temperatures

Significant dehydration does not occur during short time heating from RT to 800 °C (temperature raise of 100 °C/minute and 1 minute for the measurements at each temperature; see Section 2). Figure 6 shows high temperature behaviour of type I/II H_2O in beryl under E//c-axis (Fig. 6a) and E⊥c-axis (Fig. 6b) in the water stretching and bending regions. Spectral changes are different for each vibrational mode of type I/II H_2O: Under E//c-axis at RT, the ν_3-I and ν_1-II bands are clearly seen at 3698 and 3597 cm^{-1}, respectively. The ν_3-I band is rapidly decreased in its height with increasing temperature; for example 60 % of the band height decreases at 200 °C, compared with that at RT (Fig. 7). The rapid decreases of the type I band are also seen for the ν_1-I band and the ν_2-I related bands under E⊥c-axis (Fig. 6b). Contrary to the case for the type I bands, only 20 % of the ν_1-II and ν_2-II bands are decreased at 200 °C. Alternatively, wavenumber shifts dominantly occur for these bands. The wavenumber of the former (3597 cm^{-1} at RT) and latter bands (1628 cm^{-1} at RT) linearly shift to lower and higher, with increasing temperature (Fig. 7). The changes of the ν_3-II band (3661 cm^{-1} at RT) under E⊥c-axis are difficult to monitor because of overlapping with other bands at high temperatures. Since these changes are reversible upon heating and cooling, they are not due to dehydration but changes of the states of type I/II H_2O in the channels. Discussion is as follows.

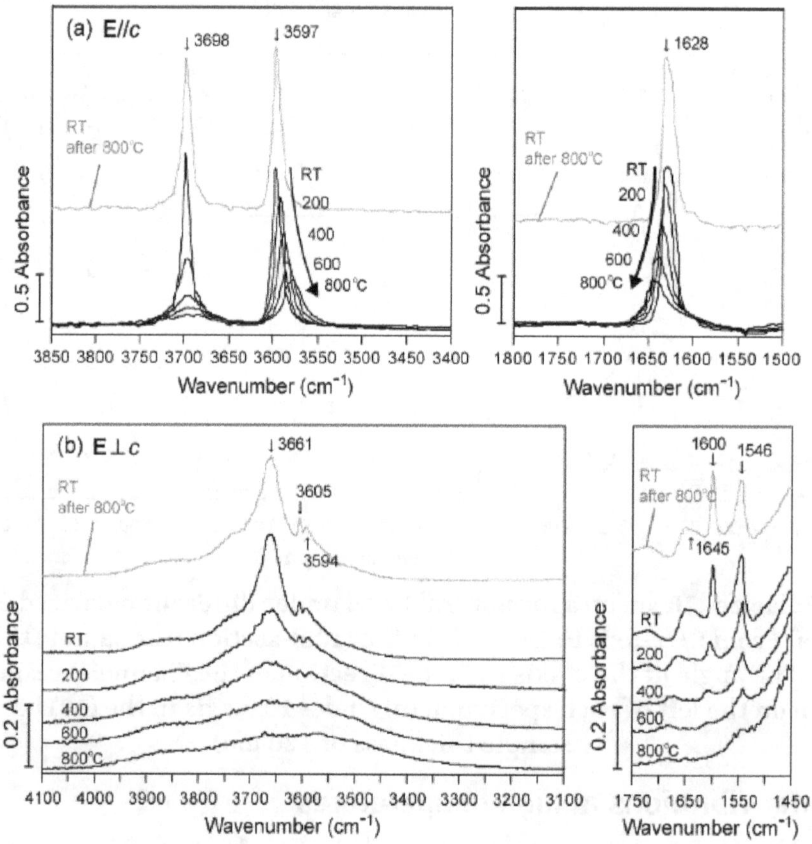

Fig. 6. High temperature behaviour of beryl from RT to 800 °C for the samples in Fig. 5 (Replotted from Fukuda & Shinoda, 2011). (a) under E//c-axis. (b) under E⊥c-axis. The RT spectra after heating at 800 °C are shown as gray lines, showing no significant dehyderation occured during heating.

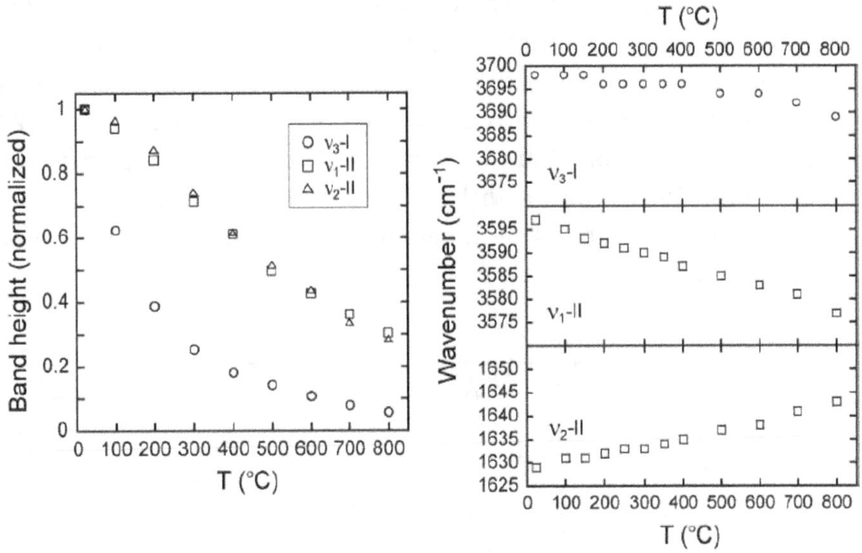

Fig. 7. Changes of band heights (left) and wavenumbers (right) of the ν_3-I, ν_1-II, and ν2-II bands with increasing temperature (Modified after Fukuda & Shinoda, 2011). Values are determined from the spectra in Fig. 6a.

These changes in band heights and wavenumbers of type I/II H_2O are interpreted mainly due to the presemce (type I) or absence of cations (type II) in the beryl channels (Fig. 4). Since type I H_2O is not coupled with a cation, its position in the channels is easy lost with increasing temperature, resulting the rapid decreases of band heights. On the other hand, Since the position of type II H_2O is fixed by a cation (mainly Na+), the decreases in band heights with increasing temperature do not significantly occur, compared with those for type I bands. Alternatively, the wavenumber shifts occur. The inverse wavenumber shifts of type II bands in stretching and bending modes would be due to modifications in vibrational constants in the thermally-expanded beryl channels (Fukuda et al., 2009b), as similar to the deviations of wavenumbers from ideal H_2O molecule in RT spectra (Section 4.2).

4.4 Dehydration behaviour

The beryl sample was heated on the heating stage at 850 °C where dehydration is enhanced (Fukuda & Shinoda, 2008). Polarized IR spectra are shown only at RT quenched from 850 °C (Fig. 8), Since in-situ high temperature IR spectra at 850 °C show broadened water bands (Fig. 6) and changes of each bands are not clearly monitored. Under E//c-axis, the ν_3-I band at 3698 cm^{-1} disappeared at heating of 12 hours, without any wavenumber changes (Fig. 8a). This trend is same with the band at 3605 cm^{-1} (ν_1-I) and three ν_2-I-related bands under E⊥c- axis (Fig. 8b). Some bands remain in the water bending region under E⊥c-axis, and they would be due to structural vibrations of beryl. The dehydration behaviour of type II bands are different with that of type I: Since the position of type II H_2O is fixed by a cation, its dehydration is obviously slower than type I H_2O. Under E//c-axis, new bands develop at 3587 and 1638 cm^{-1} with decreasing of the initial bands at 3597 (ν_1-II) and 1628 cm^{-1} (ν_2-II). The band at 3661 cm^{-1} (ν_3-II) under E⊥c-axis also shows the wavenumber shifts to the lower with decreasing its intensity. These bands are stable after 24 hours heating. The appearances of these bands are explained as follows.

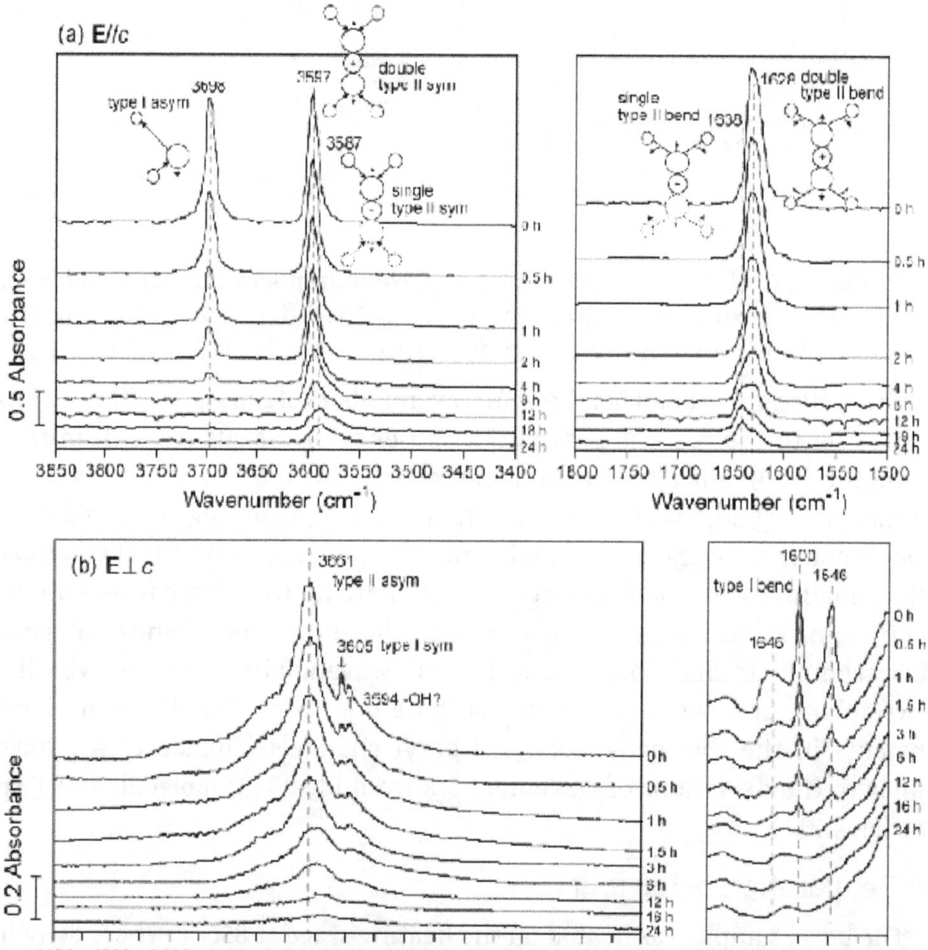

Fig. 8. IR spectra at RT quenched from heating at 850 °C, showing dehydration behaviour (Replotted from Fukuda & Shinoda, 2011). (a) under E//c-axis. (b) under E⊥c-axis. Heating times at 850 °C are shown at the right of each spectrum.

A cation is coordinated by one or two type II H_2O due to a spatial restriction of the channel (Fig. 4). Therefore, the dominant band at 3597 (ν_1-II) and 1628 cm^{-1} (ν_2-

II) before heating would be mainly due to doubly-coordinated type II to a cation, mainly Na+. Since type II H_2O dehydrates by heating, singly-coordinated type II are created. The wavenumbers of singly- and doubly-coordinated H_2O have been calculated for free H_2O molecules. According to a numerical approach for water vibrations by Bauschlicher et al. (1991), the wavenumbers of H_2O-Na+-H_2O is higher in its stretching modes and lower in the bending modes than those of Na+-H_2O in approximately 10 cm^{-1}. This is consistent with wavenumber shifts in beryl due to the formation of singly-coordinated type II at 3587 (ν_1-II) and 1638 cm^{-1} (ν_2-II).

Another possibility for the wavenumber shifts of the ν_1-II and ν_2-II bands is the presemce of Li+ in the channels. According to the calculation in Lee et al. (2004) for free H_2O molecules, the wavenumbers of Li+-H_2O is higher in its stretching modes and lower in the bending modes than those of Na+-H_2O. Also, binding energy of Li+ to H_2O is higher than that of Na+ to H_2O, which indicates the stable stability of Li+-H_2O during dehydration. If Li+ is trapped in the beryl channels, it can cause the wavenumber shifts observed in this study.

A sharp and unassigned band is seen at 3594 cm^{-1} under E⊥c-axis. This band is also more stable than that for type I bands. The wavenumber of this band is different from any vibrational modes of type I/II H_2O. Judging from the thermal stability and the wavenumber, this band may be related to Na+-OH in beryl, as its presemce has been argued in Andersson (2006).

5. IR MAPPING MEASUREMENTS FOR DEFORMED ROCKS

Rocks are deformed at shear zones in the interior of the earth. Rocks, which underwent brittle and plastic deformation at shear zones, are called as cataclasites and mylonites, respectively. Brittle deformation of continental crusts (mainly granitoids) is dominated from the ground to 10-20 km depth. Plastic deformation of rocks is dominated below that with increasing temperature and pressure. Another important factor that significantly contributes to plastic deformation of rocks is water. Water contents in ppm order dramatically promote plastic deformation of minerals, as confirmed by deformation experiments (e.g., Griggs, 1967; Jaoul et al., 1984; Post & Tullis, 1998; Dimanov et al., 1999). Also, water contributes to solution-precipitation which sometimes involves reactions among minerals (especially, feldspar and mica in granitoids) (e.g., summarized in Thompson & Rubie, 1985; Dysthe & Wogelius, 2006). Then, solution-precipitation creep may also contribute to the strength of the crusts (Wintsch & Yi, 2002; Kenis et al., 2005). Thus, water contents and distribution as well as its species are important for rock deformation.

In this section, I use IR spectroscopy to map two-dimensional water distributions as well as to consider its species in deformed granites. I especially focus on water distributions associated with solution-precipitation process of feldspar, and consider possible transport mechanisms of water.

5.1 Samples and analyses

Deformed granites were collected from outcrops in an inner shear zone of the Ryoke Metamorphic Belt in the Kishiwada district, Osaka Prefecture, SW Japan, and believed to be deformed at ~500 °C (Takagi, 1988; Imon et al., 2002; 2004). Sample thin sections of ~50 μm were at first observed under a polarized optical microscope and a back-scattered electron (BSE) image which reflects compositional differences in a scanning electron microscope (SEM). After IR mapping measurements, thin sections were again polished to suitable thickness (~20 μm) to observe Detailed microstructures under the optical microscope.

IR mapping measurements were carried out along ~1000 μm traverses with 30 μm spatial resolution (aperture size) in steps of 30 μm. See Section 2 for the instrument of IR spectroscopy. The integral absorbances of the water stretching bands in the range 3800–2750 cm^{-1} are displayed as a color-contoured image for a measured sample area. Color-contoured images can be used as a qualitative representation of the distribution of water, Since absorption coefficients tend to increase linearly with decreasing wavenumbers (Paterson, 1982; Libowitzky & Rossman, 1997). When the absolute water contents of the minerals are to be determined, Beer–Lambert law is applied, using the absorption coefficients for each mineral. For K-feldspar and plagioclase, I used the absorption coefficients of integral water stretching bands reported by Johnson & Rossman (2003) (15.3 ppm−1 cm−2), and for quartz, those reported by Kats (1962). Kats (1962) reported following relation between water content in quartz and integral absorbance of water stretching bands; C(H/106 Si)=0.812 x Aint/d, where Aint is the integral absorbance and d is the sample thickness in cm. Then, I converted H/106 Si value to a ppm H_2O unit (1 ppm = 6.67 H/106 Si), as also adopted in Gleason & DeSisto (2008). To distinguish Si-OH and H_2O contents separately, their combination bands of the stretching and bending modes should be used, as shown in Section 3. However, it is difficult for these samples, Since sample thicknesses are thin for texture observations under the optical microscope, and 1 mm thickness is needed to measure the combination bands.

5.2 Typical water distribution in deformed granite

At first, I introduce water distribution in the granite mylonite with typical microtexture (Fig. 9) (See Passchier & Trouw, 2005 for many textures of deformed rocks). As can be seen under the polarized optical microscope (Fig. 9a), quartz is recrystallized by subgrain rotation, and plastically deformed by dislocation creep. Quartz grains are elongated with the aspect ratio of ca. 3:1 and the long axis is ca. 250 μm. Feldspar is a relatively hard mineral in this deformation condition, is not plastically deformed, and behaves as rigid body sometimes with fracturing (i.e., brittle deformation). Such relatively hard minerals are called as porphyroclasts.

Under the BSE image, rims of plagioclase are replaced by K-feldspar, which would be due to solution-precipitation with or without reaction called myrmekitization (e.g., Simpson & Wintsch, 1989). Myrmekitization is the following reaction; K-feldspar + Na^{+} + CA_{2}+ = plagioclase + quartz + K+, where cations are included in circulating fluid water. Water contents in these replaced K-feldspar are difficult to determine in this region because of the limitation of its distribution, and discussed for other regions later.

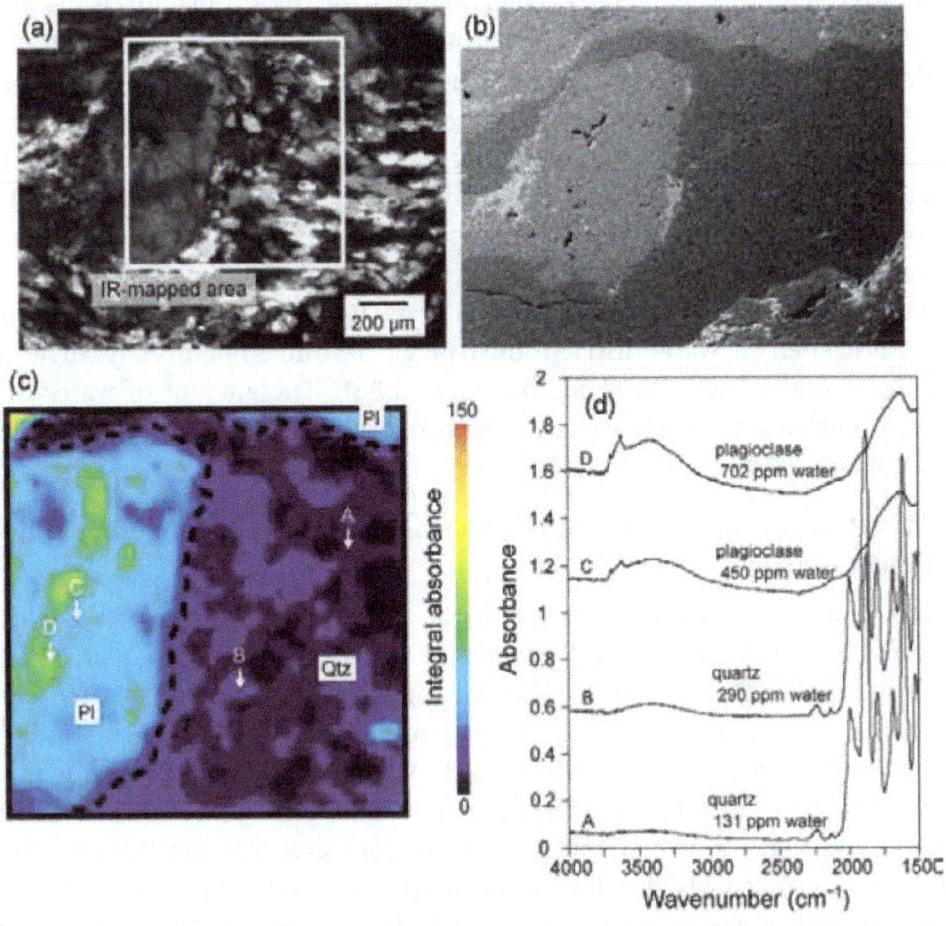

Fig. 9. (a) Optical microscopic image including the IR mapped area (bold square). (b) SEM- BSE image of the IR mapped area shown in (a): light gray; K-feldspar, medium gray; plagioclase, dark gray; quartz. (c) Water distribution mapped by integral absorbance of water stretching bands in the 3800–2750 cm–1 range of the IR spectra. The color contours from black to red approximately correspond to water contents from low to high. The boundaries of minerals are shown as dotted lines. Arrows with letters show the locations used to illustrate the various selected IR spectra in (d). Pl; plagioclase, Qtz; quartz.

The IR spectra for both plagioclase and quartz show broad bands at 3800–2750 cm^{-1}, which is due to the stretching vibration of fluid water (Fig. 9d) (See Section 3). Fluid water must be trapped as fluid inclusions within both minerals, Since spatial resolution of 30 μm (aperture size) covers intracrystalline regions, rather than intergranular regions. The IR spectra for the plagioclase porphyroclast also exhibit sharp bands at 3625 and 3700 cm^{-1}, which are due to the stretching vibrations of structural hydroxyl in plagioclase (e.g., Hofmeister & Rossman, 1985; Beran, 1987; Johnson & Rossman, 2003). The band at 1620 cm^{-1} in plagioclase is due to the bending vibrations of fluid water. Other bands in the range of 2500–1500 cm^{-1} are due to the structural vibrations of plagioclase. Six sharp bands between 2000 and 1500 cm^{-1} for quartz are due to the structural vibrations of quartz, which are same with the spectra for chalcedonic quartz (Fig. 1). Recognition of the bending vibrations of fluid water in quartz is difficult because of these sharp structural bands. Water concentration in this plagioclase porphyroclast ranges from 200 to 700 ppm, with an average of 450 ppm, consistent with values reported in the literature (Hofmeister & Rossman, 1985; Beran, 1987; Johnson & Rossman, 2003). The heterogeneity of water distribution in plagioclase does not directly correspond to textures under the optical microscope and BSE. The amount of water in the quartz is much lower than that in the plagioclase, ranging from 80 to 300 ppm, with an average of 130 ppm.

5.3 Water distribution around feldspar fine grains and possible water transportation

Water distribution was measured for an area where fine-grained K-feldspar develops around K-feldspar and plagioclase porphyroclasts (Fig. 10a). The BSE image shows that fine- grained K-feldspar regions, which are constructed by ~20 μm grains, contain patchy distribution of plagioclase (Fig. 10b). This indicates that solution-precipitation of K-feldspar, which may be accompanied with myrmekitization, occurred for the development of fine grains. The IR-mapped image shows that water contents in these regions are 220 ppm H_2O in average; low and homogeneously distributed, although fluid water must be participated in the solution-precipitation process. The features of water stretching bands of fine-grained K-feldspar and K-feldspar porphyroclasts do not show structural –OH bands, differently from plagioclase (Fig. 10d); only broad bands can be seen at 3800–2750 cm^{-1}. Water contents in K-feldspar and plagioclase porphyroclasts are 200-1150 ppm; heterogeneously distributed compared with those in fine-grained K-feldspar regions.

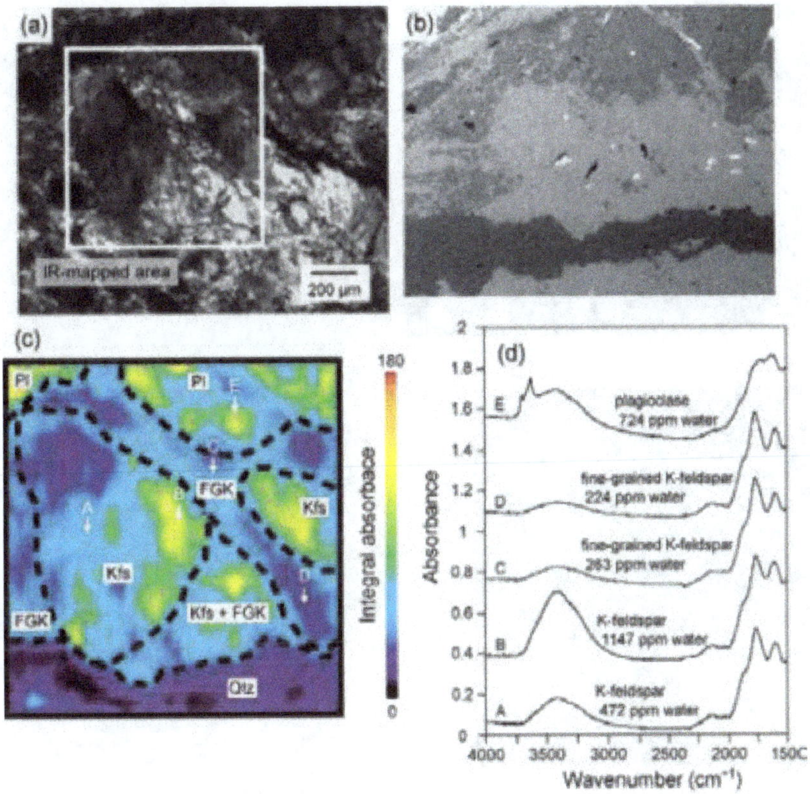

Fig. 10. (a) Optical microscopic image including the IR mapped area (bold square). (b) SEM- BSE image of the mapped area shown in (a): light gray; K-feldspar, dark gray; quartz. Fine- grained K-feldspar regions contain patchy-distributed plagioclase, indicating solution- precipitation occurred for the developments of these regions. (c) Water distribution mapped by integral absorbance of water stretching bands in the 3800–2750 cm−1 range of the IR spectra. The boundaries of minerals are shown as dotted lines. Arrows with letters show the locations used to illustrate the various selected IR spectra in (d). Pl; plagioclase, Kfs; K-feldspar, FGK; fine-grained K-feldspar, Qtz; quartz.

Figure 11 shows water distribution in an area that dominantly includes fine-grained plagioclase. The fine-grained plagioclase develops around plagioclase porphyroclasts, and some of it is closely associated with K-feldspar under the BSE image (Fig. 11b). Quartz in this region can be identified from its characteristic structural vibrations in the IR spectra (Fig. 11d), indicating that myrmekitization (K-feldspar + Na+ + CA_2+ = plagioclase + quartz + K+; see Section 5.2) occured during rocks deformation. Water contents in the area where fine-grained plagioclase grains are associated with small amounts of K-feldspar and quartz are roughly 2–4 times lower than those within plagioclase porphyroclasts, as inferred from the color contrasts in the IR mapping image (Fig. 11c). However, it is not possible to measure the absolute water contents in this area because of a mixture of plagioclase, K-feldspar, and quartz; consequently, the absorption coefficients are not clear.

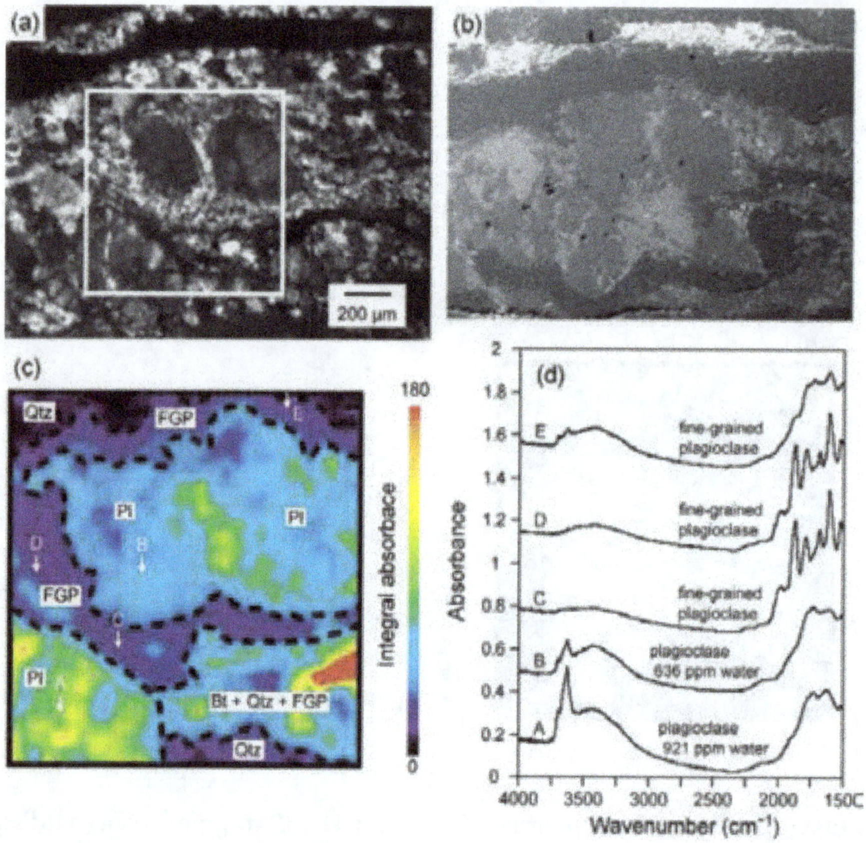

Fig. 11. (a) Optical microscopic image including the IR mapped area (bold square). (b) SEM- BSE image of the mapped area shown in (a): light gray; K-feldspar, medium gray; plagioclase, dark gray; quartz. (c) Water distribution mapped by integral absorbance of water stretching bands in the 3800–2750 cm−1 range of the IR spectra. The boundaries of minerals are shown as dotted lines. Arrows with letters show the locations used to illustrate the various selected IR spectra in (d). Pl; plagioclase, FGP; fine-grained plagioclase, Qtz; quartz. IR spectra show quartz vibrational bands in fine-grained plagioclase. Absolute water contents for fine-grained plagioclase are not determined due to the mixtures of quartz and K-feldspar.

In general, fluid water can be trapped at intergranular regions (grain boundaries) up to a few thousand ppm H_2O, as reported for quartz aggregates whose each grain size is a few tens of micrometers to a few hundred micrometers (e.g., Nakashima et al., 1995; Muto et al., 2004; O'kane et al., 2005). In this study, intergranular regions must also be covered in the measurements for fine-grained K-feldspar- and plagioclase-dominant regions. The solution-precipitation process that produces fined-grained K-feldspar and plagioclase are subsequently and/or simultaneously enhanced by the ability of fluid water along intergranular regions to carry ions in solution, especially in situations where intergranular diffusion was promoted by an increase in surface area. (e.g., Simpson & Wintch, 1989; Fitz Gerald & Stünitz, 1993; Tsurumi et al., 2003). However, contrary to the previous knowledge on water contents at intergranular regions, water contents in fine-grained K- feldspar- and plagioclase-dominant regions are low and homogeneous (av. 250 ppm). Therefore, it can be inferred that fluid water may not abundantly be trapped in newly- created grains, and rather released during and/or after the solution-precipitation process. The release of fluid water and dissolved ions, which contributed to the process, was a result of the concentration gradients formed in the context of numerous newly-created intergranular regions within the mix of fine-grained feldspar and quartz. As a consequence, the entire process results in a positive feedback to promote the solution- precipitation process.

6. CONCLUSIONS

I investigated high temperature behaviour of water in rocks and minerals. Chalcedonic quartz was used as a representative rock which contains abundant fluid water at intergranular regions and -OH in quartz crystal structures. The average coordination numbers of water molecules were degreased with increasing temperature, which causes shifts of stretching vibrations to higher wavenumbers. Dehydration of fluid water in the chalcedonic quartz was monitored by keeping at high temperatures. Fluid water was rapidly dehydrated through intergranular regions at 500 °C, and new hydroxyl bands appeared with dehydration of fluid water.

States of water molecules, which are not clustered like fluid water, were investigated for beryl, typical cyclosilicate, using high temperature polarized IR spectroscopy. The beryl channels, open cavities in the crystal structures, contain two types of water molecules which freely exist or coordinate to cations from up and/or below them. The former type of water easily looses its specific position, resulting the rapid degreases of its IR band heights without dehydration, and shows rapid dehydration at 850 °C. The latter type of water shows significant changes in its wavenumbers with increasing temperatures. There are slight modifications in the wavenumbers during dehydration due to changes of coordination to cations during dehydration.

Distribution of fluid water was measured for deformed granites. K-feldspar and plagioclase fine grains were formed around porphyroclasts by solution-precipitation process. Water contents in fine-grained K-feldspar- and plagioclase-dominant regions show low and homogeneous distribution of fluid water, while water distributions in host porphyroclasts were heterogeneous. This indicates that fluid water, which was involved in the solution-precipitation process, was released during and/or after the solution- precipitation process.

7. REFERENCES

1. Aines, R.D. & Rossman, G.R. (1984). Water in minerals? A peak in infrared. Journal of Geophysical Research, Vol. 89, pp. 4059-4071
2. Andersson, L.O. (2006). The positionof H+, Li+ and Na+ impurities in beryl. Physics and Chemistry of Minerals, Vol. 33, pp. 403-416
3. Artioli, G., Rinaldi, R., Ståhl, K. & Zanazzi, P.F. (1993). Structure refinements of beryl by single-crystal neutron and X-ray diffraction. American Mineralogist, Vol. 78, pp. 762- 768
4. Aurisicchio, C., Fioravanti, G., Grubessi, O. & Zanazzi, P.F. (1988). Reappraisal of the crystal chemistry of beryl. American Mineralogist, Vol. 73, pp. 826-837
5. Bauschlicher, C.W., Langhoff, S.R., Partridge, H., Rice, J.E. & Komornicki, A. (1991). A theoretical study of $Na(H_2O)+n$ (n = 1-4). Journal of Chemical Physics, 95, 5142-5148.
6. Beran, A., (1987). OH groups in nominally anhydrous framework structures: an infrared spectroscopic investigation of Danburite and Labradorite. Physics and Chemistry of Minerals, Vol. 14, pp. 441-445
7. Brubach, J.B., Mermet, A., Filabozzi, A., Gerschel, A. & Roy, A. (2005). Signatures of the hydrogen bonding in the infrared bands of water. Journal of Chemical Physics, Vol. 122, Article No. 184509
8. Charoy, B., de Donato, P., Barres, O. & Pinto-Coelho, C. (1996). Channel occupancy in an alkali-poor beryl from Serra Branca (Goias, Brazil): spectroscopic characterization. American Mineralogist, Vol. 81, pp. 395-403
9. Della Ventura, G., Bellatreccia, F., Cesare, B., Harley, S. & Piccinini, M. (2009). FTIR microspectroscopy and SIMS study of water-poor cordierite from El Hoyazo, Spain: application to mineral and melt devolatilization. Lithos, Vol. 113, pp. 498-506
10. Dimanov, A., Dresen, G., Xiao, X., & Wirth, R. (1999). Grain boundary diffusion creep of synthetic anorthite aggregates: the effect of water. Journal of Geophysical Research, Vol. 104, pp. 10483-10497
11. Dysthe, D.K. & Wogelius, R.A. (2006). Confined fluids in the Earth's crust – Properties and processes. Chemical Geology, Vol. 230, pp. 175-181
12. Eisenberg, D. & Kauzman, W. (1969). The Structure and Properties of Water, Oxford University Press, Oxford
13. Falk, M. (1984). The frequency of the H–O–H bending fundamental in solids and liquids.
14. Spectrochimica Acta, Vol. A40, pp. 43-48

15. Fitz Gerald, J.D. & Stünitz (1993). Deformation of granitoids at low metamorphic grade. I: reactions and grain size reduction. TectonoPhysics, Vol. 221, pp. 269-297

16. Frondel, C. (1982). Structural hydroxyl in chalcedony (Type B quartz). American Mineralogist, Vol. 67. pp. 1248-1257

17. Fukuda, J. & Shinoda, K. (2008). Coordination of water molecules with Na+ cations in a beryl channel as determined by polarized IR spectroscopy. Physics and Chemistry of Minerals, Vol. 35, pp. 347-357

18. Fukuda, J. & Nakashima, S. (2008). Water at high temperatures in a microcrystalline silica (chalcedony) by in-situ infrared spectroscopy: physicochemical states and dehydration behavior. Journal of Mineralogical and Petrological Sciences, Vol. 103, pp. 112-115

19. Fukuda, J., Peach, C.J., Spiers, C.J. & Nakashima, S. (2009a). Electrical impedance measurement of hydrous microcrystalline quartz. Journal of Mineralogical and Petrological Sciences, Vol. 104, pp. 176-181

20. Fukuda, J., Shinoda, K., Nakashima, S., Miyoshi, N. & Aikawa, N. (2009b). Polarized infrared spectroscopic study of diffusion of water molecules along structure channels in beryl. American Mineralogist, Vol. 94, pp. 981-985

21. Fukuda, J., Yokoyama, T. & Kirino, Y. (2009c). Characterization of the states and diffusivity of intergranular water in a chalcedonic quartz by high temperature in-situ infrared spectroscopy. Mineralogical Magazine, Vol. 73, pp. 825-835

22. Fukuda, J. & Shinoda, K. (2011). Water molecules in beryl and cordierite: high-temperature vibrational behavior, dehydration, and coordination to cations. Physics and Chemistry of Minerals, Vol. 38, pp. 469-481

23. Gibbs, G.V., Breck, D.W. & Meagher, E.P. (1968). Structural refinement of hydrous and anhydrous synthetic beryl, $Al_2(Be_3Si_6)O_{18}$ and emerald, $Al_{1.9}Cr_{0.1}(Be_3Si_6)O_{18}$. Lithos, Vol. 1, pp. 275-285

24. Gleason, G.C. & DeSisto, S. (2008). A natural example of crystal-plastic deformation enhancing the incorporation of water into quartz. TectonoPhysics, Vol. 446, pp. 16-30 Goldman, D.S. & Rossman, G.R. (1977). Channel constituents in cordierite. American

25. Mineralogist, Vol. 62, pp. 1144-1157

26. Graetsch, H., Flörke, O.W. & Miehe, G. (1985). The nature of water in chalcedony and opal-C from Brazilian agate geodes. Physics and Chemistry of Minerals, Vol. 12, pp. 300-306

27. Grigss, D.T. (1967). Hydrolytic weakening of quartz and other silicates. Geophysical Journal of the Royal Astronomical Society, Vol. 14, pp. 19-32
28. Hawthorne, F.C. & Černý, P. (1977). The alkali-metal positions in Cs-Li beryl. Canadian Mineralogist, Vol. 15, pp. 414-421
29. Hiraga, T., Nishikawa, O., Nagase, T. & Akizuki, M. (2001). Morphology of intergranular pores and wetting angles in pelitic schists studied by transmission electron microscopy. Contributions to Mineralogy and Petrology, Vol. 141, pp. 613-622
30. Hofmeister, A.M. & Rossman, G.R. (1985). A model for the irradiative colaration of smoky feldspar and the inhibiting influence of water. Physics and Chemistry of Minerals, Vol. 12, pp. 324-332
31. Imon, R., Okudaira, T. & Fujimoto, A. (2002). Dissolution and precipitation processes in deformed amphibolites: an example from the ductile shear zone of the Ryoke metamorphic belt, SW Japan. Journal of Metamorphic Geology, Vol. 20, pp. 297-308
32. Imon, R., Okudaira, T. & Kanagawa, K. (2004). Development of shape- and lattice-preferred orientations of amphibole grains during initial cataclastic deformation and subsequent deformation by dissolution-precipitation creep in amphibolites from the Ryoke metamorphic belt, SW Japan. Journal of Structural Geology, Vol. 26, pp. 793-805
33. Ingrin, J., Hercule, S. & Charton, T. (1995). Diffusion of hydrogen in diopside: results of dehydration experiments. Journal of Geophysical Research, Vol. 100, pp. 15489-15499
34. Jacobsen, S.D. & Van der Lee, S. (2006). Earth's Deep Water Cycle, Geophysical Monograph Series, Vol. 168, American Geophysical Union, Washington, D.C.
35. Jaoul, O. (1984). Sodium weakening of Heavitree quartzite: preliminary results. Journal of Geophysical Research, Vol. 89, pp. 4271-4280
36. Johnson, E. & Rossman, G.R. (2003). The concentration and speciation of hydrogen in feldspars using FTIR and 1H MAS NMR spectroscopy. American Mineralogist, Vol. 88, pp. 901-911
37. Kats, A. (1962). Hydrogen in alpha quartz. Philips Research Report, Vol. 17, pp. 1-31, 133–195, 201-279
38. Kenis, I., Urai, J.L., Van der Zee, W., Hilgers, C. & Sintubin, M. (2005). Rheology of fine-grained siliciclastic rocks in the middle crust-evidence from structural and numerical analysis. Earth and Planetary Science Letters, Vol. 233, pp. 351-360

39. Keppler, H. & Smyth, J.R. (2006). Water in Nominally Anhydrous Minerals, Reviews in Mineralogy and Geochemistry, Vol.62, The Mineralogical Society of America

40. Kihara, K. (2001). Moleclar dynamiCs interpretation of structural changes in quartz, Physics and Chemistry of Minerals, Vol. 28, pp. 365-376

41. Kronenberg, A.K. & Wolf, G.H. (1990). Fourier transform infrared spectroscopy determinations of intragranular water content in quartz-bearing rocks: implications for hydrolytic weakening in the laboratory and within the earth, TechtonoPhysics, Vol. 172, pp. 255-271

42. Lee, H.M., Tarakeshwar, P., Park, J., Kołaski, M.R., Yoon, Y.J., Yi, H.B., Kim, W.Y. & Kim,

43. K.S. (2004). Insights into the structures, energetiCs, and vibrations of monovalent cation-(water)1-6 clusters. Journal of Physical Chemistry A, Vol. 108, pp. 2949-2958

44. Libowitzky, E. & Rossman, G.R. (1997). An IR absorption calibration for water in minerals.

45. American Mineralogist, Vol. 82, pp. 1111-1115

46. Łodziński, M., Sitarz, M., Stec, K., Kozanecki, M., Fojud, Z. & Jurga, S. (2005). ICP, IR, Raman, NMR investigations of beryls from pegmatites of the Sudety Mts. Journal of Molecular Structure, Vol. 744, pp. 1005-1015

47. Muto, J., Nagahama, H., & Hashimoto, T. (2004). Microinfrared reflection spectroscopic mapping: application to the detection of hydrogen-related species in natural quartz. Journal of Microscopy-Oxford, Vol. 216, pp. 222-228

48. Nakahara, M., Matubayasi, N., Wakai, C. & Tsujino, Y. (2001). Structure and dynamiCs of water: from ambient to supercritical. Journal of Molecular Liquids, Vol. 90, pp. 75-83

49. Nakashima, S., Matayoshi, H., Yuko, T., Michibayashi, K., Masuda, T., Kuroki, N., Yamagishi, H., Ito, Y. & Nakamura, A. (1995). Infrared microspectroscopy analysis of water distribution in deformed and metamorphosed rocks. TectonoPhysics, Vol. 245, pp. 263-276

50. O'kane, A., Onasch, C.M. & Farver, J.R. (2007). The role of fluids in low temperature, fault- related deformation of quartz arenite. Journal of Structural Geology, Vol. 29, pp. 819- 836

51. Okumura, S. & Nakashima, S. (2004). Water diffusivity in rhyolitic glasses as determined by in situ IR spectroscopy. Physics and Chemistry of Minerals, Vol. 31, pp. 183-189

52. Passchier, C.W. & Trouw, R.A.J. (2005). MicrotectoniCs (2nd Ed). Springer-Verlag, Heidelberg

53. Paterson, M.S. (1982). The determination of hydroxyl by infrared absorption in quartz, silicate glasses and similar materials. Bulletin de Minéralogie, Vol. 105, pp. 20-29

54. Post, A. & Tullis, J. (1998). The rate of water penetration in experimentally deformed quartzite: implications for hydrolytic weakening. TectonoPhysics, Vol. 295, pp. 117- 137

55. Schwarzer, D. (2005). Energy relaxation versus spectral diffusion of the OH-stretching vibration of HOD in liquid-to-supercritical deuterated water. Journal of Chemical Physics, Vol. 123, Article No. 161105

56. Simpson, C. & Wintsch, R.P. (1989). Evidence for deformation-induced K-feldspar replacement by myrmekite. Journal of Metamorphic Geology, Vol. 7, pp. 261-275

57. Takagi, H., Mizutani, T. & Hirooka, K. (1988). Deformation of quartz in an inner shear zone of the Ryoke belt – an example in the Kishiwada area, Osaka Prefecture. Journal of the Geological Society of Japan, Vol. 94, pp. 869-886 (in Japanese with English abstract).

58. Thompson, A.B. & Rubie, D.C. (1985). Metamorphic Reactions, Kinetics, Textures and Deformation, Springer, New York

59. Tsurumi, J., Hosonuma, H. & Kanagawa, K. (2003). Strain localization due to a positive feedback of deformation and myrmekite-forming reaction in granite and aplite mylonites along the Hatagawa Shear Zone of NE Japan. Journal of Structural Geology, Vol. 25, pp. 557-574

60. Wintsch, R.P. & Yi, R. (2002). Dissolution and replacement creep: a significant deformation mechanism in mid-crustal rocks. Journal of Structural Geology, Vol. 24, pp. 1179-1193

61. Wood, D.L. & Nassau, K. (1967). Infrared spectra of foreign molecules in beryl. Journal of Chemical Physics, Vol. 47, pp. 2220-2228

62. Yamagishi, H., Nakashima, S. & Ito, Y. (1997). High temperature infrared spectra of hydrous microcrystalline quartz, Physics and Chemistry of Minerals, Vol. 24, pp. 66-74

5 Chapter

ATTENUATED TOTAL REFLECTION –INFRARED SPECTROSCOPY APPLIED TO THE STUDY OF MINERAL – AQUEOUS ELECTROLYTE SOLUTION INTERFACES

A GENERAL OVERVIEW AND A CASE STUDY

1. INTRODUCTION

The present chapter gives an overview of the application of Attenuated total reflection - Infrared spectroscopy (ATR-IR) to the environmentally important mineral - aqueous electrolyte interface. At these interfaces the important adsorption processes occur that limit the availability of potentially toxic solutes. These retention processes may retard for example the migration of solutes in aquifer systems or even immobilize them on the aquifer material, which is usually a natural mineral. Selected solutes may also via a preliminary adsorption process, which weakens bonds, enhance both dissolution kinetics and the equilibrium solubility of a given mineral.

In the context of retardation (oxy)(hydr)oxide minerals are of major importance. At the surface of these minerals surface functional groups exist that are able to bind metal ions and organic ligands as well as they may promote the formation of so-called ternary surface complexes involving both metal ions and some ligand. To be able to quantify these retention phenomena in porous media (such as aquifers or soils) a physical model of solvent movement is coupled to a (chemical) adsorption model (usually some variant of the surface complexation approach). The intent in the chemical part of the model is to invoke as much understanding of the adsorption process as possible. Thus it turns out to be important whether an adsorption process results in monodentate or multidentate surface complexes. This can have profound consequences in the use of a surface complexation model under different conditions (Kulik et al., 2010; Kallay et al., 2011). Evaluating a surface complexation model based on macroscopic adsorption data alone usually is not unambiguous. Consequently, it is required to study the adsorption process at the molecular level. Various spectroscopic approaches have been used to resolve the adsorption mechanism, one being ATR-IR.

We give an introduction to the approach and an overview of its possible applicability (and in this context its use in contributing to the understanding of the acid-base chemistry of (oxy)(hydr)oxide mineral surfaces, the adsorption of anions and cations like the uranyl-ion, and the formation of ternary surface complexes can be mentioned in general). Our contribution focuses on a review on the interaction of small organic molecules with oxidic surfaces and we highlight previous studies and point to some controversary issues in selected studies that continue to exist despite extensive research. Obviously such studies relate to other vibrational spectroscopies like Raman or sum frequency generation vibrational spectroscopies.

Finally we discuss results from an experimental study on the mineral gibbsite ($Al(OH)_3$) in the presemce of 5-sulfosalicylic acid (5-SSA). We show how ideally such a study should be designed, starting from the study of the gibbsite-electrolyte solution system (i.e. in absence of 5-SSA) and that of 5-SSA in aqueous solution (i.e. in the absence of gibbsite). Furthermore, we show that it is necessary to study in aqueous solution the interaction of 5-SSA with dissolved aluminium, Since the pH – dependent solubility of gibbsite will ultimately cause the appearance of aluminium ions in solution. The system involving gibbsite and 5-SSA is discussed in more detail. We relate the data to calculations of the species distribution for the solution systems, which indicate the dominant aqueous species thus facilitating the assignment of bands.

2. REVIEW OF USE OF ATR IN STUDIES ABOUT ADSORPTION OF SELECTED SMALL ORGANIC MOLECULES

2.1 Principles of ATR

The Attenuated Total Reflection effect is based on the existence of an evanescent wave in a medium of lower index of refraction in contact with an optically denser medium in which the infrared beam is sent. This evanescent field decays exponentially in the less dense medium according to equation (1).

$$E = E_0 \exp\left[-\frac{2\pi}{\lambda_1}\left(\sin^2\theta - n_{21}^2\right)^{1/2} Z\right] \qquad (1)$$

where $l_1 = l / n_1$ is the wavelength of the radiation in the denser medium, l the wavelength in free space, q the angle of incidence with respect to the normal. The parameter n_{21} is defined as the ratio of the refractive indices, i.e. $n_{21} = n_2 / n_1$, where n_1 and n_2 are respectively, the refractive indices of the optically denser and less dense media, and Z is the distance from the surface (Mirabella, 1993) (see Fig. 1).

From the ATR element, the infrared beam probes only the first few micrometers of the sample medium. From equation (1), different parameters can be defined to characterize the depth of penetration. A first definition was the depth at which the electric field amplitude falls to half its value at the interface (Z = 0.69 / g). Another definition of the depth of penetration (d_p) is given by Z = 1 / g, i.e. a decay of the electric field of 63 %. Moreover, this value is lower than the actual depth sampled (d_s), which is about three times d_p (decay of the electric field of 95%) (Mirabella, 1993; Tickanen et al., 1991). Equation (1) can be used to obtain the value of d_p in a homogeneous solution, but the determination of the penetration across oxy-hydroxide films is more complex. The depth of penetration, d_p, is expressed as (Coates, 1993)):

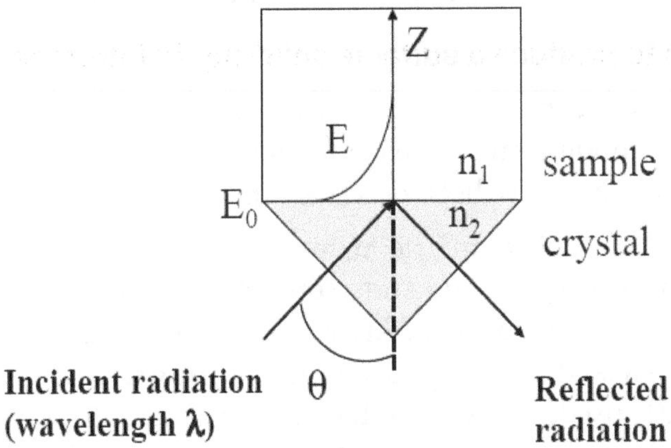

Fig. 1. Schematic diagram of the attenuated total reflection of the infrared beam in a monoreflection ATR accessory.

$$d_p = \frac{\lambda_1}{2\pi}\left(\sin^2\theta - n_{21}^2\right)^{-1/2} \quad (2)$$

or, with n, the wavenumbers (cm^{-1}):

$$d_p = \frac{10000}{2\pi v n_1}\left(\sin^2\theta - n_{21}^2\right)^{-1/2} \quad (3)$$

In studies on the adsorption of ions onto layers of particles deposited on ATR crystals, it is important that the whole layer be probed. Otherwise sorption which takes place in the top of the layer (i.e. further away from the crystal) does not significantly contribute to the observed signal. To take into account the presemce of a layer of particles (pores filled with solution) formula (1) can be used with a volume-weighted average of the refractive index of the particle material and the aqueous solution (Hug and Sulzberger, 1994):

$$d_p = F_v \times n_{par} + (1 - F_v) \times n_{water} \quad (4)$$

where F_v is the volume fraction of solid and n_{par} the refractive index of the pure solid. A volume fraction between 0.30 and 0.40 was estimated for TiO_2 (n_{par} = 2.6), leading to a maximum d_p of 2.6 µm at 1100 cm^{-1}. Thus, the actual depth sampled would be ca. 7 µm (dS = 3´d_p), indicating that the deposited layer should be thinner than this value.

2.2 Experimental

Using an accessory allowing to record infrared spectra in ATR mode is the first requirement to get in situ signals of the solid/solution interface. However, the way to prepare this interface and even the choice of the accessory is not straightforward.

2.2.1 Protocols to produce a suitable solid-liquid interface

The first step in ATR-related studies involves the formation of a suitable solid-liquid interface. To obtain a metal oxide / solution interface which can be probed by ATR, several methods have been described in literature.

The first one, described in the pioneering work by Tejedor-Tejedor and collaborators (Tejedor-Tejedor and Anderson, 1986; Tejedor-Tejedor and Anderson, 1990; Tickanen et al., 1991) consisted in a cylindrical internal reflection cell (a rod-shaped crystal of ZnSe) dipped in a suspension of 100 g/L goethite. This method is now less frequently used, to the advantage of horizontal ATR crystals. Using such instrumentation, Hug and Sulzberger (1994) have developed a method which has become standard. The approach consists in coating the ATR crystal by colloidal particles to form a film. As a typical protocol, a mixture of solid and ethanol is spread over the ATR crystal, then dried using a nitrogen flux. After drying, the layer is rinsed with water or with an electrolyte solution. More details are given in articles by Hug (1997) or Peak et al. (1999).

In another method, the equilibrium of the system solid/solution is reached by a classical batch experiment, using diluted suspensions of the solid. Then the suspensions are centrifuged to obtain a higher mass/volume ratio, for example 100-1000 g/L, or even a paste. The sample is then spread on the ATR crystal using a spatula (Villalobos and Leckie, 2001).

A final possibility is to use the surface of the crystal as the sample itself. Either the surface of the crystal is used as received, as ZnSe on which sodium dodecyl sulfate (Gao and Chorover, 2010) or Ge on which heptyl xanthate (Larsson et al., 2004) formed a monolayer, or the surface was chemically modified and is different from the bulk. Thus, Asay and Kim (2005) studied the adsorption of water molecules on the native layer of silica present on a silicium ATR crystal, or Wang et al. (2006) studied the adsorption of hexane and ethylbenzene from the vapor phase on a layer of zeolite grown directly on the surface of a silicium ATR crystal. Frederiksson and Holmgren (2008) have formed a PbS film on a ZnS ATR crystal by a chemical bath deposition process in order to study the adsorption of heptyl xanthate. In these latter studies, the system is very close to a film obtained by drying of a suspension, but the optical properties are expected to be better. Couzis and Gulari (1993) have deposited 600 Å of alumina by sputtering on a ZnSe crystal.

The advantages and drawbacks of the three methods to prepare the solid/ solution interface discussed above are listed in table 1. As of today, the most common method is to prepare a dry layer, even though it is simpler to use a paste. However, using a paste has a major drawback Since the contact between particles and the ATR crystal is not optimal, the sensitivity is low and depends on the suspension structure (which in general is pH- dependent). On the other hand, using the results obtained with a dry layer to interpret macroscopic data obtained in well-dispersed suspensions can be tricky, Since effects due to the confinement of the solutions are ignored. The last advantage of using a film is the possibility to perform experiments with a flow cell. This set-up allows recording spectra, while varying the composition of the solution, e.g. by modifying pH, or the concentration of adsorbing species.

Table 1. Summary of characteristics of the three methods of preparation of the solid/solution interface probed by ATR-IR: (++) strong advantage, (+) advantage, (-) drawback.

Method	Paste	Dried Layer	Film growth or crystal only
Ease of preparation	++	+	-
Variety of solids to best studied	++	+	-
Optical quality (sensitivity)	-	+	++
Quantitative evaluation of spectra	-	+	++
Representativity/suspension	++	+	-
Flow cell	-	+	++

2.2.2 Limitations in the wavenumber range

Once the procedure to prepare the interface has been chosen, the wavenumber range covered by the measurement is another important experimental aspect. Indeed, a number of interferences may occur between bands of adsorbed species and the experimental set- up.

The main limitation may arise from the ATR element itself. Each material has a transmission threshold, which may be located at a high wavenumber, such as silicium. Other materials with a low transmission threshold may be too reactive towards solutions. Thus, ZnSe can be attacked by acid or zinc-complexing species. A usual choice made by ATR-elements-dealers is an element made in ZnSe, but covered by a thin layer of diamond to increase its chemical resistance. This possibility exists only for small ATR crystals, allowing only few reflections of the infrared beam. To increase sensitivity, large ATR crystals are used. For example 40 mm × 10 mm crystals with a thickness of around 1 mm allow dozens of reflections. Such crystals usually consist of a pure material.

Besides the above limitations due to the ATR element, two gases present in the ambient atmosphere lead to absorption bands in IR spectra: carbon dioxide, and water. The main bands (Fig. 2) consist in a doublet at 2361 and 2339 cm^{-1} (CO_2), and numerous narrow peaks in the range 2000 – 1300 (H_2O bending) and 4000 – 3400 (H_2O stretching). Generally, the band of CO_2 does not interfere with bands of adsorbates, but H_2O bending can interfere with adsorbed organic molecules. Several methods exist to solve this problem. In fact, the presemce of CO_2 and water in the atmosphere of the spectrometer is not the actual problem Since it is taken into account in the background spectrum. It is rather the evolution of their concentrations (or partial pressures) during the subsequent spectra collection that leads to the presemce of bands, which varies with time. The less concentrated these gases are, the lower are the bands, Since the signal comes from the fluctuation of the partial pressures. Thus, some spectrometers are evacuated to enhance sensitivity and stability. Other spectrometers are purged with inert gas or with compressed dried Air. Another possible solution to the problem consists in the use of spectrometers which are sealed and equipped with desiccant powder. In all cases, if bands of atmospheric compounds remain, they can be tentatively removed by subtracting the atmosphere spectra.

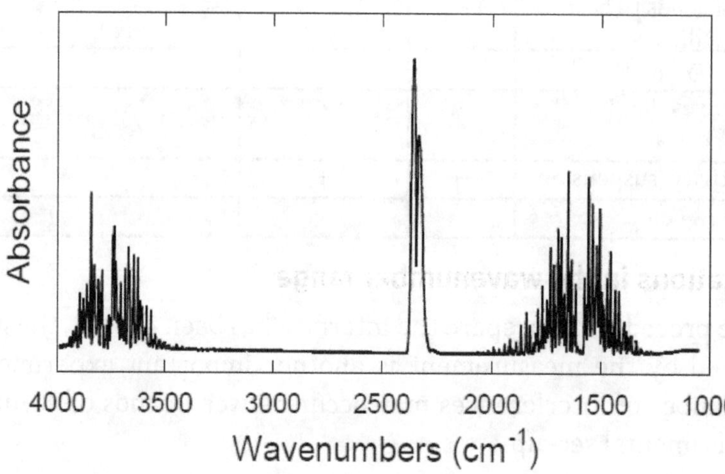

Fig. 2. Spectrum of the atmosphere showing the contributions as discussed in the text.

Since the studies generally consist in probing the species adsorbed on a solid deposited on the ATR crystal, it is important to take into account bands from the solid itself. For metal oxides, the absorption bands are generally located at low wavenumbers, which does not cause interferences with adsorbed species. Exceptions exist with light metals as SiO_2 (around 1060 cm^{-1}). For metal hydroxides, stretching of M-OH can lead to the presemce of bands above 800 cm^{-1} as is the case with goethite (900 and 800 cm^{-1}) or gibbsite (around 1000 cm^{-1}).

Ideally, if the layer formed by particles is stable, the signal coming from the solid can be subtracted from the final spectra, and the presemce of these bands does not hamper the detection and interpretation of bands from adsorbed species. However, in practice subtraction is often difficult due to the evolution of the signal of the solid with time or solution composition. Phenomena such as re-entrainment of particles by flowing solution, or swelling/shrinkage due to the change in surface potential can explain this problem.

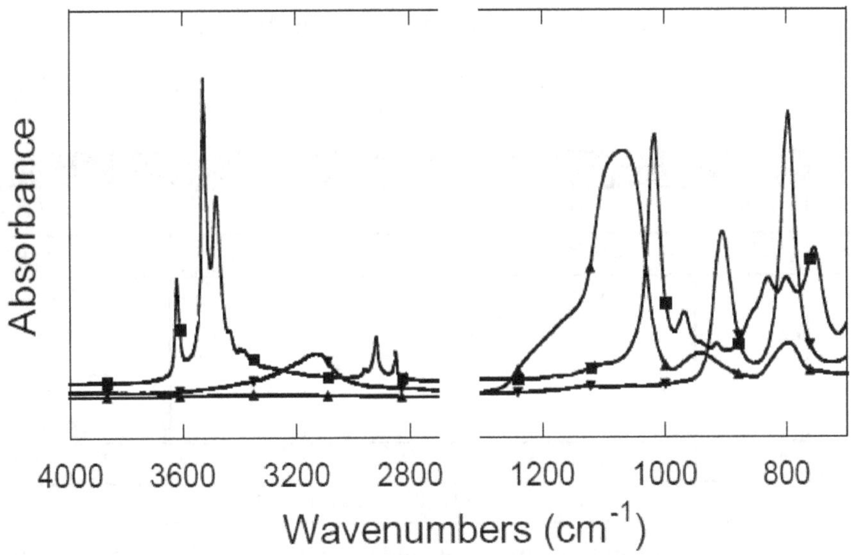

Fig. 3. Spectra of solids as dried layer on an ATR element: silica ( ), gibbsite (n), goethite (,)

Water is the most common solvent in environmental studies and its absorption bands can be a problem too. Stretching of H_2O occurs around 3000-3600 cm^{-1} and interferes with stretching of surface hydroxyl groups. Bending takes place at 1643 cm^{-1} (Venyaminov and Prendergast, 1997), close to the stretching of C=O groups (see below). This can complicate the accurate measurement of nC=O maxima. Finally, water absorption is very strong below ca. 900 cm^{-1}, and this can prevent the measurement of any bands in the lowest wavenumber range. In fact the actual threshold appears to depend on the number of reflections in the ATR system. For a monoreflection accessory, a measurement can be made down to 650 cm^{-1} without large absorption of H_2O, while for a 25-reflection crystal, the signal becomes noisy below 890 cm^{-1} (Lefèvre et al., 2006). To be able to record spectra at lower wavenumbers, heavy water (D2O) can be used because the absorption bands are shifted by a factor of ca. 1.4 to lower wavenumbers. Thus, a good signal can be obtained for bands located between 850 and 950 cm^{-1} (Lefèvre et al., 2008) using the same 25-reflection crystal. It can be useful to avoid interferences with bands around 1650 cm^{-1} Since D2O bending is located at 1209 cm^{-1} (Venyaminov and Prendergast, 1997).

2.3 Review of adsorption of carboxylic acids onto metal (hydr)oxides by ATR-IR

2.3.1 Monoacids: Formic, acetic, benzoic, lauric

A number of monoacids are discussed in the context of this review. Table 2 gives some information on the monoacids both in solution and at the interface. The systems are discussed in detail in the remainder of the section.

Table 2. Characteristics of carboxylic acid (R-C(O)OH). DCOO=nas(COO)-ns(COO)

Acid	pKa	R	DCOO (cm^{-1}) in solution	DCOO (cm^{-1}) adsorbed
Formic	3.75 *	–H	230	192 (TiO_2)
Acetic	4.76 *	–CH3	137	90 (TiO_2)
Benzoic	4.19 *	–C6H5	154	109 (TiO_2) 117 (TA_2O5) 122 (goethite) 141 (ZrO_2)
Lauric	4.90 **	$-CH_2-(CH_2)_9-CH_3$	136	185 (alumina)
from *Lide (1998), ** Dean (1999)				

Several surface complexes can be formed with monoacids, such as monodentate, or bidentates (Fig. 4). Monodentate surface complexes can be distinguished from bidentates based on the occurrence or not of the free C=O group band, with a stretching frequency at about 1700 cm^{-1}.

The ATR-FTIR spectrum of 1M of formate ion is characterized by bands located at 1350, 1383 and 1580 cm^{-1}, assigned to ns(COO), d(HCO) and n_{as}(COO), respectively (Rotzinger et al., 2004). Spectra of formate adsorbed on TiO_2 at pH 5.0, up to 30 mM display the presemce of bands of formate ions and a new peak at 1540 cm^{-1}, assigned to n_{as}(COO) of species interacting with the surface. Spectra in D2O confirmed this assignment Since only a small shift (8 cm^{-1}) of this band was observed, which precludes the vibration of a protonated/deuterated species. A decrease of pH from 9 to 3 leads to the decrease of the peak area. A series of experiments where the adsorption of formic acid as a gas has been studied has shown the presemce of bands of formic acid, formate, and a peak at ca. 1540 cm^{-1}. In support of this, molecular calculations have been performed for the three hypothetical surface complexes (Fig. 4), leading to calculated frequencies. Calculations on the stability of the surface complexes were found to support the binuclear bidentate coordination.

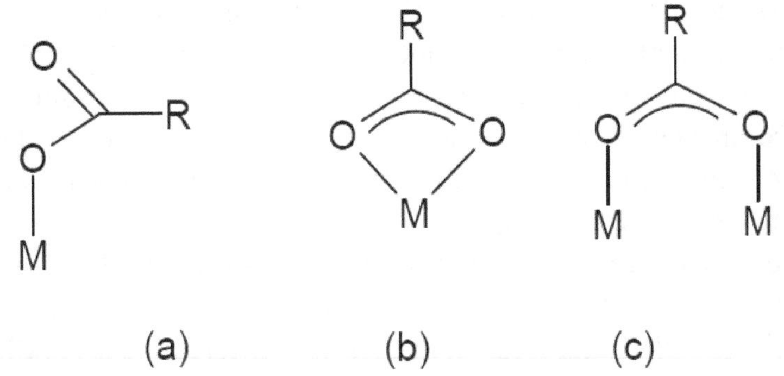

Fig. 4. Surface complexes between a monoacid and a metal oxide: (a) monodentate, (b) mononuclear bidentate and (c) binuclear bidentate

Sorption of acetate ions has been studied by ATR on rutile (Rotzinger et al., 2004) and several other minerals (Kubicki et al., 1999). In solution, the acetate ion is characterized by bands at 1348-1349, 1415-1422 and 1552-1555 cm^{-1} (Rotzinger et al., 2004; Kubicki et al., 1999) assigned to d(CH3), ns(COO), and n_{as}(COO), respectively. Acetic acid is characterized by bands at 1279-1283, 1370-1371, 1392-1397(dCH3), 1642-1650 and 1711-1717 (nC=O) (Rotzinger et al., 2004; Kubicki et al., 1999). Spectra of adsorbed species have been recorded at pH 5.0 (ca. 1:1 mixtures of the acetate ion and acetic acid in solution Since pH is close to pKa), and at total acetate concentrations up to 25 mM on TiO_2 (Rotzinger et al., 2004), and at pH 3 and 6 in the presemce of 2 M acetate on quartz, albite, illite, kaolinite and montmorillonite (Kubicki et al., 1999). On TiO_2, bands of acetate are present with a new band at 1512 cm^{-1} assigned to n_{as}(COO) shifted due to the adsorption. The absence of a band at ca. 1700 cm^{-1} indicates that the C=O group is not present in the surface species. On several minerals (Kubicki et al., 1999), spectra recorded at pH 3 and pH 6 are similar to spectra of solution species. With acetic acid adsorbed on quartz, two bands are seen around 1720 cm^{-1} (at 1709 and 1732), suggesting two different bonding environments. For the other minerals, the authors conclude that chemisorption is below the detection limit of the spectroscopy. This might be due to the low specific surface area of the minerals used in the study.

Adsorption of benzoic acid on minerals was studied by several authors on quartz, albite, illite, kaolinite and montmorillonite (Kubicki et al., 1999), goethite (Tejedor-Tejedor et al., 1990), TiO_2 (Tunesi and Anderson, 1992; Dobson and McQuillan, 1999), as well as Al_2O_3, ZrO_2 and TA_2O5 (Dobson and McQuillan, 1999). Aqueous benzoate is characterized by bands at 1542 cm^{-1} (n_{as}(COO)), 1388 cm^{-1} (ns(COO)) and 1593 cm^{-1} (nC=C). At pH < pKa, spectra are characterized by bands at 1705 cm^{-1} (nC=O), 1319

cm^{-1} ($n_{CO}H$), 1279 cm^{-1} (dCOH), and bands associated with C=C and C-H vibrations (1603, 1494, 1452, 1178, 1073 and 1026 cm^{-1}). Benzoic acid adsorbed on quartz displays bands of the aqueous species with two new peaks (at 1604 and 1569 cm^{-1}). The lower frequency was found by calculation to correspond to a monodentate complex, and the higher one to an outer-sphere complex. On albite at pH 3, no peaks above 1700 cm^{-1} were observed, indicating that the C=O group is absent from the surface complex even in the pH range where the acid species predominates over the benzoate anion. This result is a direct evidence of the formation of a bidentate complex, stable over a wide range of pH. On goethite at pD 3.9 (Tejedor-Tejedor et al., 1990) and on TiO_2 at pH 3.6 (Tunesi and Anderson, 1992), the nC=O mode is also absent. Another interesting point is that the asymmetric / symmetric carboxylate group stretching ratio decreases when benzoate interacts with Fe(III), which can be explained by the increase of co- planarity between the benzene ring and the n_{as}(COO). These observations are consistent with the formation of a bidentate complex. c.f. Fig. 5 (Tejedor-Tejedor et al., 1990).

On TiO_2, the difference between n_{as}(COO) et ns(COO) for the adsorbed species is lower by 45 cm^{-1} compared to the corresponding difference for the solute species. It is believed that a lower value is indicative of a bidentate complex, and that such a large value indicates a chelate structure with a single centre (Fig. 5) (Tunesi and Anderson, 1992). On goethite, the difference was lower by 32 cm^{-1}, consistent with a bridging complex (Tunesi and Anderson, 1992).

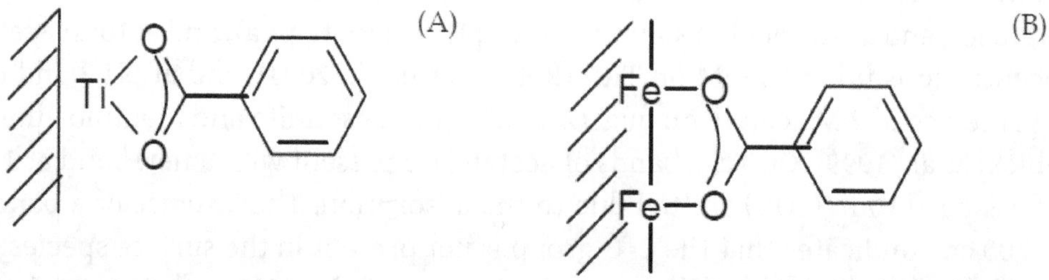

Fig. 5. Proposed surface complexes of benzoate on (A) TiO2 and (B) goethite.

The bands pertaining to n_{as}(COO) and ns(COO) modes of laurate in solution are located at 1547 and 1411 cm^{-1}, respectively. Between 2850 and 3000 cm^{-1}, several bands are reported corresponding to hydrocarbon stretching. Laurate anions were adsorbed onto alumina, which had been deposited on the ATR element by a sputtering technique the thickness of the film being 600 Å (Couzis and Gulari, 1993). The recorded spectra depended on contact time and pH. At pH 8, up to 20 minutes after initiation of the solid-liquid contact, the observed peaks mainly corresponded to the solute species and the authors inferred the presemce of an outer-sphere surface complex, Since the surface is positively charged at this pH. For longer times of exposure, a new band appeared at 1597, along with the increase of the band at 1412 cm^{-1}. This new band was assigned to n_{as}(COO) of the adsorbed species. The difference between n_{as}(COO) and ns(COO) for the surface species was higher than the value obtained for the solute species. This behaviour, contrary to that observed for carboxylic acids with a shorter alkyl chain (Table 3) has been interpreted as a different, i.e. monodentate, surface coordination. From the evolution of the spectra recorded in the hydrocarbon stretching range (2750 – 3000 cm^{-1}), a chain-chain interaction is inferred after adsorption of laurate for short contact times, suggesting the association of the aliphatic chains at low surface coverage. For longer contact times, corresponding to a higher surface coverage, the chain-chain interactions become negligible.

2.3.2 Saturated and unsaturated diacids

This section in a similar way as the previous one summarizes a number of studies on the adsorption saturated and unsaturated diacids to (oxy)(hydr)oxide minerals. The chemical speciation (in terms of the number of species in solution) becomes more complex for these compounds, which concomitantly enhances the possibilities of the diacids to form surface complexes of different stoichiometries in terms of bonding and proton balances. The diacids addressed are summarized in tables 3 (saturated diacids) and 4 (unsaturated diacids). The remainder of the section discusses in some detail published findings from ATR-FTIR spectroscopy.

Table 3. Characteristics of dicarbocylic acids (HO(O)C–R–C(O)OH).

Acid	pKA_1, pKA_2 Lide (1998)	R
Oxalic	1.23, 4.19	N.A.
Malonic	2.83, 5.69	$-CH_2-$
Succinic	4.16, 5.61	$-(CH_2)_2-$
Glutaric	4.31, 5.41	$-(CH_2)_3-$
Adipic	4.43, 5.51	$-(CH_2)_4-$

Table 4. Characteristics of unsaturated dicarbocylic acids.

Acid	pK_{a1}, pK_{a2} from Lide (1998)	Formula
Trans-Fumaric	3.03,4.44	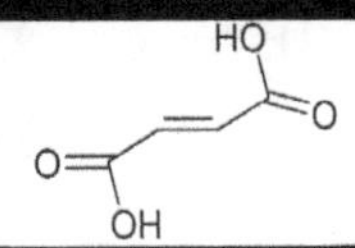
Maleic	1.83,6.07	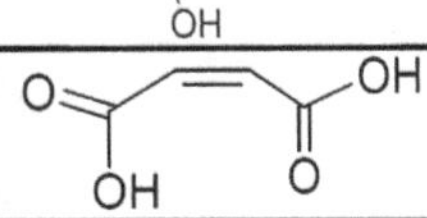
o-Phtalic	2.89,5.51	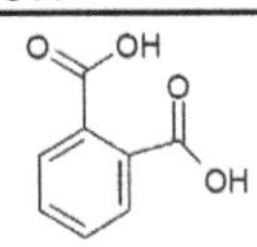

Oxalic acid is the simplest polyacid molecule (COOH)2 and its adsorption is the most common subject of study by ATR-IR on oxy-hydroxides of aluminum (Axe and Persson, 2001; Johnson et al., 2004; Rosenqvist et al., 2003; Yoon et al., 2004;

Dobson and McQuillan, 1999), iron (Borda et al., 2003; Duckworth and Martin, 2001; Persson and Axe, 2001), chromium (Degenhardt and McQuillan, 1999; Garcia Rodenas et al., 1997), titanium (Hug and Sulzberger, 1994; Weisz et al., 2001; Weisz et al., 2002; Dobson and McQuillan, 1999), silicon (Kubicki et al., 1999), tantalum (Dobson and McQuillan, 1999) and zirconium (Dobson and McQuillan, 1999). The spectra of species in solution, i.e. (COOH)2, HOOCCOO- and (COO-)2 were reported in several studies. The oxalate ion is characterized by two peaks at 1307 and 1571 cm^{-1}, respectively, which are assigned to n_{as}(COO) and ns(COO) modes. The spectrum of the oxalic acid species in solution is dominated by C=O stretching at 1735 cm^{-1} and C-OH stretching at 1227 cm^{-1}. These features of oxalate and oxalic acid are consistent with theoretical frequency calculations (Axe and Persson, 2001). The spectra of hydrogen oxalate shows three peaks assigned to C=O stretching (1725 cm^{-1}), n_{as}(COO) (1620 cm^{-1}), and C-OH stretching (1240 cm^{-1}) (Degenhardt and McQuillan, 1999).

Sorption of oxalate on boehmite was studied as a function of oxalate concentration and pH (Axe and Persson, 2001). Two different complexes were identified: an outer-sphere complex characterized by a spectrum similar to that of dissolved oxalate (two bands at 1577 and 1308 cm^{-1}), and an inner-sphere complex. The assignment of this latter was based on the comparison of the spectra of the boehmite surface after sorption of oxalate (characterized by strong bands at 1722, 1702, 1413, 1288 cm^{-1}) with the spectra of dissolved [Al(Ox)(H_2O)4]+ (1725, 1706, 1412, 1281 cm^{-1}). The very close resemblance suggests a mononuclear five-membered chelate geometry. The possibility of a symmetric bridging coordination to two equivalent Al(III) ions was ruled out by Raman spectra of the surface species. Indeed, the comparison of Raman spectra of [Al(Ox)(H_2O)4]+ with theoretical frequency calculations have indicated that the intensity of Raman bands can be used to distinguish a ring chelate from a bridging structure.

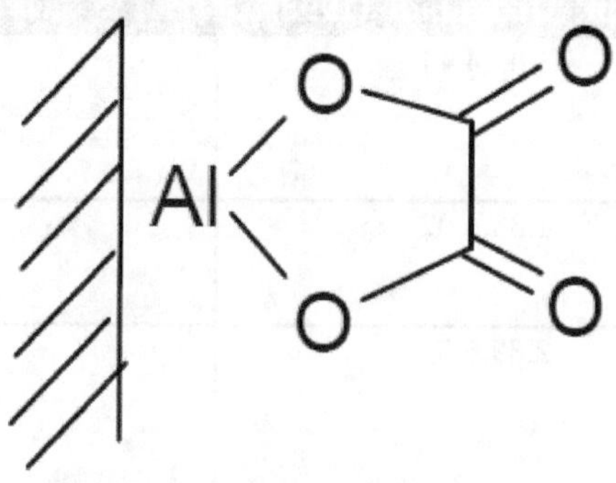

Fig. 6. Ring chelate of oxalate on alumina (Axe and Persson, 2001)

This interpretation has been supported by a study of oxalate sorption on corundum modelled by the CD-MUSIC model involving ATR-IR spectroscopy

(Johnson et al., 2004). A mononuclear bidentate complex was found up to 14 μmol/m2, whereupon oxalate additionally adsorbed as an outer-sphere complex. Sorption of oxalate has also been studied on boehmite and corundum by Yoon et al. (2004) The peaks assigned to the inner-sphere complex in previous works (near 1286, 1418, 1700 and 1720 cm^{-1}) were claimed to arise from the presemce of several species. Evidence for this phenomenon comes from the observation that peaks at 1286 and 1418 cm^{-1} are shifted to 1297 and 1408 cm^{-1} as the oxalate surface coverage increases. The authors finally postulated the existence of two species: species "A" at 1286 and 1418 cm^{-1}, and species "B" at 1297 and 1408 cm^{-1}, respectively, which were assigned to an inner-sphere surface complex on boehmite and to dissolved oxalate coordinated to aqueous Al(III). This assumption is supported by oxalate promoted dissolution of the aluminum oxy-hydroxide arising from the complexation reactions of dissolved Al(III) cations. Quantum calculations of infrared vibrational frequencies of possible surface complexes were carried out on aluminium oxide clusters (Al18O12 and Al14O22) including monodentate, bidentates with 4- and 5-membered ring, and bridging bidentates. They showed that the bidentate 5-membered ring most closely matched the experimental observations (within 15 cm^{-1}), while the simulation results for the other models showed deviations between 17 and 102 cm^{-1}.

On hematite (Duckworth and Martin, 2001), the spectra of sorbed oxalate are similar to the above discussed surface complex on an aluminum oxy-hydroxide and consequently a 5- member bidentate complex was proposed. The effect of pH on the sorption of oxalate on goethite has also been studied (Persson and Axe, 2001). An outer-sphere surface complex and a 5-member ring inner-sphere surface complex were inferred from spectra of the goethite/oxalate system and the aqueous Fe(III)-oxalate complex. At low pH, the presemce of outer-sphere surface complexes (COOH)2 was ruled out because of the absence of a band corresponding to these species in aqueous solutions (around 1735 and 1233 cm^{-1}).

Oxalate sorption on chromium oxide (Degenhardt and McQuillan, 1999) is characterized by bands at 1708, 1680, 1405 and 1269 cm^{-1}, which was interpreted as a side-on surface complex (both carboxylic groups interact with the surface), but without distinguishing between a 5- or a 7-member ring. An additional weakly bound oxalate ion was detected (bands at 1620- 1580 and around 1306 cm^{-1}). The absence of absorption at 1725 cm^{-1} (corresponding to C=O group) eliminates the singly protonated oxalate species. However, an upward shift of n_{as}(COO) is observed (from 1571 cm^{-1} in Ox2-), suggesting hydrogen bonding with the surface. On chromium oxide (Garcia Rodenas et al., 1997), spectra were recorded after exposure to 0.1 M oxalate solution at pH 3.6 followed by washing with pure water. The resulting spectra were compared to the spectra of Cr(Ox)33- species in solution. As in the work by Degenhardt and McQuillan (1999), an inner-sphere surface complex was inferred from the bands at 1710, 1680, 1410 and 1260 cm^{-1}. The remaining shoulder at 1620 cm^{-1} and the peak at 1310 cm^{-1} were attributed to uncoordinated oxalate ions. Since the solid has been washed after contacting with oxalate solution, the solute species are expected to be removed, and these data could be reinterpreted as a species involving hydrogen bonding.

Oxalate sorption onto TiO_2 was amongst the first in situ adsorption studies involving ATR- IR spectroscopy at the solid/solution interface (Hug and Sulzberger, 1994). Hug and Sulzberger (1994) have focused their study on the measurement of adsorbed oxalate to plot an isotherm curve. The isotherm at constant pH (3) was fitted by three Langmuir components, correlated with the three possible solute species (H_2Ox/HOx-/Ox2-). Weisz et al. (2001, 2002) have used the same protocol, and have measured three Langmuir stability constants. Dobson and McQuillan (1999) have recorded the spectra of $NA_2[TiO(Ox)2]2.3H_2O(S)$, where the oxalate ion forms a µ2-oxo bridged Ti dimeric complex. Its spectrum is close to those obtained with oxalate adsorbed onto TiO_2 and the comparison would lead to the interpretation of the spectroscopic results in terms of a bidentate-bridging surface complex. However, this interpretation disagrees with observations by Scott et al. (1973) who have shown on oxalato-Co(III) complexes that bidentate-chelating and bidentate- bridging oxalato ligands are characterized by nearly identical spectra.

ATR-FTIR studies of the sorption of other saturated diacids HO(O)C-(CH_2)n-C(O)OH with n=1 to 4 have been reported, as malonic (Dobson and McQuillan, 1999; Dolamic and Bürgi, 2006; Duckworth and Martin, 2001; Rosenqvist et al., 2003), succinic (Dobson and McQuillan, 1999; Duckworth and Martin, 2001), glutaric (Duckworth and Martin, 2001) and adipic (Dobson and McQuillan, 1999; Duckworth and Martin, 2001) acids. In solution, the dicarboxylate ions are characterized by $n_{as}(CO_2)$ around 1560-1550 cm^{-1} and ns(CO_2) around 1410-1350 cm^{-1} (Dobson and McQuillan, 1999). The value of $n_{as}(CO_2)$ is close for all values of n (and equal to frequency in oxalate), but ns(CO_2) increases with n, from ca. 1360 cm^{-1} (1310 cm^{-1} in oxalate) to 1400 cm^{-1}. Dicarboxylic acids are characterized by n(C=O) at 1718 cm^{-1} (malonic) and n(C-O) at 1328 cm^{-1} (malonic). The $-CH_2-$ bending frequencies are located at 1410-1440 cm^{-1} and 1300-1250 cm^{-1}. Malonate adsorbed on hematite at pH 5 is characterized by peaks at 1260 (d($-CH_2-$)), 1349 (ns(CO_2)), 1439 (d($-CH_2-$)) and 1631 (n(C=O)) cm^{-1}. ns(CO_2) remains at 1349 cm^{-1} without shift. In comparison with solution spectra, the intensity of - CH_2- bending is enhanced, and the CO_2 asymmetric stretching is replaced by n(C=O) (70 cm^{-1} higher). Such an assignment is consistent with a single-bonded surface complex, a structure similar to adsorbed oxalate. The $-CH_2-$ bending band enhancement indicates a change of the dipole moment, which may be an indication of a strained surface structure with increased bond angles (Dobson and McQuillan, 1999). On other metallic oxides, as TiO_2 (Dobson and McQuillan, 1999; Dolamic and Bürgi, 2006), ZrO_2 (Dobson and McQuillan, 1999), Al_2O_3 (Dobson and McQuillan, 1999; Rosenqvist et al., 2003) and TA_2O5 (Dobson and McQuillan, 1999), the spectra of sorbed malonate species are similar: d(-CH_2-) at 1430-1450 and 1270- 1280 cm^{-1}, n(C=O) at 1580-1600 cm^{-1}, and unshifted ns(CO_2) at 1360-1380 cm^{-1}. On gibbsite (Rosenqvist et al., 2003), the study of the evolution of the spectra with pH has shown two independent species: an inner-sphere complex (corresponding to a peak at 1438 cm^{-1}) and an outer-sphere complex (corresponding to the unshifted ns(CO_2)). Since the other authors (Dobson and McQuillan, 1999; Dolamic and Bürgi, 2006; Duckworth and Martin, 2001) have not studied the effect of pH, it is not possible to rule out the presemce of an outer-sphere complex in their studies, even if they have not mentioned this possibility.

Duckworth and Martin (2001) and Dobson and McQuillan (1999) have studied the effect of longer carbon chains on the adsorption of dicarboxylic acids. Dobson and McQuillan (1999) found surface structures similar to malonate (bridging bidentate via a loop) consistent with greater molecular flexibility. On the contrary, Duckworth and Martin (2001) have found either a bridging bidentate via a loop for malonate and glutarate, or a monodentate complex for succinate and adipate. This interpretation is supported by the behaviour of these ions in the dissolution of hematite: oxalate, malonate and glutarate promote the dissolution whereas succinate and adipate show less of an effect. However, the spectral evidence for this difference in the geometry of complexes is the presemce of a peak at 1550-1540 cm^{-1}, i.e. at a similar location as ($n_{as}(CO_2)$) of the solution species, whereas this stretching mode was absent in malonate and glutarate surface species.

Three unsaturated dicarboxylic acids have been studied by ATR: fumaric (Dobson and McQuillan, 1999; Rosenqvist et al., 2003), maleic (Dobson and McQuillan, 1999; Borda et al., 2003; Rosenqvist et al., 2003, Johnson et al., 2004), (respectively trans- and cis- butendioic acids), and phtalic (Boily et al., 2000; Dobson and McQuillan, 1999; Hwang et al., 2007; Klug and Forsling, 1999; Kubicki et al., 1999; Nordin et al., 1997; Rosenqvist et al., 2003; Tunesi and Anderson, 1992) acid. On gibbsite, the spectra of these three species after sorption are very similar to spectra of the carboxylate ion in solution. Consequently, Rosenqvist et al. (2003) have concluded that an outer-sphere complex was formed. On Al_2O_3, the same behaviour was shown for maleate and fumarate (Dobson and McQuillan, 1999). On corundum, high resolution spectra of the adsorbed maleate ion were recorded (Johnson et al., 2004), and only small differences between bands of dissolved and adsorbed species have been detected, supporting outer-sphere complexation. For phtalate, inner-sphere and outer- sphere complexes were found on ferric oxy-hydroxides (Boily et al., 2000; Hwang et al., 2007).

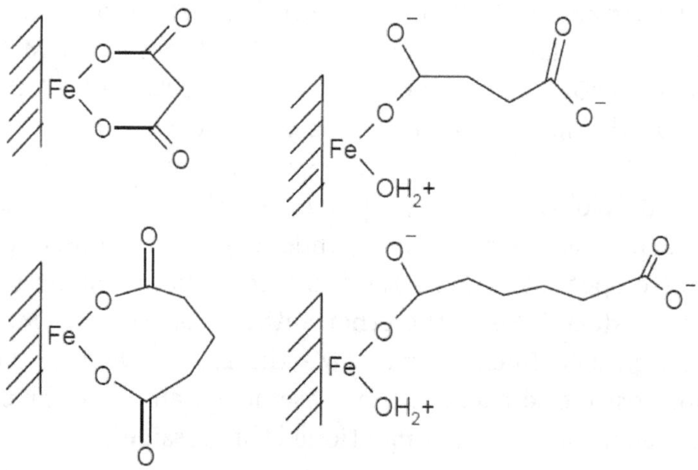

Fig. 7. Proposed structures of diacids adsorbed on hematite (Duckworth and Martin, 2001)

3. ADSORPTION OF 5-SULFOSALICYLIC ACID ONTO GIBBSITE

3.1 Introduction

The system 5-sulfosalicylic acid (5-SSA)/gibbsite is part of a broader study involving Cm adsorption onto gibbsite (Huittinen et al., 2009). Interaction of dissolved Cm with 5-SSA (Panak, 1996) and acid-base equilibria of gibbsite (Adekola et al., 2011) have been studied separately. Gibbsite as used in this study has a platelike morphology. Most of its surface relates to the basal plane. A very complex issue is the interfacial behaviour of the bare gibbsite. While the basal plane from a conventional point of view is considered quite inert, there are a number of indications that rather show that the basal plane can be quite reactive (Rosenquist et al., 2002; Gan and Franks, 2006). The present view is that a number of effects can be of importance, such as the precise conditions for the preparation of the gibbsite particles. This has been further studied by mimicking the basal plane by individual single crystals of sapphire, which are structurally very similar to the ideal basal plane of gibbsite (Lützenkirchen et al., 2010). Here we present some results on the interaction of 5-SSA with the gibbsite particles used in the Cm adsorption and basic charging studies mentioned above to lay a foundation to the study of the ternary Cm- SSA-gibbsite system.

Adsorption of 5-SSA to related minerals (i.e. alumina) has been previously studied previously by Jiang et al. (2002) by both a set of electrokinetic data and batch adsorption. Furthermore they reported IR data which they interpreted in terms of 5-SSA forming a bidentate surface complex involving the carboxylate group and the phenol group.

3.2 Experimental

3.2.1 Chemicals

The gibbsite particles were synthesized by the following procedure: 1 mol.dm-3 aluminum chloride solution was titrated with 4 mol.dm-3 NaOH until pH reached a value of about 4.6. Dialysis was carried out at 70 °C during four months, with initially one change of water per day. Subsequently, water was changed two to three times a week. The gibbsite was stored as a suspension at a concentration of 41.9 g.dm-3. The radius of the particles was determined by several methods, including AFM and field flow fractionation yielding an average width of the platelets of about 200 nm and a thickness of about 10 nm.

The suspension and solution were prepared with de-ionized water (conductivity ~ 18.2 MΩ cm). All solutions and suspensions were prepared in plastic containers. The following chemical reagents were used: NaCl (p.a., Merck), HCl (0.1 mol. dm–3, titrival, Merck), NaOH (0.1 mol.dm–3, titrival, Merck), 5-sulfosalicylic acid (5-Sulfosalicylic acid.2H_2O, Sigma Aldrich), AlCl3.6H_2O (Merck p.a.), standard buffers (pH = 3, 5, 7, 9 and 11).

Speciation of 5-SSA in solution results in four species. The occurrence of these species depends on pH (Fig. 8).

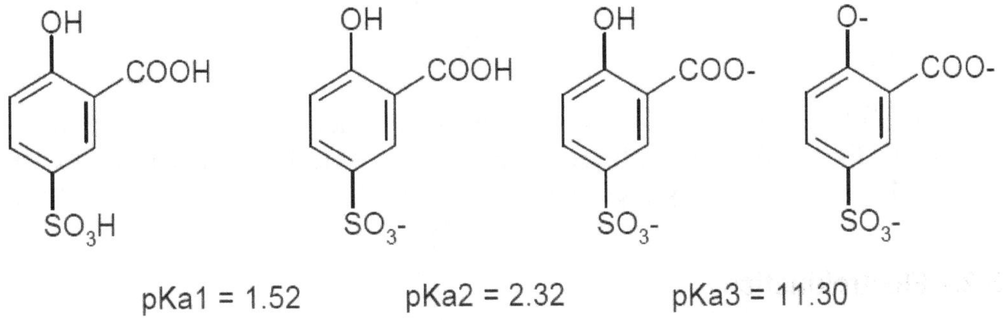

Fig. 8. Equilibrium constants for the deprotonation of 5-SSA (Panak, 1996).

In presemce of aluminium ions, a number of Al(5-SSA)x species (with x=1 to 3) can be formed. The speciation of 5-SSA and Al as a function of pH for 5′10-2 mol. dm-3 of each component is plotted in Fig. 9. The relevant Al-species are shown in the upper part, indicating that Al is preferably bound to 5-SSA up to pH 8. Above pH 8, the tetra-hydroxo species of aluminium dominates the speciation under the given conditions. The lower part of Figure 9 shows the distribution of 5-SSA for the equimolar solution. 5-SSA is predominantly bound in Al-complexes up to pH 11.

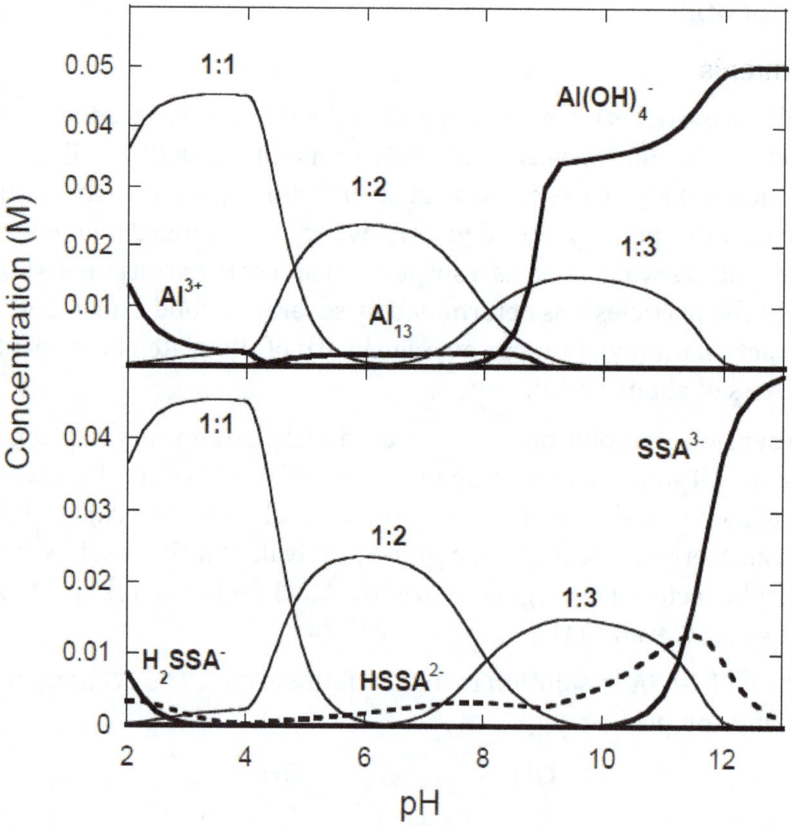

Fig. 9. Calculated speciation for an equimolar solution (5´10-2 M) of Al(III) (top) and 5-SSA (bottom). Uncomplexed species are in thin lines, and complex Al:5-SSA are in bold lines. Only species whose concentration are higher than 5% are represented.

3.2.2 Electrokinetics

The electrokinetic (zeta) potential of gibbsite particles was measured after adsorption of 5- SSA on gibbsite surfaces by means of a ZetaPals (Brookhaven Instruments). The mass concentration of gibbsite particles was 0.1 g.dm-3 and 5-SSA concentration was 10–3 mol.dm-3. The experiments were performed at two different ionic strength values (Ic = 10-1 mol.dm-3 and 10-2 mol.dm-3). The results are shown in Figure 10 together with zeta potential of gibbsite particles in the absence of 5-SSA (Adekola et al., 2011).

For both ionic strengths studied, the electrokinetic data show (i) a strong shift of the isoelectric point and (ii) negative zeta-potentials over a wide pH range, indicating adsorption of 5-SSA and transfer of negative charges to the gibbsite surface. The speciation diagram (Fig. 9) actually shows that the speciation of 5-SSA is dominated both in the absence and the presemce of Al by negatively charged aqueous species.

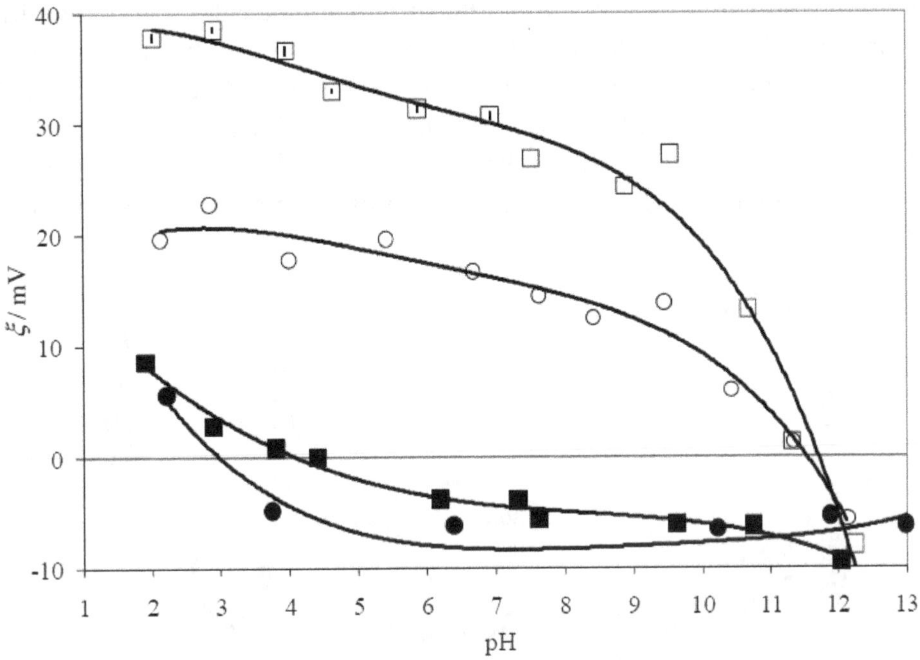

Fig. 10. Zeta-potentials of gibbsite particles in the absence (□,○) and in the presemce (■,●) of 5-SSA. Ionic strength was controlled by NaCl: 10-1 mol.dm-3 (circles) and 10-2 mol.dm-3 (squares). Temperature was 25 ºC.

3.2.3 Spectroscopy

The IR-ATR spectra of gibbsite in absence and presemce of 5-SSA, as well as spectra of the aqueous solutions of the 5-SSA and 5-SSA/Al^{3+} were obtained using a Bruker spectrometer (IFS 55). A ZnSe crystal (multiBounce) was used and for each measurement 1024 scans were recorded with a resolution of 4 cm^{-1}. All measurements were made under dry argon atmosphere. The effect of pH, which has an influence on gibbsite surface charge as well on the speciation of 5-SSA in solution was examined. For spectra of the gibbsite layer in contact with solution of 5-SSA, the gibbsite layer was prepared by drying an aliquot of a suspension.

3.3 Results and discussion

3.3.1 Spectral characterization of 5-SSA in solution

Solution spectra of 5-SSA, with and without aluminium ions have been recorded by ATR-IR. For pure 5-SSA solution at pH 2, 5 and 12, the spectra are shown in Fig. 11. At pH 2, the spectrum consists in several bands which have been assigned following previous works by Varghese et al. (2007) and Jiang et al. (2002) as shown in Table 5. From pH 2 to 5, some differences can be seen: bands around 1200 cm^{-1} decrease, bands at 1270 and 1306 cm^{-1} appear, while bands at 1370 cm^{-1}, 1430 cm^{-1} increase. Around 1200 cm^{-1}, bands corresponding to vibration of the sulfate group are expected, so this evolution would be in agreement with the change in the molecule from –SO3H to –SO3- at pH 1.5. From pH 5 to 12, the shift of several bands is visible, the most important is 1292 cm^{-1} to 1306 cm^{-1}. According to Jiang et al. (2002), this band corresponds to stretching of Ph-OH, which is deprotonated at pH > 11.3. Thus, the difference between these spectra would correspond to the change of speciation from pH 5, where only $HSSA_2$- is present to pH 12, dominated by 83% of SSA3-.

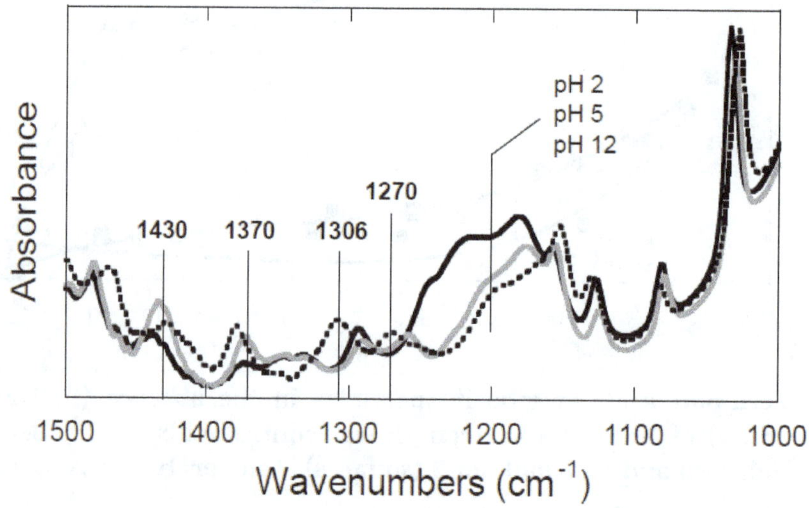

Fig. 11. Spectra of 5-SSA at pH 2 (black line), pH 5 (grey line), pH 12 (dotted line)

Table 5. Assignments of bands in spectra of 5-SSA in solution at different pH values.

	wavenumbers (cm^{-1})		
Assignments from literature	pH 2	pH 5	pH 12
nPh *	1480	1479	1468
nPh,dCOHc *	1439	1431	1425
nsCOc **	1371	1373	1379
n_{as}SO2 *	1346-1330	1346-1330	1306
nPh-OH **	1292	1294	1269
nsSO2 *	1242sh		
	1218sh	1201sh	1198sh
nC-OH, dSOH *	1180	1175	
dCH *	1159	1159	1151
nC-COOH *	1082	1080	1084
nS-OH *	1034	1030	1028
Varghese et al. (2007), ** Jiang et al. (2002), sh: shoulder, c:carboxyl			

The main impact would be on the vibration band of hydroxyl group which is deprotonated, but a consequence on vibration of other groups would be possible.

3.3.2 Spectral characterization of 5-SSA in presemce of aluminium ions or adsorbed onto gibbsite

Spectra of solutions of 5-SSA in the absence and presemce of aluminium ions have been recorded. The results are shown as dotted and grey lines in Figure 12. The main difference is the presemce of bands at 1330 and 1270 cm^{-1} at pH 2 and pH 5. At pH 12, the pure 5-SSA presents the same band too, yet a difference between the spectra is rather observed in the shift of bands at 1308, 1428, 1470 cm^{-1} to 1326, 1431, 1480 cm^{-1} respectively. Speciation calculation suggests predominance of Al-SSA at pH 2, a mixture of Al-SSA and Al-SSA_2 at pH 5 and no Al-SSA complex at pH 12.

From the study of the effect of pH on pure SSA, it was concluded that the band visible at 1270 cm^{-1} indicates the presemce of Ph-O-, while a band at 1294 cm^{-1} wich indicates the presemce of Ph-OH is absent in spectra at pH 5 in presemce of aluminium ion, or less intense at pH 2. At both pH values, the spectra would indicate that the presemce of aluminium promotes the deprotonation of the phenol group, which leads to a bond between 5-SSA and the aluminium ion. The band corresponding to carboxyl groups at 1370 cm^{-1} is not shifted in presemce of aluminium. However, a hypothetical carboxyl-Al group could perturb also the intramolecular hydrogen bond, as suggested for salicylate-iron interaction, which makes difficult to use this method to evaluate the interaction with carboxyl groups. We may conclude that the affinity of carboxyl groups for aluminium ion is well known, and an interaction is expected, despite any direct experimental evidence. At pH 12, the speciation calculation indicates the favourable formation of Al(OH)4- and the absence of any Al-SSA complexes. However, the spectra evolves in presemce of aluminium ions, mainly by the shift of bands at 1425 and 1468 cm^{-1} towards values close to the spectra of pure SSA at pH 2. This result remains unexplained but suggests an interaction between both species.

As the last step, the spectra of 5-SSA in the presemce of gibbsite have been recorded. The results are shown as black lines in Figure 12. In the 1500-1000 cm^{-1} range, no difference between spectra of pure 5-SSA solution and gibbsite-SSA was observed at pH 2 and 12. At pH 2, this behaviour is different from that of aluminium in solution. This might be consistent with an extent of 5-SSA adsorption that is too low to be detectable by the ATR measurements.

A comparison of the spectroscopic results to the zeta-potential measurements shown in Fig. 10 suggests the following picture. The differences in zeta potential of gibbsite particles with and without 5-SSA being present are substantial over the complete pH range investigated, except for pH 12. At pH 12, no adsorption is expected according to the electrokinetic results, but the spectra of 5-SSA in the presemce of gibbsite do show a slight shift of 1425 cm^{-1} and 1468 cm^{-1} bands, similar to that observed with aluminium in solution. At pH 5, the spectra in presemce of gibbsite are similar to that of 5-SSA in presemce of aluminium ions, mainly due to the presemce of a peak near 1270 cm^{-1}. The effect of 5-SSA on zeta potential is strong, so it can be concluded that a complex between 5-SSA and aluminium atom occurs at the gibbsite surface at pH 5, similar to what has been observed with aluminium ion. In such a geometry (Fig. 13), the negative charge observed in the electrokinetic data would be consistent with the free sulfonate group, deprotonated in the whole pH range. At very low pH, the zeta- potential turns positive even in the presemce of 5-SSA, which can be explained by the speciation of the organic molecule itself, which would on average be less negative than at higher pH.

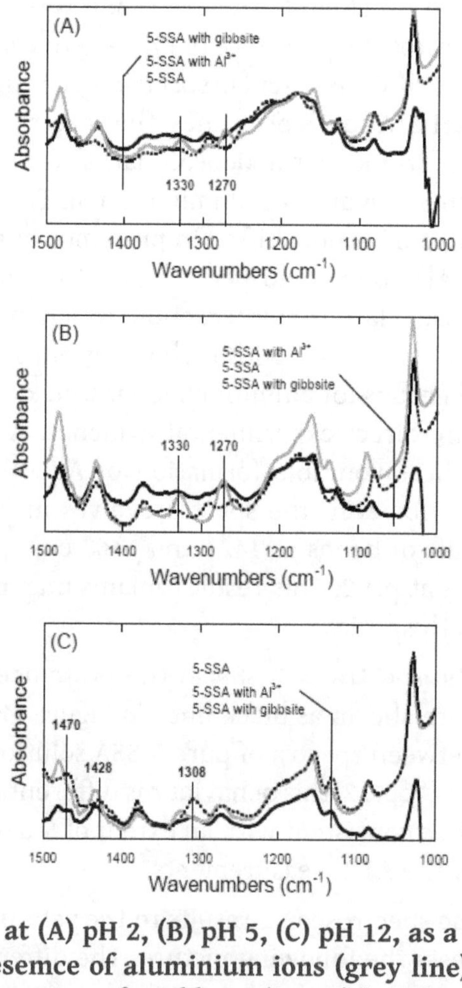

Fig. 12. Spectra of SSA at (A) pH 2, (B) pH 5, (C) pH 12, as a pure solution (dotted line), a solution in presemce of aluminium ions (grey line), a solution after the deposition of a gibbsite layer (black line).

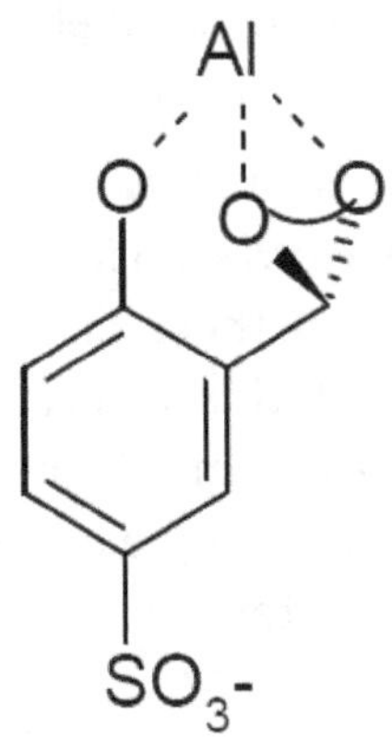

Fig. 13. Structure of the Al-SSA complex. Al is either an ion in solution, or bound to surface in surface complexes.

At pH 12, there might still be adsorption of 5-SSA, Since the spectra of 5-SSA/gibbsite system is close to those of 5-SSA/Al(III). However, it would be difficult to detect by the electrokinetic method, Since the gibbsite itself turns negative at pH > 11.3. The fact that the solution spectra of 5-SSA are different with or without Al(III) species indicates the possibility that there is some interaction. Again, the importance of spectroscopic studies in pinpointing interactions which are not detectable by macroscopic methods can be highlighted here.

4. CONCLUSIONS

In the first part of this chapter, the principles and experimental protocols of the use of ATR- IR to probe solid/solution interfaces have been described. A review of the literature has shown that ATR-IR can be useful to get information about the surface speciation of adsorbed carboxylic acids.

The second part was devoted to the comprehensive study of the adsorption of 5-SSA onto gibbsite platelets, and it was suggested that this phenomenon occurs via a dominant tridentate surface complex involving the phenolic and carboxylic groups. The ring structure of this surface complex had been previously postulated by Jiang et al. (2002) based on a limited set of spectra. The fact that the sulphate group is not involved in co-ordination to the surface, and therefore oriented towards the solution side of the interface, agrees with the negative charge of the gibbsite in the presemce of 5-SSA.

The solution study strongly indicates that the present speciation schemes for the 5-SSA- dissolved aluminium system are incomplete. The spectra show differences between the absence and presemce of Al(III) in solution at high pH, where within the current speciation scheme no complexes are expected. At high pH in the 5-SSA-gibbsite-system, the spectroscopic data also suggest interaction, which is undetectable by the electrokinetic method, Since the gibbsite surface has an overall negative charge in that pH-range.

A general conclusion of our experimental study would be that comprehensive studies of solid-solution interfaces via ATR-IR may contribute to solving the structure of surface complexes. Furthermore, such studies may help finding previously unidentified species in solution speciation schemes. Overall the interplay of surface complexation and solution complexation in this system indicates that the adsorption process may be very complex and that multi-method approaches are best suited to gain deeper understanding.

5. REFERENCES

1. Adekola, F., Fédoroff, M., Geckeis, H., Kupcik, T., Lefèvre, G., Lützenkirchen, J., Plaschke, M., Preocanin, T., Rabung, T., & Schild, D. (2011). Characterization of acid-base properties of two gibbsite samples in the context of literature results, J. Colloid Interface Sci. Vol. 354, 306-317, 0021-9797.
2. Asay, D. B. , & Kim, S.H. (2005). Evolution of the Adsorbed Water Layer Structure on Silicon Oxide at Room Temperature, J. Phys. Chem. B Vol. 109, 16760-16763, 1089- 5647.
3. Axe, K., & Persson, P. (2001). Time-dependent surface speciation of oxalate at the water- boehmite (g-AlOOH) interface: implications for dissolution, Geochim. Cosmochim. Acta Vol. 65, 4481-4492, 0016-7037.
4. Boily, J.-F., Persson, P., & Sjöberg, S. (2000). Benzenecarboxylate surface complexation at the goethite (a-FeOOH)/water interface: II. Linking IR spectroscopic observations to mechanistic surface complexation models for phthalate, trimellitate, and pyromellitate, Geochim. Cosmochim. Acta Vol. 64, No. 20, pp. 3453-3470, 0016-7037.
5. Borda, M. J. , Strongin, D. R., & Schoonen, M. A. (2003). A novel vertical attenuated total reßectance photochemical ßow-through reaction cell for Fourier transform infrared spectroscopy, Spectrochim. Acta A Vol. 59, 1103-1106, 0584-8539.
6. Coates, J.P. (1993). The Industrial Applications of Infared Internal Reflection Spectroscopy, In: Internal Reflection Spectroscopy, Dekker, New York, Mirabella, F. M. (Ed.), pp.53- 96.
7. Couzist, A., & Gulari E. (1993). Adsorption of sodium laurate from its aqueous solution onto an alumina surface. A dynamic study of the surface-surfactant interaction using attenuated total reflection Fourier transform infrared spectroscopy, Langmuir Vol. 9, 3414-3421, 0743-7463.
8. Dean, J.A. (1999) Lange's Handbook of Chemistry (15th ed.), McGraw-Hill.

9. Degenhardt, J., & McQuillan, A.J. (1999) Mechanism of oxalate ion adsorption on chromium oxide-hydroxide from pH dependance and time evolution of ATR-IR spectra, Chem. Phys. Lett. Vol. 31, 179-184, 0009-2614.

10. Dobson, K. D., & McQuillan, A. J. (1999). In situ infrared spectroscopic analysis of the adsorption of aliphatic carboxylic acids to TiO_2, ZrO_2, Al_2O_3, and TA_2O5 from aqueous solutions, Spectrochim. Acta A Vol. 55, 1395-1405, 0584-8539.

11. Dolamic, I., & Bürgi, T. (2006). Photoassisted decomposition of malonic acid on TiO_2 studied by in situ attenuated total reflection infrared spectroscopy, J. Phys. Chem. B Vol. 110, 14898-14904, 1089-5647.

12. Duckworth, O.W., & Martin, S.T. (2001), Surface complexation and dissolution of hematite by C1-C6 dicarboxylic acids at pH = 5.0, Geochimi. Cosmochim. Acta Vol. 65, 4289- 4301, 0016-7037.

13. Fredriksson, A., & Holmgren, A. (2008). An in situ ATR-FTIR investigation of adsorption and orientation of heptyl xanthate at the lead sulphide/aqueous solution interface. Miner. Engineer. Vol. 21, 1000-1004, 0892-6875.

14. Gan, Y., & Franks, G.V. (2006). Charging Behavior of the Gibbsite Basal (001) Surface in NaCl Solution Investigated by AFM Colloidal Probe Technique. Langmuir Vol. 22, pp. 6087-6092.

15. Gao, X., & Chorover, J. (2010). Adsorption of sodium dodecyl sulfate (SDS) at ZnSe and α- Fe2O3 surfaces: Combining infrared spectroscopy and batch uptake studies. J. Colloid Interface Sci. Vol. 348, pp. 167-176, 0021-9797.

16. Garcia Rodenas, L. A., Iglesias, A. M., Weisz, A. D., Morando, P. J., & Blesa, M. A. (1997). Surface complexation description of the dissolution of chromium(III) hydrous oxides by oxalic acid, Inorg. Chem. Vol. 36, 6423-6430, 0020-1669.

17. Hwang, Y. S., Liu, J, Lenhart, J. J. , & Hadad, C. M. (2007). Surface complexes of phthalic acid at the hematite/water interface, J. Colloid Interface Sci. Vol. 307, 124–134, 0021-9797.

18. Hug, S.J., & Sulzberger, B. (1994). In situ Fourier transform infrared spectroscopic evidence for the formation of several different surface complexes of oxalate on TiO_2 in the aqueous phase, Langmuir Vol. 10, pp. 3587-3597, 0743-7463.

19. Hug, S. H. (1997). In situ Fourier transform infrared measurements of sulfate adsorption on hematite in aqueous solutions. J. Colloid Interface Sci. Vol. 188, pp. 415-422, 0021- 9797.

20. Jiang, L. , Gao, L. , & Liu, Y. (2002). Adsorption of salicylic acid, 5-sulfosalicylic acid and Tiron at the alumina–water interface, Colloids Surf. A. Vol. 211, 165-172, 0927- 7757.

21. Johnson, S. B. , Yoon, T. H. , Slowey, A. J. , & Brown, G. E. (2004). Adsorption of organic matter at mineral/water interfaces: 3. Implications of surface dissolution for adsorption of oxalate, Langmuir Vol. 20, 11480-11492, 0743-7463.

22. Johnson, S. B. , Yoon, Kocar, B.D. , & Brown, G.E. (2004). Adsorption of organic matter at mineral/water interfaces. 2. outer-sphere adsorption of maleate and implications for dissolution processes, Langmuir Vol. 20, 4996-5006 , 0743-7463.

23. Kallay, N., Preočanin, T., Kovačević, D., Lützenkirchen, J., & Villalobos, M. (2011), ThermodynamiCs of the Reactions at Solid/Liquid Interfaces, Croat. Chem. Acta Vol. 84, 1, pp. 1-10, 0011-1643.

24. Klug, O., & Forsling, W. (1999). A spectroscopic study of phthalate adsorption on g- aluminum oxide, Langmuir Vol. 15, 6961-6968, 0743-7463.

25. Kubicki, J.D., Schroeter, L.M., Itoh, M.J., Nguyen, B.N., & Apitz, S.E. (1999). Attenuated total Reflectance Fourier-transform infrared spectroscopy of carboxylic acids adsorbed onto mineral surfaces, Geochim. Cosmochim. Acta Vol. 63, 2709-2725, 0016-7037.

26. Kulik, D. A., Luetzenkirchen, J., & Payne, T. (2010). Consistent treatment of 'denticity' in surface complexation models. Geochim. Cosmochim Acta Vol. 74, 12 , pp. A544-A544, 0016-7037.

27. Larsson, M.L., Fredriksson, A., & Holmgren, A. (2004). Direct observation of a self- assembled monolayer of heptyl xanthate at the germanium/water interface: a polarized FTIR study. J. Colloid Interface Sci. Vol.273, 345–349, 0021-9797.

28. Lefèvre, G., Noinville, S., & Fédoroff, M. (2006). Study of uranyl sorption onto hematite by in situ attenuated total reflection - infrared spectroscopy, J. Colloid Interface Sci. Vol. 296, 608-613, 0021-9797.

29. Lefèvre, G., Kneppers, J., & Fédoroff, M. (2008). Sorption of uranyl ions on titanium oxide studied by ATR-IR spectroscopy, J. Colloid Interface Sci. 327, 15-20, 0021-9797.

30. Lide, D.R. (1998). Handbook of Chemistry and Physics 79th.; Lide, D.R., Ed.; CRC Press: Boca Raton.

31. Huittinen, N., Rabung, Th., Lützenkirchen, J., Mitchell, S.C., Bickmore, B.R., Lehto, J., & Geckeis, H. (2009). Sorption of Cm(III) and Gd(III) onto gibbsite, alpha-Al(OH)3: A batch and TRLFS study, J. Colloid Interface Sci. Vol. 332, 158-164, 0021-9797.

32. Lützenkirchen, J., Kupcik, T., Fuss, M., Walther, C., Sarpola, A., & Sundman, O. (2010). Adsorption of Al(13)-Keggin clusters to sapphire c-plane single crystals: Kinetic observations by streaming current measurements, Appl. Surf. Sci. Vol. 256, 5406- 5411, 0169-4332.

33. Mirabella, F.M. (1993). Principles, Theory, and Practice of Internal Reflection Spectroscopy, In: Internal Reflection Spectroscopy, Dekker, New York, Mirabella, F. M. (Ed.), pp.17- 52.

34. Nordin, J. , Persson, P., Laiti, E., & Sjöberg, S. (1997). Adsorption of o-phthalate at the water- boehmite (g-AlOOH) interface: evidence for two coordination modes, Langmuir Vol. 13, 4085-4093, 0743-7463.

35. Panak, P. (1996). Untersuchung von intramolekularen Energietransferprozessen in Cm(III)- und Tb(III) Komplexen mit organischen Liganden mit Hilfe der zeitaufgelösten Laserfluoreszencspektroskopie. Ph.D. Thesis, Technische Universität München, 236 p.

36. Peak, D., Ford, R.G., & Sparks, D.L. (1999). An in-situ FTIR-ATR investigation of sulfate bonding mechanisms on goethite. J. Colloid Interface Sci.Vol. 218, pp. 289-299, 0021- 9797.

37. Persson, P., & Axe, K. (2005). Adsorption of oxalate and malonate at the water-goethite interface: molecular surface speciation from IR spectroscopy, Geochim. Cosmochim. Acta Vol. 69, No. 3, pp. 541–552, 0016-7037.

38. Rosenqvist, J., Persson, P., & Sjöberg, S. (2002). Protonation and Charging of Nanosized Gibbsite (a-Al(OH)3) Particles in Aqueous Suspension. Langmuir Vol. 18, pp. 4598- 4604.

39. Rosenqvist, J., Axe, K., Sjöberg, S., & Persson P. (2003). Adsorption of dicarboxylates on nano-sized gibbsite particles effects of ligand structure on bonding mechanisms, Colloids Surf. A Vol. 220, 91-104, 0927-7757.

40. Rotzinger, F.P., Kesselman-Truttmann, J.M., Hug, S.J., Shklover, V. , & Grätzel, M. (2004). Structure and vibrational spectrum of formate and acetate adsorbed from aqueous

41. solution onto the TiO_2 rutile (110) surface, J. Phys. Chem. B Vol. 108, 5004-5017, 1089- 5647.

42. Scott, K.L., Wieghardt, K., & Sykes, A.G. (1973). Mu-oxalato-cobalt(III) complexes, Inorg.

43. Chem. Vol. 12, 655-663, 0020-1669.

44. Tejedor -Tejedor, M.I., & Anderson, M.A. (1986). "In situ" attenuated total reflection Fourier transform infrared studies at the goethite (a-FeOOH)-aqueous solution interface. Langmuir Vol. 2, pp. 203-210, 0743-7463.

45. Tejedor -Tejedor, M.I., & Anderson, M.A. (1990). Protonation of phosphate on the surface of goethite as studied by CIR-FTIR and electrophoretic mobility. Langmuir Vol. 6, pp. 602-611, 0743-7463.

46. Tejedor-Tejedor, M.I., Yost, E.C., & Anderson, M.A. (1990), Characterization of benzoic and phenolic complexes at the goethite/aqueous solution interface using cylindrical internal reflection Fourier transform infrared spectroscopy. Part 1. Methodology. Langmuir Vol. 6, 979-987, 0743-7463.

47. Tickanen, L.D., Tejedor-Tejedor, M.I., & Anderson, M.A. (1991). Quantitative characterization of aqueous suspensions using attenuated total reflection Fourier transform infrared spectroscopy: influence of internal reflection element-particle interactions on spectral absorbance values. Langmuir Vol. 7, pp. 451-456, 0743- 7463.

48. Tunesi, S., & Anderson, M.A. (1992). Surface effects in photochemistry: an in situ cylindrical internal reflection-Fourier transform infrared investigation of the effect of ring substituents on chemisorption onto TiO_2 ceramic membranes, Langmuir Vol. 8, 487- 495, 0743-7463.

49. Varghese, H.T., Panicker, C. Y., & Philip, D. (2007). IR, Raman and SERS spectra of 5- sulphosalicylic acid dihydrate, J. Raman Spectr. Vol. 38, 309-315, 1097-4555.

50. Venyaminov, S.Y., & Prendergast, F.G. (1997). Water (H_2O and D2O) molar absorptivity in the 1000–4000 cm^{-1} range and quantitative infrared spectroscopy of aqueous solutions. Anal. Biochem. Vol. 248, 234-245, 0003-2697.

51. Villalobos, M., & Leckie, J. O. (2001). Surface complexation modeling and FTIR study of carbonate adsorption to goethite. J. Colloid Interface Sci. Vol. 235, pp. 15-32, 0021- 9797.

52. Wang, Z., Grahn, M., Larsson, M.L., Holmgren, A., Sterte, J., & Hedlund, J. (2006). Zeolite coated ATR crystal probes, Sensors Actuators B. Vol. 115, 685-690, 0925-4005.

53. Weisz, A.D., Regazzoni, A.E., & Blesa, M.A. (2001). ATR-FTIR study of the stability trends of carboxylate complexes formed on the surface of titanium dioxide particles immersed in water, Solid State IoniCs Vol. 143, 125-130, 0167-2738.

54. Weisz, A.D., Garcia Rodenas, L., Morando, P.J., Regazzoni, A.E., & Blesa, M.A. (2002). FTIR study of the adorption of single pollutants and mixtures of pollutants onto titanium dioxide in water: oxalic and salycylic acids, Catal. Today Vol. 76, 103-112, 0920-5861.

55. Yoon, T. H., Johnson, S. B., Musgrave, C. B., & Brown, G. E (2004). Adsorption of organic matter at mineral/water interfaces: I. ATR-FTIR spectroscopic and quantum chemical study of oxalate adsorbed at boehmite/water and corundum/water interfaces, Geochim. Cosmochim. Acta Vol. 68, No. 22, pp. 4505-4518, 0016-7037.

56. Yost, E.C., Tejedor-Tejedor, M.I., & Anderson, M.A. (1990). In situ CIR-FTIR characterization of salicylate complexes at the goethite/aqueous solution interface, Environ. Sci. Technol. Vol. 24, pp. 822-828, 0013-936X.

6 Chapter

RESEARCH OF CALCIUM PHOSPHATES USING FOURIER TRANSFORM INFRARED SPECTROSCOPY

1. INTRODUCTION

In the biomaterial research field, nowadays a great attention is driven onto calcium phosphates synthesis and obtaining of ceramics that can be used in orthopedics and dentistry, in the form of coatings, granules, porous or solid blocks, as well as in the form of various composite materials. The most frequently studied, clinically tested and used synthetic materials based on calcium phosphate (CaP) are hydroxyapatite [HAp - $CA_{1}0(PO_{4})6(OH)_{2}$], β-tricalcium phosphate [β-TCP - $Ca_{3}(PO_{4})2$] and biphasic HAp/β-TCP mixture. CaP ceramic demonstrates high biocompatibility and bioactivity while it contacts bone cells, builds a direct chemical connection between bone tissues and ceramic implant. As practice shows, purchased materials, most often commercial CaP materials, not always have properties and qualities defined by the manufacturer. Frequently, the manufacturer's information about the offered product is not complete or precise, by it troubling usage of raw CaP material for development of implants. The most common imperfections of CaP materials are - unpredictable properties after the high temperature treatment (composition and clarity of crystal phases, chemical composition, thermal stability, etc.) which become clear only after the high temperature treatment of the ready implant material has occurred. For several years, Riga Biomaterial Innovation and Development Centre (RBIDC) of Riga Technical University perform a wide range of property studies of various commercial CaP raw materials.

Properties of bioceramic implants obtained from various commercial and laboratory synthesized calcium phosphate precursors are different, Since behavior of those precursors is different within the thermal treatment processes, which are a significant stage of obtaining ceramics.

CaP synthesis methods and their technological parameters can significantly impact stoichiometry of the synthesis product, its grade of crystallization, particle size, bioceramic phase composition, thermal stability, microstructure and mechanical properties. The important technologic parameters that impact properties of calcium phosphate synthesis product and then also of bioceramic, are temperature of synthesis, pH of synthesis environment, reagent type and concentration, as well as selection of raw materials, their purity and quality. All of the above mentioned also brings a significant impact on the tissue response of these bioceramic implants.

Fourier transform infrared spectroscopy (FTIR) is one of the methods which, systematically monitoring variations of structural characteristic groups and vibrations bonds, can provide an indirect evaluation of the synthesized Ca/P implant materials from TCP up to HAp and bioceramics, obtained from these materials.

FTIR spectroscopy has numerous advantages when used for chemical analysis of CaP products. First of all, an obtained spectrogram provides useful information about location of peaks, their intensity, width and shape in the required wave number range. Secondly, FTIR is also a very sensitive technique for determining phase composition. In the third place, FTIR is a comparatively quick and easy everyday approach.

During recent years, many authors' attention is turned onto synthesis of CaP and research of structure of the synthesized products depending on their technological parameters, with various methods, including an X-ray diffraction (XRD) and FTIR methods. However, the data of the literary sources is often incomplete or sometimes even contradictory. Studying various literary sources and analyzing the taken spectra of laboratory synthesized and commercial CaP products was aimed onto creating summary IR spectrum tables for the characteristic calcium phosphate chemical groups absorption bands. HAp stoichiometry is very important if the material has undergone a high temperature treatment.

A minor misbalance of synthesis product in a stoichiometric ratio (standard molar ratio of Ca/P is 1.67) during high temperature treatment can lead to composition of b-, a-TCP, or other phases. Thermally treating the stoichiometric calcium phosphates, it is possible to obtain stable phases at temperatures up to 1300 oC. One of the main non-stoichiometry reasons is inclusion of impurities, often substitutions of CA_2+ or interpenetration of other ions in the crystal lattice. In total, biological calcium phosphates are defined as calcium hydroxyapatites with deficient of calcium, $CA_{10-x}(PO_4)_{6-x}(HPO_4)x(OH)_{2-x}$ (0<x<2), including the substituting atoms or groups, as, for example, Mg^{2+}, Na^+, K^+, Sr^{2+}, or BA_2^+ substitute CA^{2+}, CO_3^{2}-, H_2PO_4-, HPO_4^{2}-; SO_4^{2}- substitute PO_4^{3}-; F- , Cl- , CO_3^{2}- , PO_4^{3}- substitute OH-.

The main target of our work was to perform a FTIR spectroscopy analysis of the CaP products synthesized in RBIDC laboratory and make summarizing conclusions about chemical groups of calcium phosphates, their variations under impact of synthesis parameters and further thermal treatment, as well as creating summary tables for:

- CaP powders synthesized in the laboratory with a chemical solution precipitation method with different synthesis parameters (temperature, final suspension pH, maturation time) as-synthesized and then thermally treated at various temperatures from 200oC up to 1400oC;
- Commercial CaP products;
- CaO containing materials of various origins (marble, eggshells, land Snail shells) and FTIR spectra of obtained products;

2. CALCIUM PHOSPHATES AND FTIR ABSORPTION BANDS OF THEIR CHEMICAL GROUPS

2.1 Calcium phosphates

Calcium phosphates as chemical compounds arise interest of the numerous fields of Science, like geology, chemistry, biology and medicine. Many forms of calcium phosphates are determined by their Ca/P molar ratio. From the point of view of chemistry, they are formed by three main elements: calcium, phosphorus and oxygen. Many calcium phosphates also contain hydrogen in an acidic phosphate anion (for example, HPO_4^{2}-), hydroxyl groups (for example, $CA_10(PO_4)_6(OH)_2$) or in a form of bonded water (for example, $CaHPO_4{\cdot}2H_2O$). Majority of compounds of this class are poorly soluble in the water and non-soluble in alkaline solutions, but all of them easily dissolve in acids. Chemically pure calcium orthophosphates are white crystals with an average hardness, while natural materials are always of some other color which depends on the type and amount of impurities. Biologic calcium phosphates are main mineral components in the calcified tissues of the vertebrates (Dorozhkin, 2009a).

Main components of the natural bone tissues are calcium phosphates which, along with the other elements (Na, K, F, and Cl) form ~ 70% of the bone tissue mass. Also, bone tissues contain water (10% of mass) and collagen along with the other organic materials in small amounts. In living system CaP are found in the form of crystalline hydroxyapatite (HAp) and in the amorphous calcium phosphate (ACP) form.

As bone substitution materials, calcium orthophosphates are researched for more than 80 years. The most significant characteristics of calcium phosphates are their bioresorbtion and bioactivity. They are non-toxic and biocompatible. Bioactivity shows as an ability to create a physical chemical bond between an implant and a bone. This process is called ostheointegration (Dorozhkin, 2009b).

Depending on the calcium/phosphorus (Ca/P) molar ratio and solubility of the compound, it is possible to obtain numerous calcium phosphates of different composition. Molar Ca/P ratio and solubility are connected with the pH of the solution. Majority of materials of this class are resorbable and dissolve when inserted in a physical environment. Calcium phosphates that are most frequently used in the biomaterial field are demonstrated in Table 1 (Dorozhkin, 2009c; El Kady, 2009; Shi, 2006).

For biomedical application, the following calcium phosphates are most frequently used: HAp (Ca/P = 1.67) and β-TCP (Ca/P=1.5), as well as biphasic calcium phosphate which mainly consists of HAp and β-TCP mixture in various ratios.

2.2 FTIR absorption bands of the synthesized HAp

Hydroxyapatite $CA_10(PO_4)_0(OH)_2$ is dominating and the most significant mineral phase in the solid tissues of the vertebrates. It consists of the same ions that form mineral part of teeth and bones.

A biological HAp usually has a calcium deficient; it is always substituted with a carbonate. Two types of carbonate substitution are possible: (1) direct substitution of OH- with CO_3^{2-} (A-type substitution $(CO_3)_2^-$ ↔ 2OH-) and (2) necessity after charge compensation, PO_4^{3-} substituting a tetrahedral group with CO_3^{2-} (B-type substitution). Substitution groups may provoke characteristic changes in the lattice parameters, crystallinity, crystal symmetry, thermal stability, morphology, and solubility, physical, chemical and biological characteristics (Shi, 2006).

The most characteristic chemical groups in the FTIR spectrum of synthesized HAp are PO_4^{3-}, OH-, CO_3^{2-}, as well as HPO_4^{2-} that characterize non-stoichiometric HAp.

Table 1. Calcium phosphates used in the biomaterial field.

Name	Abbreviation	Chemical formula	Ca/P
Amorphous calcium phosphate	ACP	$Ca_xH_y(PO_4)z\,.\,nH_2O$	1.2-2.2
Dicalcium phosphate anhydride	DCPA	$CaHPO_4$	1.00
Dicalcium phosphate dehydrate	DCPD	$CaHPO_4\,.\,2H_2O$	1.00
Octacalcium phosphate	OCP	$Ca8(HPO_4)2(PO_4)4\,.\,5H_2O$	1.33
β-tricalcium phosphate	β-TCP	$Ca_3(PO_4)2$	1.50
α-tricalcium phosphate	α-TCP	$Ca_3(PO_4)2$	1.50
Hydroxyapatite with calcium deficient	CDHA	$CA_{10-x}(HPO_4)_x(PO_4)_{6-x}(OH)_{2-x}$ $0 \leq x \leq 1$	1.5-1.67
Hydroxyapatite	HAp	$CA_{10}(PO_4)_6(OH)2$	1.67
Tetra calcium phosphate	TTCP (TetCP)	$Ca_4(PO_4)_2O$	2.00
β-Ca pyrophosphate	CPP	$CA_2P_2O_7$	<1.5
Oxyapatite	OAp	$CA_{10}(PO_4)_6O$	1.67

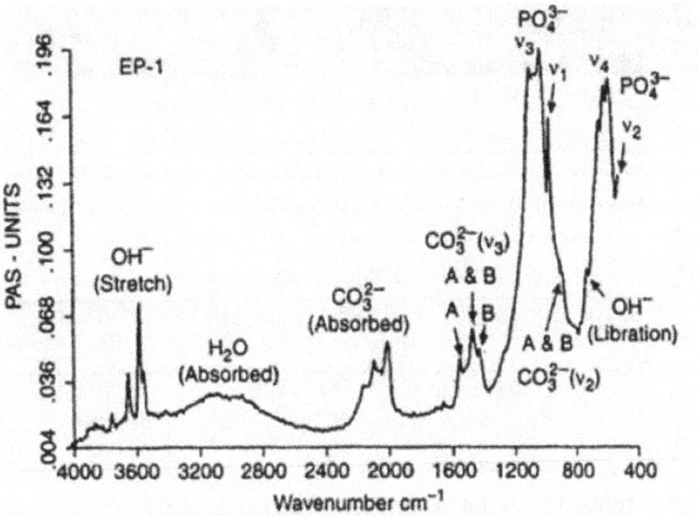

Fig. 1. A typical FTIR spectrum of hydroxyapatite (Ratner, 2004)

PO_4^{3-} group forms intensive IR absorption bands at 560 and 600 cm^{-1} and at 1000 - 1100 cm^{-1}. Adsorbed water band is relatively wide, from 3600 to 2600 cm^{-1}, with an explicit peak at 3570 cm^{-1}, a weaker peak is formed at 630 cm^{-1}. CO_3^{2-} group forms weak peaks between 870 and 880 cm^{-1} and more intensive peaks between 1460 and 1530 cm^{-1}. Absorption bands of chemical bonds of the synthesized HAp spectrum are summarized in Table 2.

Table 2. FTIR absorption bands of synthesized HAp chemical groups.

Chemical groups	Absorption bands, (cm^{-1})	Description
CO_3^{2-}	873; 1450; 1640 (Meejoo, et al., 2006) 1650 (Raynaud, et al., 2002); 870 and 880; 1460 and 1530 (Ratner, 2004)	Substitutes phosphate ion, B-type HAp is formed (Meejoo, et al., 2006)
OH-	3500 (Meejoo, et al., 2006) 630 and 3540 (Destainville, et al., 2003), (Raynaud, et al., 2002); 3570 and 3420 (Han J-K., et al., 2006); 1650 (Raynaud, et al., 2002)	OH- ions prove presemce of HAp
Adsorbed water	2600 – 3600 (Meejoo, et al., 2006)	Under influence of thermal treatment, absorption band becomes narrower
HPO_4^{2-}	875 (Destainville, et al., 2003), (Raynaud, et al., 2002); 880 (Kwon, et al.,2003)	Characterizes HAp with deficient of calcium. (Raynaud, et al., 2002); Refers to non-stoichiometric HAp (Kwon, et al.,2003);

Chemical groups	Absorption bands, (cm^{-1})	Description
PO_4^{3-}	460 (Destainville, et al., 2003); (Raynaud, et al., 2002);	n_2 (Destainville, et al., 2003);
	560 - 600 (Destainville, et al., 2003), (Raynaud, et al., 2002), (Mobasherpour & Heshajin, 2007); 602 un 555 (Han J-K., et al., 2006)	n_4 (Destainville, et al., 2003); bending mode (Han J-K., et al., 2006)
	960 (Destainville, et al., 2003), (Raynaud, et al., 2002)	n_1 (Destainville, et al., 2003);
	1020 -1120 (Destainville, et al., 2003), (Raynaud, et al., 2002); 1040 (Han J-K., et al., 2006); 1000 - 1100 (Mobasherpour & Heshajin, 2007);	n_3 (Destainville, et al., 2003); bending mode (Han J-K., et al., 2006);
NO_3^-	820 and 1380 (Destainville, et al., 2003); (Raynaud, et al., 2002)	Synthesis residue that disappears during the calcifying process (Destainville, et al., 2003)

2.3 FTIR absorption bands of thermally treated calcium phosphates

As a biomaterial, HAp is mainly used in its ceramic form that was obtained by sintering the powder at 1000 - 1350 °C or as a coating on the implant surface. During the process of thermal decomposition of HAp, sintering of ceramic or obtaining the coating, physical, chemical, mechanical and, most important, biomedical properties may be negatively affected. Thus, HAp and other CaP materials should be thoroughly studied while thermally treated.

Table 3. FTIR absorption bands of thermally treated CaP chemical bonds.

Temperature, °C	Chemical groups and phases	Absorption bands, (cm^{-1})	Description
250	H_2O		Molecules of adsorbed water disappear (Mobasherpour & Heshajin, 2007);
600	NO_3^-	820 and 1380 (Raynaud, et al., 2002)	Synthesis impurities disappear (Raynaud, et al., 2002);
700	CO_3^{2-}	1450 (Meejoo, et al., 2006)	Intensity decreases;

Temperature, °C	Chemical groups and phases	Absorption bands, (cm^{-1})	Description
	H_2O, OH-	3500 (Meejoo, et al., 2006) 630 and 3540 (Destainville, et al., 2003);	Adsorbed water band becomes narrower (Meejoo, et al., 2006); Refers to variations of OH- (Destainville, et al., 2003);
800	CO_3^{2-}	1450 (Meejoo, et al., 2006)	Disappears (Meejoo, et al., 2006)
900	OH-		Disappears (Meejoo, et al., 2006)
	β- TCP	947, 974 and 1120(Meejoo, et al., 2006)	β-TCP shoulders begin to show up (Meejoo, et al., 2006);
	PO_4^{3-}	603 and 565; 1094 and 1032 (Meejoo, et al., 2006)	Shifts position at 1200 0C (Meejoo, et al., 2006);
1200	β- TCP		Can see the characteristic peaks better (Ratner, 2004);
	PO_4^{3-}	601 and 571; 1090 and 1046 (Meejoo, et al., 2006)	Indicates that under influence of temperature, phosphates decompose and β-TCP shoulders become wider (Meejoo, et al., 2006).
1200 - 1400			β- TCP transforms onto α-TCP (Mobasherpour & Heshajin, 2007);
1400	α-TCP	551; 585; 597; 613; 984; 1025; 1055 (Han J-K., et al., 2006)	

During the thermal treatment, behavior of CaP is affected by various factors, like, atmosphere of sintering, ratio of Ca/P, method and conditions of powder synthesis, type and amount of impurities, sample size, particle size, etc.

During thermal treating HAp undergoes the following processes:

- dehydration (separation of adsorbed water);
- dehydroxylation (separation of structured water), forming oxy-hydroxyapatite (OHAp) and oxyapatite (OAp);
- HAp decomposition with formation of other phases.

In Tables 3-5, the data obtained from literary sources about the FTIR absorption bands of thermally treated CaP chemical groups, is summarized.

Table 4. FTIR absorption bands of thermally treated TCP chemical groups.

Temperature, oC	Chemical groups and phases	Absorption bands, (cm^{-1})	Description
650	TCP;		Synthesis residue is taken away, but the initial TCP remains unchanged (Destainville, et al., 2003);
750	TCP→β-TCP (Destainville, et al., 2003);		Agglutination begins (Destainville, et al., 2003);
950 - 1000			Maximum speed of compaction, comparing with HAp, β-TCP sintering occurs at a lower temperature. (Destainville, et al., 2003);
	OH-	630 (Destainville, et al., 2003);	Disappeared; spectrum is similar β-TCP (Kwon, et al., 2003)
	P_2O_74-	727 and 1200 (Destainville, et al., 2003)	Lack of P_2O_74- proves that there is no CPP phase and spectrum is similar to pure β-TCP (Destainville, et al., 2003);
1200	β-TCP→αTCP		(Destainville, et al., 2003);

Losing the adsorbed water do not impact lattice parameters. The water adsorbed on the surface discharges under temperature of less than 250°C, when the moisture is discharged from pores up to 500°C. With temperature rising, wide water bands at 3540 cm^{-1} become narrower and gradually disappear, but the sharp narrow peaks at 630 and 3570 cm^{-1} refer to variations of structural OH- groups, which is characteristic to structure of HAp. Depending on the synthesis condition, a carbonate containing apatite is often obtained. Then, it should be considered that CO_2 is discharged from the sample between 450-950°C.

Thermal stability is characterized by the decomposition temperature of HAp sample. The decomposition occurs when a critical dehydration point is achieved. In the temperatures less than the critical point, crystal structure of HAp remains unchanged in spite of the stage of dehydration. Achieving the critical point, a complete and irreversible dehydroxillation occurs, which results damage of HAp structure, decomposing onto tricalcium phosphate (β-TCP under 1200 °C and α-TCP in higher temperatures) and tetracalcium phosphate (TTCP).

At 900oC, β-TCP shoulders begin to show up at 947, 974 and 1120 cm^{-1}, but during heating at higher temperature as, for example, 1200oC, β-TCP phase becomes more visible and PO_4^{3-} peaks shift from 603 and 565 cm^{-1} to 601 and 571 cm^{-1}, also from 1094 and 1032 to 1090 and 1046 cm^{-1}.

Table 5. Analysis of FTIR absorption bands of thermally treated biphasic calcium phosphate (HAp/β-TCP) chemical groups

Temperature, oC	Chemical groups	Absorption bands, (cm^{-1})	Description
350	HPO_4^{2-}		Begins showing up (Raynaud, et al., 2002);
350-720	HPO_4^{2-}	875 (Raynaud, et al., 2002);	As a result of condensation, P2O7 is formed (Raynaud, et al., 2002);
400	$P_2O_7$4-	720 (Raynaud, et al., 2002);	Begins showing up if 1.5<Ca/P<1.677 (Raynaud, et al., 2002);
600	NO_3-	820 and 1380 (Raynaud, et al., 2002);	Disappears (Raynaud, et al., 2002)
750	OH-	630 and 3540 (Destainville, et al., 2003);	Proves presemce of HAp (Destainville, et al., 2003);
800	$P_2O_7$4-	715 (Meejoo, et al., 2006);	Forming of pyrophosphate groups (Meejoo, et al., 2006);
700-900			HAp with deficient of calcium, by decomposing, forms HAp and β - TCP(Raynaud, et al., 2002);
1000	$P_2O_7$4-	720 (Raynaud, et al., 2002);	Disappears above 1000 oC, if 1.5<Ca/P<1.677 (Raynaud, et al., 2002);

For stoichiometric HAp, HPO_4^{2-} group is not detected, even though it can appear from the synthesis impurities (NO_3-, NH_4^+).

Various studies show that in the result of HAp (OAp) decomposition, apart from TCP and TTCP, also other calcium compounds may form, like calcium pyrophosphate (CPP, β- CA_2P2O7) and calcium oxide (CaO).

Apatitic TCP (ap-TCP) Ca9(HPO_4)(PO_4)5(OH) is a calcium orthophosphate which, during thermal treatment at temperature higher than 750oC, transforms onto β-tricalcium phosphate Ca_3(PO_4)2.

During the synthesis of TCP, the most important controllable parameters are temperature and pH. According to the literary sources, pH is almost neutral or slightly acidic, and is synthesized at lower temperatures.

A pure stoichiometric β-TCP with a molar ratio Ca/P=1.500, is formed in the result of temperature treatment. If Ca/P > 1.500, HAp is formed as the second phase. When Ca/P ratio is changed for 1%, HAp is formed for 10 wt%. If Ca/P < 1.500, then DCPA is formed, this is proven by presemce of calcium pyrophosphate CA_2P2O_7.

In order to control speed of biodegradation, a biphasic calcium phosphate (BCP) bioceramic is developed, containing both HAp and TCP. By variation HAp/TCP ratio, it is possible to control bioactivity and biodegradation of implant. Since β-TCP is more soluble and HAp allows a biological precipitation of apatites, solubility of BCP depends on the ratio of HAp/TCP. Osteoconductivity among BCP, HAp and TCP does not significantly differ.

BCP is formed, if 1.500 < Ca/P < 1.667, which refers to a hydroxyapatite with calcium deficient, its chemical formula is $CA_10\text{-x}(PO_4)6\text{-x}\ (HPO_4)x(OH)2\text{-x}$ (0< x <2). Ca/P ratio of synthesis sedimentary is not directly connected with the initial Ca/P ratio. At the constant pH, molar ratio of calcium phosphates may be varied by changing the temperature during the synthesis process.

3. MATERIALS AND METHODS

3.1 Synthesis of calcium phosphates

In order to achieve higher assay of the obtained material, its thermal stability and predictability of other properties, the calcium phosphates were synthesized in a laboratory using a wet chemical precipitation method from CaO (or calcium hydroxide Ca(OH)2) of various origins (commercial, marble, eggshells, land Snails shells) as precursors and orthophosphoric acid H_3PO_4.

Main advantages of this method are a simple synthesis process, a relatively quick obtaining of end product, possibility to obtain large quantities of end product, relatively cheap raw materials and the only by-product it gives is water. It is also important that calcium phosphates with nanometric crystal size can be obtained at a low process temperature (from room to water boiling temperature).

During synthesis, technological parameters like final pH of calcium phosphate suspension and synthesis temperature (T,°C) were changed. Final suspension pH was stabilized in the range of 5-11, using solution of acid. Synthesis temperature was changed in the range from room temperature (21°C) up to 70°C, the following parameters were controlled: acid solution adding speed (ml/min), stirring speed (rpm), synthesis temperature (T,°C), final suspension pH, stabilization time (h), maturity time (τ, h), drying temperature and time (T,°C; h), calcifying temperature and time (T,°C; h).

For further obtaining HAp or biphasic bioceramic synthesized under impact of various parameters (final pH and synthesis temperature), powder is thermally treated. Samples are heated in different environments (Air, vacuum and water vapor), variating thermal treatment temperature in range 200-1400°C and processing time.

3.2 Analysis methods and sample preparation

In order to determine raw materials, structure of synthesized and heated powder, phase composition and functional groups, two important methods are used, complementing each other: Fourier transform infrared spectroscopy (FTIR) and X-ray difractometry (XRD). XRD method is widely used for apatite characterization, for it provides data about the crystal structure of material and its phase composition, however, it is not convenient to determine amount of [OH] or [CO_3] groups in hydroxyapatite. FTIR method, in many cases is more sensitive than XRD when determining presemce of new phases. Using FTIR, CaP can be characterized, considering three spectrum parameters:

- Location of absorption maximum indicates material composition, even slight variations of the composition influence energy of material bondings and, as follows, frequency of variations;
- Peak width shows degree of the atoms' order in the apatite elementary cell.
- Considering the absorption maximum of [OH] vibrations, presemce of HAp and its thermal stability can be determined, as well as hydroxyl group concentration in the sample.

During the research process, X"Pert PRO X-ray diffractometer has been used (PANalitical, the Netherlands). Samples were measured in a spinning mode, in the 2θ angle range from 20-90º, with a scanning step 0,0334o, with a $Cu_{K\alpha}$ radiation. Ratio of HAp/TCP is determined using a XRD semi-quantitative method after calibration line.

Calcium phosphate spectra are measured and analyzed with a FTIR spectrometer "Varian 800" of Scimitar Series, with a wave length range from 400 - 4000 cm^{-1}, with precision of 4 cm^{-1} and RESOLUTION software.

Samples are prepared by mixing powder with KBr and pressing the pellet. This method of sample preparation has some complications and requires a certain experience, in order to obtain a good quality spectrum in everyday work routine. Special factors should be considered in order to perform invariable sample analysis by using a KBr method, and these include pellet thickness, particle dispersion, ensuring vacuum state during the pressing, pressure influence, ion exchange, etc.

In a prepared KBr pellet, there should be material concentration of 2-10% from the total weight. For preparing a 300g KBr pellet, from 1 to 5mg of the sample is required, and the pellet size will be 13 mm.

Powder grain size should be ~150 μm. The analyzed sample is crashed in a powder and thoroughly mixed with the KBr powder. A powder mixer „Pulverisette 23" was used for crashing and mixing of the powder. Prepared powder was located in a specific SPECAC (d = 13 mm) mould, and the required pressure was achieved by applying a uniaxial press (required pressure is ~5.103 kg/cm^2, pressing time 1 min).

KBr attracts water molecules from the environment and they create wide water bands in the spectrum, so they are hard to or even impossible to analyze. KBr powder should be of the highest assay, Riedel-de-Haen KBr (Lot 51520) brand was used with an assay in the range of 99.5-100.5%. Usually, absorption bands of the main impurities in the KBr are: OH- groups and H_2O molecules (3500 cm^{-1} and 1630 cm^{-1}), NO2 (1390 cm^{-1}), SO_4^{2-} (1160-1140 cm^{-1}). Spectrum of the KBr used in our research is demonstrated on the Fig. 2. Considering that the powder is hygroscopic, KBr powder is dried at 105oC and kept in special hermetic containers. Prepared pellets with the analyzed material are dried Once again for 24 h. The obtained pellets should be transparent and equally colored. Weak bands connected with water can also be compensated by using a KBr pellet of the same thickness, but not containing the analyzed material, for background spectrum measuring.

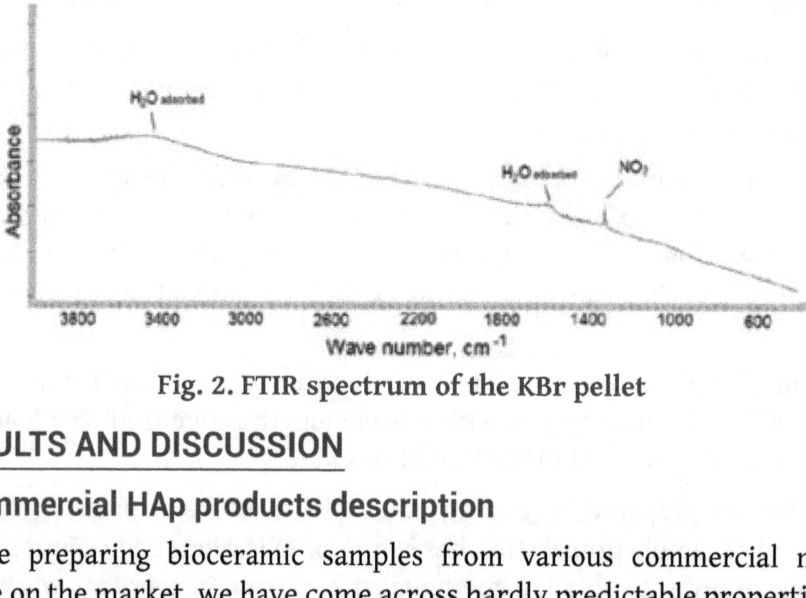

Fig. 2. FTIR spectrum of the KBr pellet

4. RESULTS AND DISCUSSION

4.1 Commercial HAp products description

While preparing bioceramic samples from various commercial materials available on the market, we have come across hardly predictable properties of the end product, like crystallinity degree, phase composition and, following, bioactivity and mechanical characteristics. One of the disadvantages while purchasing commercial calcium phosphate powders or commercial calcium phosphate ceramic materials is insufficient information about synthesis conditions of these calcium phosphates, raw materials and in which proportions these materials are taken.

It is significant to know if the purchased powder is thermally treated, and in which temperature range this thermal treatment was performed. Exactly temperature, at which the sample was obtained and processed, is one of the conditions that influences outcome of ceramic and phase composition.

For example, in spectra of several overviewed commercial hydroxyapatites, OH- and PO_4^{3-} groups are observed, but band shape, width and intensity are different. Differences in spectra are also observed at CO_3^{2-} and HPO_4^{2-}-groups location and intensity. On Fig. 3, there are three spectra from different commercial HAp compared: "Fluka"(F), "Riedel-de Haën®" (R-dH) and „Sigma Aldrich" (S-A). IR spectrum of commercial S-A product has a lower CO_3^{2-} and HPO_4^{2-} intensity that could mean a higher assay degree than the other materials have. These spectra also demonstrate that the products have a low crystallinity degree and they were not thermally treated.

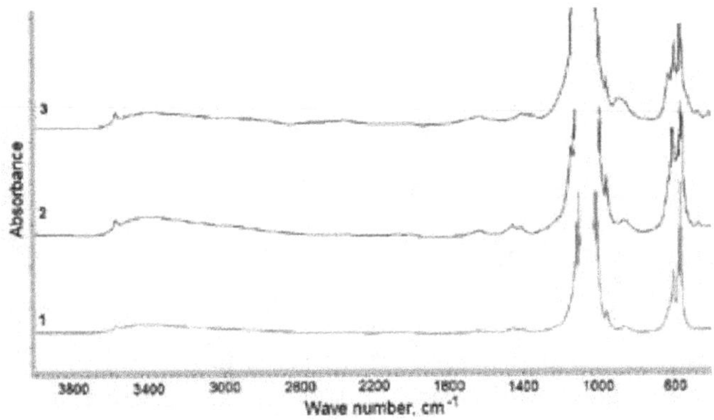

Fig. 3. FTIR spectra of commercial HAp products (1 - S-A; 2 - R-dH; 3 - Fluka)

Characteristic chemical groups of the commercial synthesized products are summarized in table 6.

Table 6. Characteristic chemical groups of commercial HAp FTIR absorption bands

Commercial HAp	Fluka	R-dH	S-A
Chemical groups	Absorption bands, cm^{-1}		
CO_3^{2-}	1386; 1411; 1635; 1997; (2359 C≡C)	1382; 1413; 1457; 1634; 1997	1382; 1417; 2457; 1639; 1990; 2359
H_2O adsorbed	3100 - 3600	3000 -3600	3200 - 3600
OH-	635; 3568	3568	630; 3569;
HPO_4^{2-} ; CO_3^{2-}	891; 875	870	874
PO_4^{3-}	470; 553 - 600; 964; 1000 - 1156	470; 553 - 610; 964; 1000 - 1150	471; 561; 601; 605; 964; 1013 - 1120

In the FTIR spectra of commercial β-TCP products, PO_4^{3-} groups are observed, which is characteristic to β-TCP (Fig. 4). In HAp IR spectrum, OH- group peaks are observed (at 630 and 3570 cm^{-1}), but there are no such in the IR spectrum of commercial β-TCP, which means that there is no HAp phase in this β-TCP product.

In addition, at 725 cm-1, presemce of $P_2O_7$4- group can be observed, which is characteristic to calcium pyrophosphate phase.

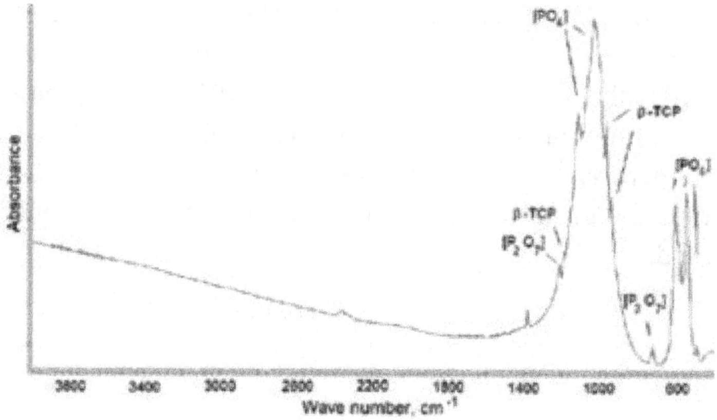

Fig. 4. FTIR spectra of commercial product (Fluka β-TCP)

4.2 Phase composition control during synthesis reaction

As above mentioned, CaP material synthesis was performed in our laboratory, variating synthesis and further sample heating parameters, searching for their optimal combination depending on the required properties of the obtained material.

In order to control reaction process and possible appearance of by-products, control samples were taken every 5 minutes and analyzed by measuring their spectra. After taking a sample from reactor with a plastic dropper, it was inserted in a glass bottle, hermetically sealed and frozen by putting it in the mixture of dry ice and acetone. After full freezing, the temperature is supported by storing the bottle in the dry ice. Before inserting in the cryogenic drying device, bottles are covered with a perforated plastic film. From the spectra of samples prepared this way (Fig.5), synthesis with a final pH=9,3 and synthesis temperature 45oC, it can be seen that the synthesized material is formed with an apatitic HAp structure with a slight Ca deficient, which is proven by presemce of the CO_3^{2-} group, amount of which is constantly reducing along with reaction approaching its end. No by-products were detected, during the reaction, CaO has reacted fully which is proven by an OH- peak disappearing at 3642 cm^{-1} and forming OH- peaks at 3571 cm^{-1} and 631 cm^{-1} which is characteristic for HAp phase.

Such control is very important for scaling the synthesis, relatively increasing amount of synthesis and amount of obtained CaP.

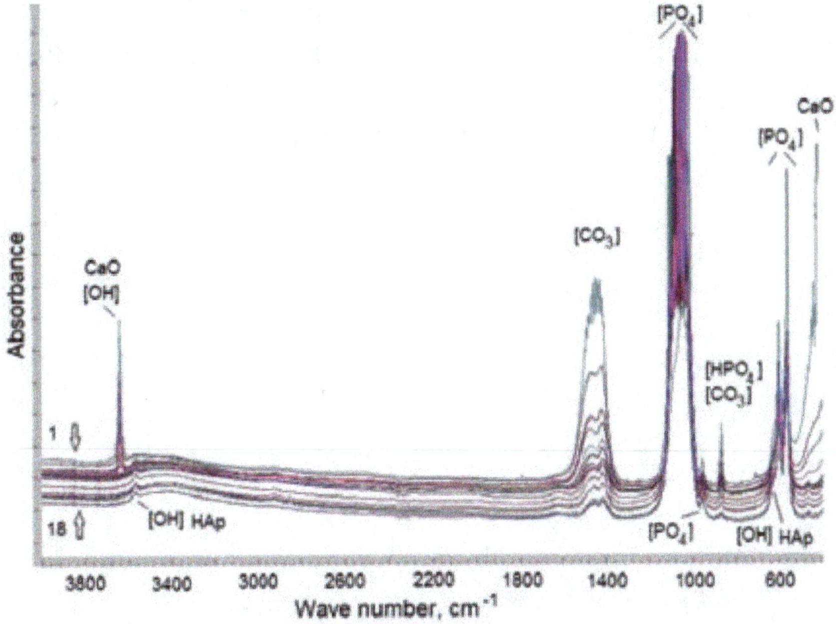

Fig. 5. FTIR spectra of the samples from the synthesis series with final pH=9,3 at temperature 45oC, depending on the reaction occurrence time (from 1 to18).

4.3 Selection of CaO containing materials of various origins (marble, eggshells, Snail shells) and its influence on the CaP product properties

Scientific literature contains very few information about research of how raw materials' (for example, CaO, Ca(OH)2) quality (chemical and physical properties) impacts properties of the obtained bioceramics.

Before starting synthesis of CaP products, selection of CaO containing raw material and complex research were performed. For „Ca" precursors, two commercial available synthetic CaO powders from "Riedel-de Haën®" (CaOR) and "Fluka" (CaOF) and two materials of biogenic origin, widespread in the nature - egg shells and land Snail (Arianta arbustorum) shells, were chosen. These materials were selected as raw materials for obtaining CaO and further usage in the synthesis process of CaP.

In composition of commercial CaO, presemce of Ca(OH)2 phase, small amount of MgO phase, as well as small amount of polymorphous $CaCO_3$ modification – calcite phase (Fig. 6), is detected. Presemce of $CaCO_3$ is undesirable in CaO, for during the process of suspension obtaining, Ca(OH)2 creates an error in preparation of precise amount of the reagent and, as follows, product reproduction, also preventing obtaining a homogenous Ca(OH)2 suspension which is required for further synthesis of CaP. After heating biogenic and commercial CaO at 1000ºC, X-ray diagrams of all the synthesis materials demonstrated a CaO crystal phase with a small amount of MgO phase, along with Ca(OH)2 that formed in the result of CaO contact with Air moisture; considering that CaO is hygroscopic (Siva Rama Krishna, et al., 2007).

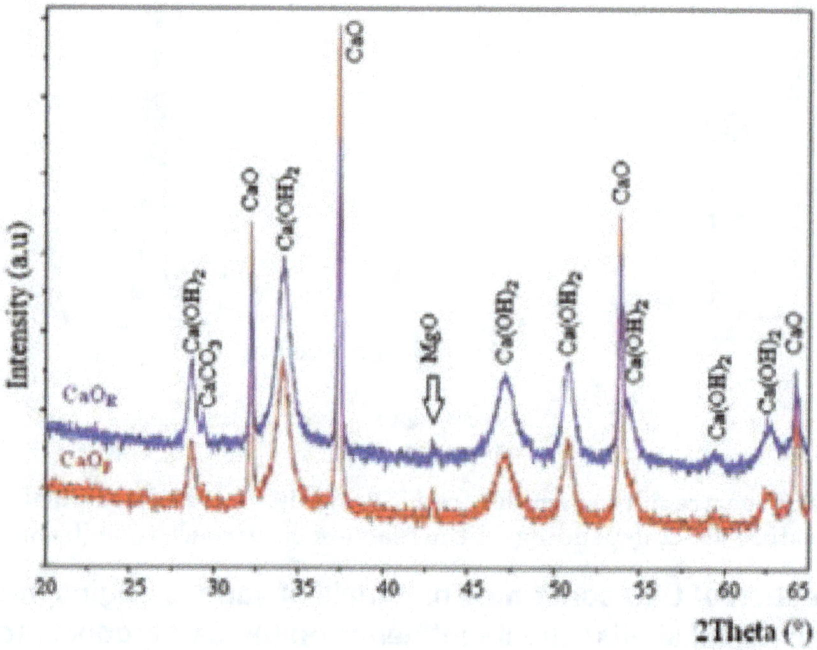

Fig. 6. Commercial CaO X-ray diffractograms before calcifying

FTIR spectra were also taken for the thermally treated CaO samples, and their compositions were similar (Fig. 7.). In all the IR spectra, an explicit absorption peak is visible at 3642 cm^{-1}, that indicates stretching variations of Ca(OH)2 (Ji, et al., 2009) and [OH] groups (Siva Rama Krishna, et al., 2007). Presemce of CaO is also proven by a wide intensive absorption band of [Ca-O] group, which is centered at ~ 400 cm^{-1} (Ji, et al., 2009). Absorption peaks at 874 cm^{-1}, 1080 cm^{-1} [113], as well as at 1420 cm^{-1} prove presemce of the [CO_3] groups in the samples which, therefore, shows a slight carboxilation of Ca(OH)2 from CO_2 of the atmosphere. Absorption band from 3430-3550 cm^{-1} proves presemce of adsorbed water molecules in the samples.

FTIR spectra of the synthesized powders demonstrate absorption bands of the chemical functional groups, characteristic to HAp phase (Fig. 8.). Number of [CO_3] groups in those is different, but after thermal treatment at 1100ºC for 1 h, the spectra become very similar (Fig. 9).

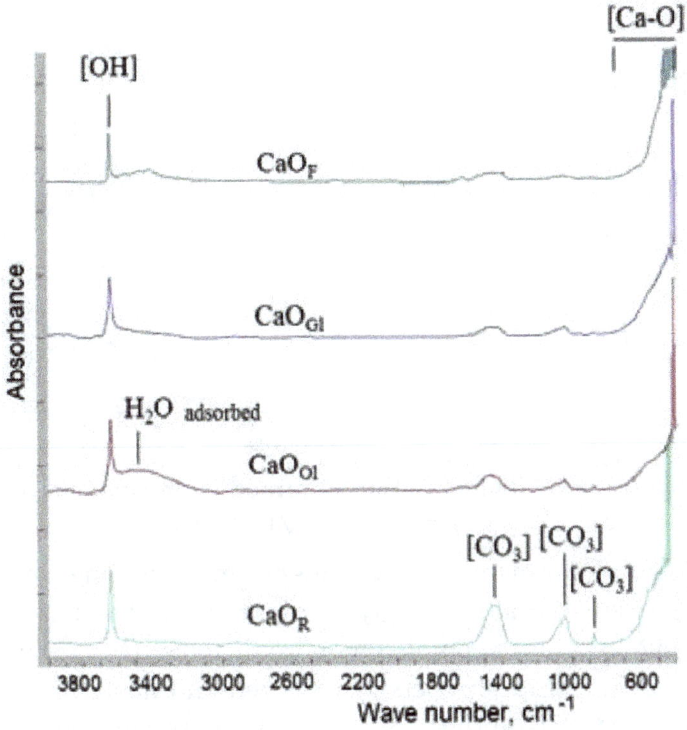

Fig. 7. FTIR spectra of biogenic and commercial CaO after heating at 1000ºC for 1 h (F - "Fluka", Gl - land Snail shells, Ol - egg shells and R - "Riedel-de Haën®")

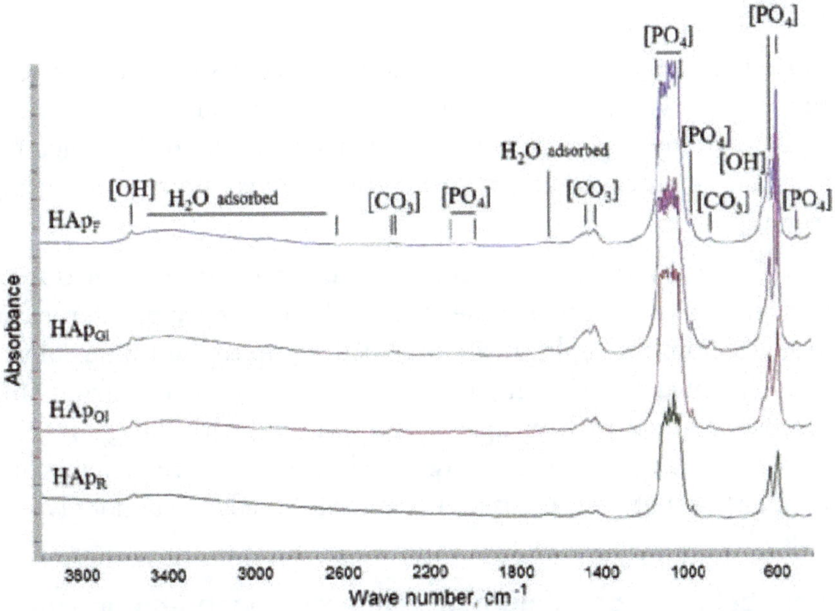

Fig. 8. FTIR spectra of HAp products as-synthesized after drying at 105ºC for ~ 20 h from various CaO (F - "Fluka", Gl - land Snail shells, Ol - egg shells and R - "Riedel-de Haën®").

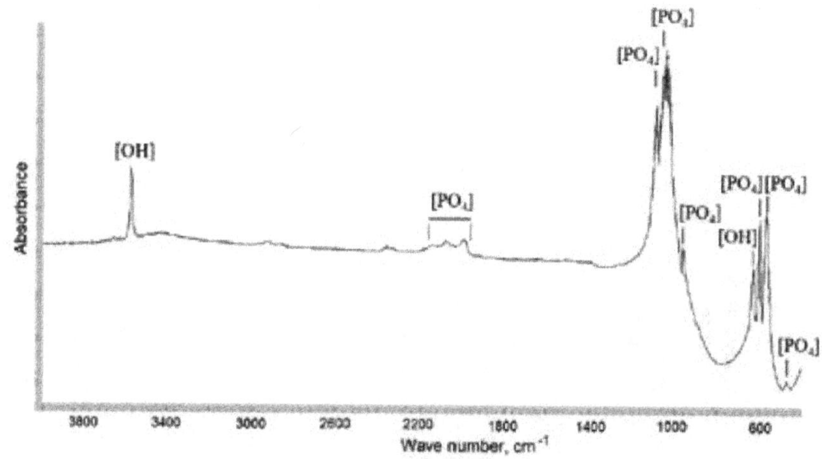

Fig. 9. FTIR spectra of HAp bioceramics obtained from various CaO after thermal treatment at 1100ºC for 1 h.

Using CaO containing materials of various origins, including the commercial CaO available on the market that seemingly correspond with the assay specification, it is still impossible to obtain a completely reproductable HAp bioceramic materials. XRD and FTIR of the HAp products synthesized and heated at 1100ºC, produce similar pictures, and post-synthesis morphology, phase composition and molecular structure of these bioceramics are identical. However, analyzing it with FE-SEM micrographs, it can be observed that obtained HAp products demonstrate different microstructures.

A homogenous, fine-grained microstructure with a small grain size - about 150-200 nm, is observed at the HAp bioceramic from commercial reagents, and non-homogenic grainy structure with irregular grain size in the range from 200 nm up to 1 mcm is observed at the ceramic which was synthesized using Ca of natural origin.

Therefore, the samples from commercial reagents demonstrated color changing, from white in the synthesized and just dried powder to average aquamarine, at ceramic samples heated at 1000 °C, up to light blue color at 1300 °C. It can be explained by oxidizing of the manganese impurities from Mn_2^+ up to Mn5+ and substitution of hydroxyapatite (PO_4^{3-}) group with (MnO_4^{3-}) (Ślósarczyk, et al., 2010). Color change can be also explained by other microelements or defects in the crystal lattice, but, in this case, FTIR and XRD analysis cannot give a precise answer to this.

4.4 Thermal behavior of calcium phosphates depending on synthesis parameters

Sample structure, phase composition and thermal stability after heating depend on synthesis parameters, especially from final pH and synthesis temperature. Also, connection between pH value and temperature (with temperature increasing, pH decreases); so, by combining and analyzing those parameters, both pure HAp and TCP materials with a good thermal stability are obtained and biphasic materials with various Ca/P ratios and various percentages of phase composition.

Thermal behavior data of laboratory synthesed calcium phosphates with various synthesis parameters were summarized. Final reaction pH was variated in the range from 5.0 up to 10.7 and synthesis temperature was chosen as room (22oC), 45oC and 70oC (Table 7).

Table 7. Variable parameters of some calcium phosphate synthesis and calcium phosphate products phase composition after thermal treatment at 1100ºC for 1 h

Nr. of synthesis	Final pH of synthesis	Synthesis temperature, oC	Phase composition by XRD after thermal treatment at 1100oC (1h)
1	5,0	room	TCP+$CaHPO_4$
2	5,1	70	HAp/TCP (60/40)
3	5,3	45	HAp/TCP (35/65)
4	5,9	room	TCP
5	7	45	HAp/TCP (80/20)
6	9,3	45	HAp
7	10,7	45	HAp/CaO

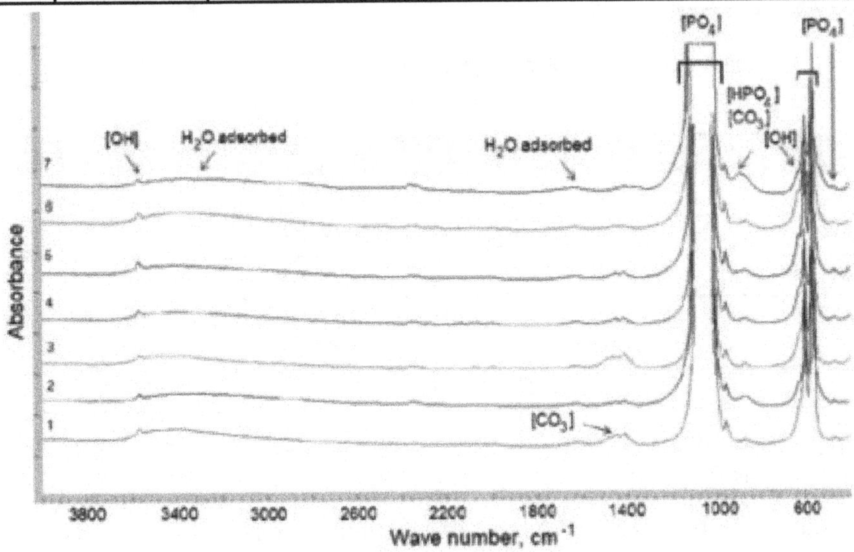

Fig. 10. FTIR spectrum of as-synthesized and drying at 105ºC (20 h) CaP products of 1-7 synthesis

It can be ascertained that, in spite of variable parameters of synthesis, all the spectra of samples and XRD diffraction diagrams, are similar, all the functional groups correspond with non-stoichiometric apatitic HAp structure with a low crystallinity degree (Fig. 10). The only slight differences that can be observed between spectra, are intensities of CO_3^{2-} group bands, however, no significant differences in the number of OH- groups, depending on the synthesis parameters at the non-calcified samples, are detected (Table 8).

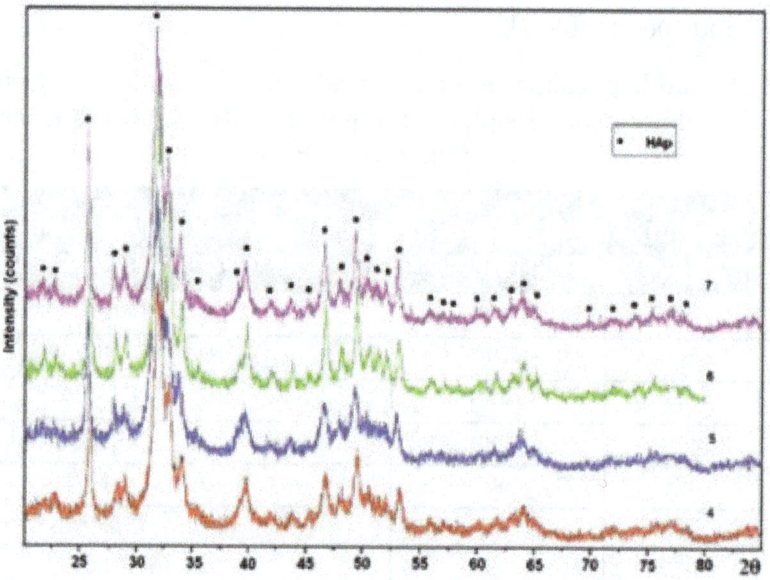

Fig. 11. XRD patterns of as-synthesized CaP products from 4, 5, 6, 7 synthesis.

Table 8. Absorption bands of as-synthesized and drying at 105ºC (20 h) CaP products

Chemical groups	Absorption bands, cm^{-1}
PO_4^{3-}	472; 570; 602; 963; 1000 - 1140;
H_2O adsorbed	3100 - 3600;
OH-	631; 3570;
HPO_4^{2-}	875 (identifies HAp with deficient of calcium and non- stoichiometric structure)
CO_3^{2-}	875; 1418; 1458; 1632 and 1650; 1994

Absorption bands of chemical functional groups characteristic to HAp phase can be defined as follows:

- Absorption bands at 3570 cm^{-1} and 631 cm^{-1} are referable to structural [OH] groups (O- H) stretching and libration modes at the HAp crystallite surface or at the crystallites.
- Presemce of [PO_4] groups, characteristic to tetrahedral apatite structure, is proven by absorption bands at 472 cm^{-1} , which is characteristic to [PO_4]

ν2 group (ν2 O-P-O) bending variations; double band at 570 cm^{-1} and 602 cm^{-1} with a high resolution is referable to asymmetric and symmetric deformation modes of $[PO_4]\nu4$ group (ν_4 O-P- O); absorption band at 963 cm^{-1} corresponds to a symmetric stretching mode; intensive absorption band in the range of 1040-1090 cm^{-1} corresponds to a band characteristic to $[PO_4]\nu_3$ groups (ν_3 P-O) at 1040 cm^{-1} and 1090 cm^{-1} asymmetrical stretching mode, which, as explicit maximums, can be observed after thermal treatment of the samples;

- An absorption band of weak intensity within the range between 1950- 2100 cm^{-1} is connected with combinations of $[PO_4]\nu_3$, ν_1 modes.

These locations of absorption bands of the functional groups explicitly indicate forming of a typical HAp structure in the synthesized samples (Nilen & Richter, 2008; Siva Rama Krishna, et al., 2007; Kothapalli, et al., 2004; Landi, et al., 2000; Lioua, et al., 2004).

Weak absorption bands at 2365 cm^{-1} and 2344 cm^{-1} appeared due to attraction of CO_2 from the atmosphere.

Presemce of $[CO_3]$ bands can be identified with explicitly visible absorption bands within the range between 1600-1400 cm^{-1} and at 875 cm^{-1}, which are observed in the spectra of synthesized samples. Absorption bands of weak intensity, centered at 1418 cm^{-1} and 1458 cm^{-1} correspond to symmetrical and asymmetrical stretching modes of the $[CO_3]\nu_3$ groups (C-O). Absorption band at ~875 cm^{-1} can prove presemce of $[CO_3]\nu2$ stretching mode (at ~872 cm^{-1}), intensity of which is approximately 1/5 share of $[CO_3]\nu_3$, or presemce of $[HPO_4]$ group absorption maximum (Siva Rama Krishna, et al., 2007; Landi, et al., 2000). Considering that the $[HPO_4]$ group band partially covers $[CO_3]\nu_2$, it is complicated to detect, of which group is this band. Presemce of this absorption band ascertains solution of atmosphere CO_2 in the suspension, if synthesis occurs in the alkaline environment.

Combination of absorption bands – absorption bands of $[CO_3]\nu_3$ groups at 1418 cm^{-1} and 1458 cm^{-1}, as well as at 875 cm^{-1}, proves substitution of „B-type" $[PO_4]$ groups with $[CO_3]$ groups in the HAp crystal lattice (Barinov, et al., 2006; Siva Rama Krishna, et al., 2007).

Absorption maximum of $[CO_3]$ group at 875 cm^{-1} can also prove that „AB-type" ($[PO_4]$ and [OH] groups) substitution in the structure of HAp, as well as a weak absorption band at 3571 cm^{-1} in the synthesized HAp samples can mean the „AB-type" substitution (Barinov, et al., 2006).

A wide absorption band within the range from ~3600 cm^{-1} up to 3100 cm^{-1} points on ν_3 and ν_1 with H_2O molecules bonded with hydrogen for stretching modes and an absorption band at 1629 cm^{-1} is referable onto deformation mode ν2 of H_2O

molecules (Siva Rama Krishna, et al., 2007), that proves presemce of physically adsorbed water in the synthesized samples.

Processing the samples in the temperature range between 200 ºC to 1400 ºC in the Air atmosphere, a similar sample behavior, even up to characteristic bioceramic sintering temperatures, is observed in all the syntheses. Dehydratation of the samples occurs up to 500oC, and at 600oC, the spectra demonstrate that adsorbed water band disappears from the spectrum, adsorbed and capillary water is eliminated from CaP. Within the temperature range between 500oC and 800oC, amount of CO_3^{2-} groups in the samples also reduces, and bands of CO_3^{2-} group fully disappear at 900 oC. FTIR spectra (Fig. 12) and XRD diffractograms (Fig. 13) of the thermally treated at 1100ºC samples considerably differ, comparing with the samples that just have been synthesized. A restructurization of functional [PO_4] groups have occurred, and sample phase composition is considerably different from the combination of the initial synthesis parameters. It can be concluded that phases with a high crystallinity degree were formed.

A pure, stoichiometric and stable HAp in a wide temperature range is obtained using the following synthesis parameters: final pH=9,3 and synthesis temperature Ts=45oC (synthesis 6). Sample thermal treatment is performed in the Air atmosphere, in the temperature range 200oC – 1400oC for 1 h. HAp spectra at various heating temperatures are demonstrated on the Fig. 14 and 15.

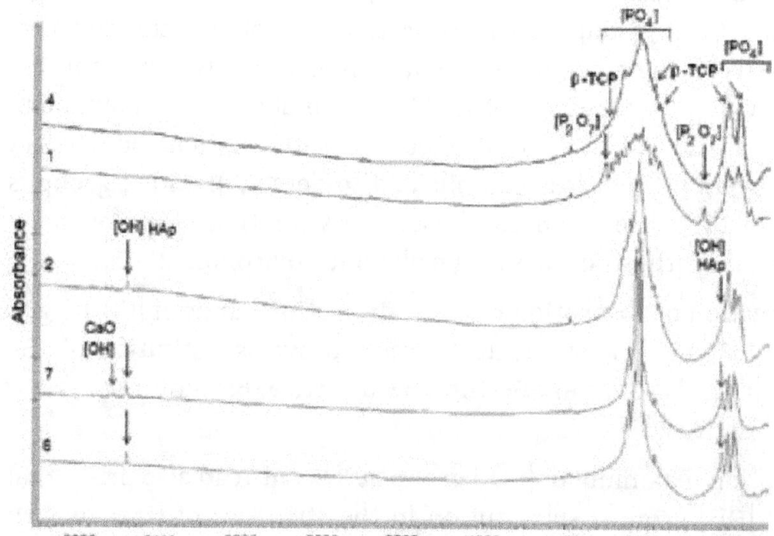

Fig. 12. FTIR spectra of calcium phosphates thermally treated at 1100oC for 1 h, depending on synthesis parameters (1, 2, 4, 6 and 7 synthesis)

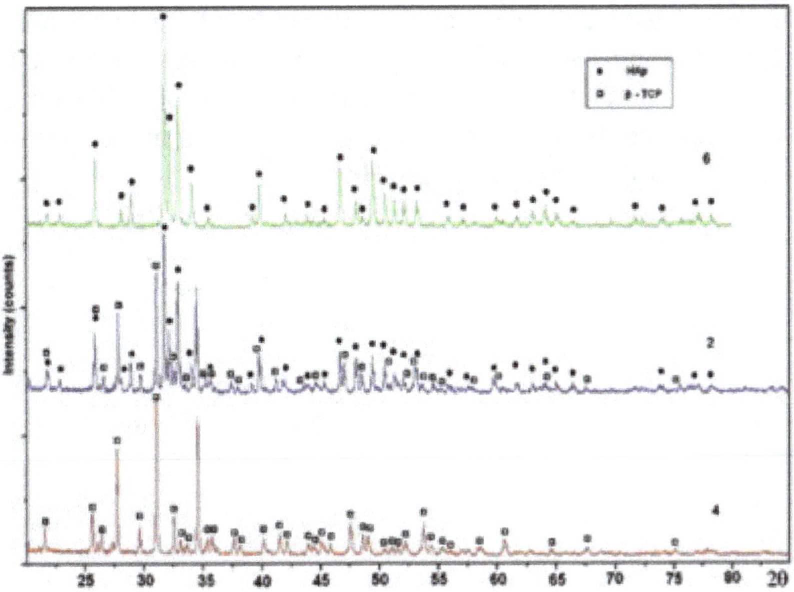

Fig. 13. X-ray diffraction patterns of different calcium phosphate of calcium phosphates thermally treated at 1100oC for 1 h with different phase compositions for synthesis 6 – pure HAp, 4 – pure b-TCP and 2 – biphasic mixture of HAp/ b-TCP

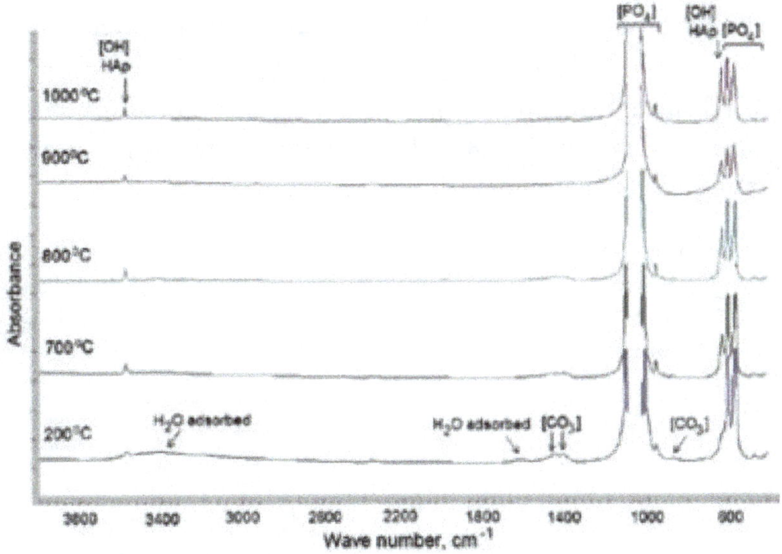

Fig. 14. FTIR spectra of HAp (synthesis 6) depending of thermal treatment temperature in range 200 – 1000oC for 1 h

As can be seen after FTIR analysis which is often more sensitive than XRD, HAp begins to decompose when a sample is heated for 1 h at more than 1200°C, forming TCP shoulders at 948 cm^{-1}, 975 cm^{-1} and poorly intensive band at 432 cm^{-1}, which proves dehydroxilation of HAp and forming of α-TCP or OHAp phases. Therefore, XRD analysis shows that the decomposition occurs only at 1400°C, and is resulted with mixture of α-tricalcium phosphate (α-TCP, $Ca_3(PO_4)2$), tetracalcium phosphate (TTCP, $Ca4(PO_4)2O$) and HAp. By heating and then cooling a HAp sample in the Air atmosphere, a complete HAp decomposition cannot be achieved, for during cooling [OH] groups get back in the structure from the Air and therefore a partial reversible α-TCP and TTCPtransform into HAp.

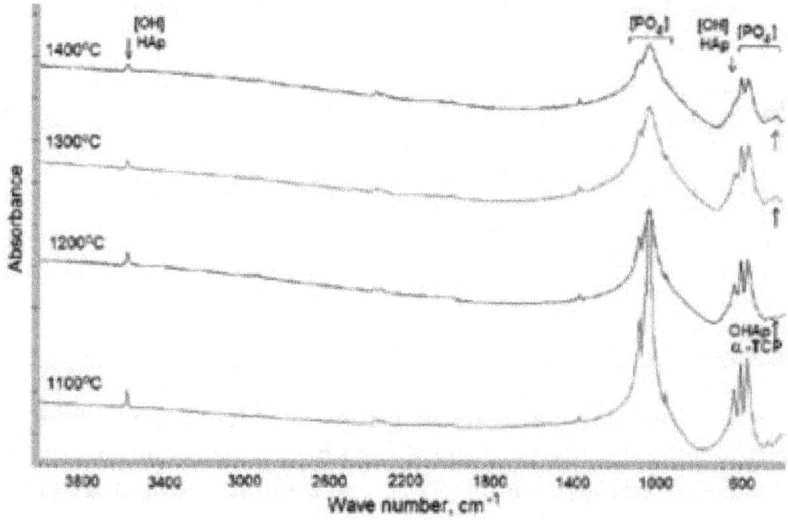

Fig. 15. FTIR spectra of HAp (synthesis 6) depending of thermal treatment temperature in range 1100 – 1400oC

A pure β-tricalcium phosphate phase $Ca_3(PO_4)2$, obtained using the following synthesis parameters: final pH=5,8 and T= 22 oC (synthesis 4), it is synthesized at low temperatures and pH is acidic.

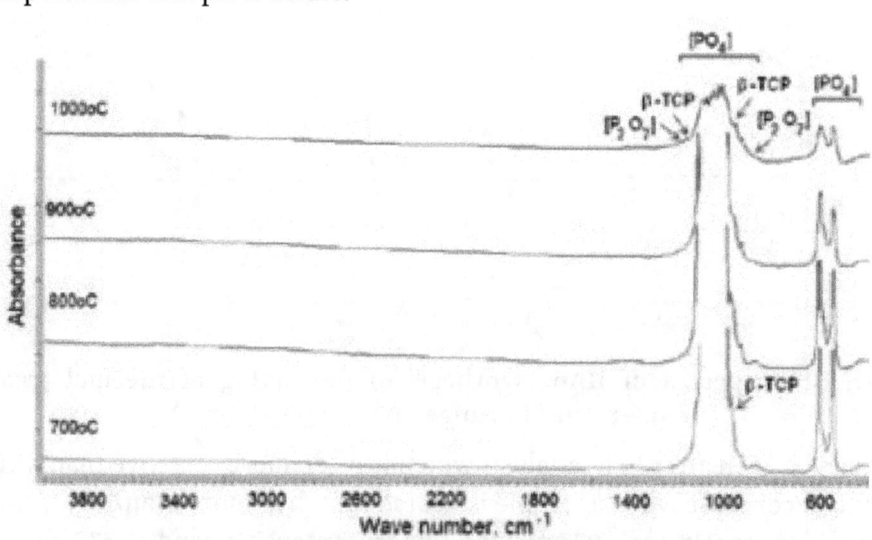

Fig. 16. FTIR spectra of β-TCP (synthesis 4) depending of thermal treatment temperature in range 700 – 1000oC for 1 h

OH- group bands completely disappear at 700oC with a wave length 3569 cm^{-1} and 631 cm^{-1}, which is characteristical for HAp. It can also be observed that at 900oC, bands that correspond to a CO_3^{2-} functional group, disappear. At 700oC, a β-TCP phase begins to form, which is shown by characteristic shoulders which become more sharply explicit with increasing of temperature. Until 1100oC, also poorly intensive pyrophosphate (CPP) group bands at 727 and 1212 cm^{-1}, that disappear at higher temperatures.

Since the sample does not contain CPP and HAp any longer, a pure TCP is obtained in the result, and it is stable up to 1400 oC (Fig. 16 and 17). A β-TCP phase can be identified in the spectrum by appearance of characteristic bands at 947 cm^{-1} and 975 cm^{-1}.

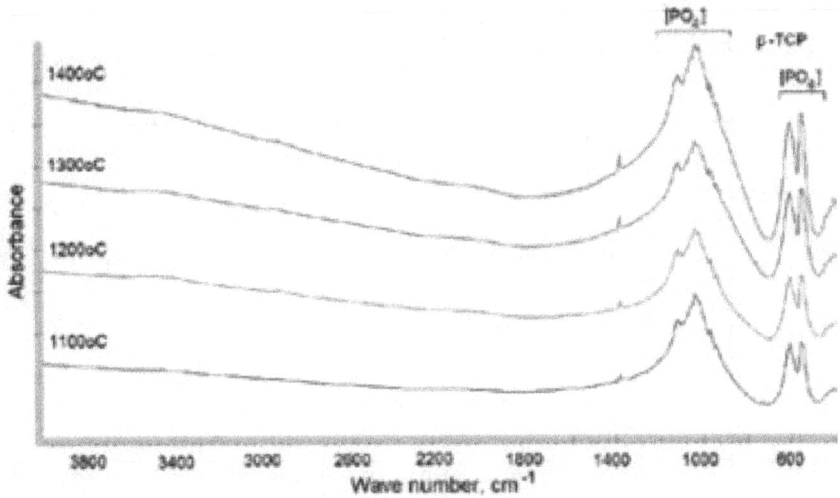

Fig. 17. FTIR spectra of β-TCP (synthesis 4) depending of thermal treatment temperature in range 1100 – 1400oC for 1 h

In the CaP syntheses with final pH=7, T=45oC (synthesis 5), final pH=5,1 and T=70oC (synthesis 2), final pH=5,3 and T=45oC (synthesis 3), spectra demonstrate that a biphasic mixture with various HAp/TCP proportions, respectively 80/20, 60/40 and 35/65, measured at 1100 oC with an XRD semi-quantitative method, is formed in the synthesized samples.

In the synthesis with final pH=5,00 and T=22 oC (synthesis 1), a complicated phase composition is formed from β-TCP, $CaHPO_4$ and TTCP phases, and it remains stable up to 1200 oC. $CaHPO_4$ phase can be recognized after a HPO_4^{2-} group band at 897 cm^{-1}, which is visible in the spectra already at 200 oC.

In the synthesis with a final pH=10,74 and T=45oC (synthesis 7), a HAp structure is formed but at 1000oC, and additional OH- peak appears at 3644 cm^{-1}, which means that there is a unreacted CaO left. The reaction did not occur completely and this synthesis, as well as the above mentioned synthesis 1, cannot be used in practice.

Table 9. FTIR absorption bands (cm^{-1}) of thermally treated laboratory synthesized HAp and b -TCP (synthesis 6 and 4) chemical groups.

T, oC Chemical groups and phases	800	900	1000	1100	1200	1300	1400
CO_3^{2-}	875; 1418; 1456; 1466; 1636	875; 1384; 1418; 1457; 1636	875; 1385; 1419; 1457	-	-	-	-
HPO_4^{2-}	875	875	875	-	-	-	-
H_2O adsorbed	3100 - 3600	3250-3600	3330 - 3595	-	-	-	-
OH-	3572; 634	3572; 634	3572; 633	3572; 632	3570	3570	3570
PO_4^{3-} b -TCP (OAp)	-	-	-	435;	436;	434	434
PO_4^{3-}	1000 -1120	1006 - 1120	1046; 1091	1046; 1091	1046; 1091	1045; 1094;	1037; 1045; 1090
PO_4^{3-} HAp	473; 550 - 640; 963	471; 556-604; 963	473; 554-601; 963	472; 570; 602; 963	472; 569; 602; 962	472; 568; 602; 961	472; 568 601; 961
PO_4^{3-} b -TCP	946; 975	946; 975	946; 975; 1127	944; 971 1127	944; 970; 1121	944; 970; 1121	943; 970; 1120
$P_2O_7$4-	725; 1211	725; 1212	725;1210	-	-	-	-

4.5 Thermal stability of HAp in various environments

Dehydroxilation of HAp under temperature influence is proven by reduction of [OH] absorption maximums intensity. Absorption band of [OH] bonding at 632 cm^{-1} is very sensitive to temperature changes and at higher heating temperatures (over 1300°C) is only shown as shoulder of [PO_4] absorption band at 601 cm^{-1}, when [OH] bonding at 3572 cm^{-1} is more stable in the higher temperatures (Fig. 18).

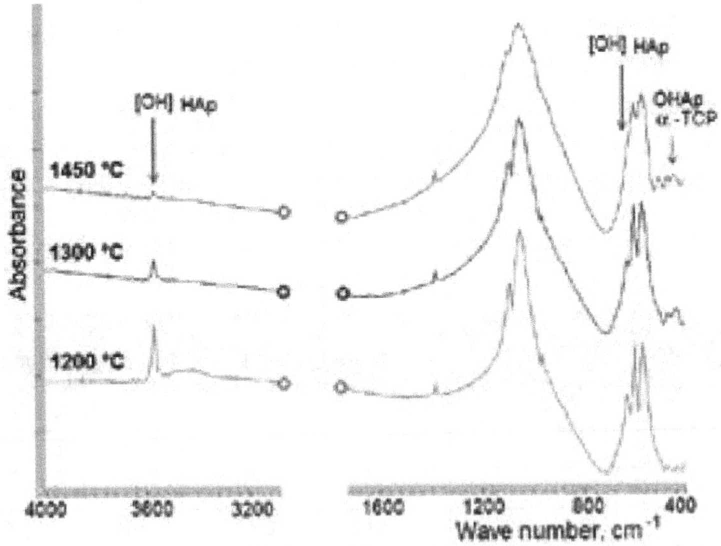

Fig. 18. FTIR spectra of HAp heated in the Air at 1200, 1300, 1450°C (1h)

Table 10. Phase composition after HAp thermal treatment in the Air

Temperature, °C	Heating time, h	Phase composition	
		After XRD	after FTIR
1000	8	HAp	HAp
1100	1	HAp	HAp
1200	1	HAp	HAp + TCP
1300	1	HAp	HAp + TCP
	12	HAp	HAp + TCP
1400	1	HAp + α-TCP + TTCP	HAp + TCP
	3	HAp + α-TCP + TTCP	HAp + TCP +TTCP
	1	HAp + α-TCP + TTCP	HAp + TCP +TTCP
1450	3	HAp + α-TCP + TTCP	HAp + TCP +TTCP
	6	HAp + α-TCP + TTCP	HAp + TCP +TTCP
1500	1	HAp + α-TCP + TTCP	HAp + TCP +TTCP

Stability of hydroxyapatite phase depends on the partial pressure of the water in the atmosphere, so, in an environment with no presemce of water, by addition of sufficient amount of energy, HAp will turn into more stable calcium phosphates in the waterless environment. After thermally treating the sample in vacuum, there is an absorption band detected at 948 cm^{-1} in the FTIR spectrum already at 600 °C. It appears in the result of [PO_4] group fluctuations and usually points onto TCP phase, however the literary sources mention that the absorption band at this wave length could also be characteristic to oxyapatite phase. In the result of thermal treatment, in vacuum at 1300 °C, HAp phase has completely decomposed and absorption bands, characteristic to TCP and TTCP phases, are visible in the FTIR spectrum (overlapping of numerous [PO_4] groups vibrations occurs, so it is not possible to distinguish, which absorption bands belong to TCP, and which to TTCP phase).

Table 11. Phase composition after thermal treatment of HAp in the vacuum oven.

Temperature, °C	Processing time, h	Phase composition	
		after XRD	after FTIR
600	1	HAp	HAp + OAp/TCP
800	1	HAp	HAp + OAp/TCP
900	1	HAp	HAp + OAp/TCP
1000	1	HAp + α-TCP	HAp + OAp/TCP/TTCP
1100	1	HAp + α-TCP + TTCP	HAp + TCP + TTCP
1200	1	HAp + α-TCP + TTCP	TCP + TTCP
1300	1	α-TCP + TTCP	TCP + TTCP

According to XRD data, supplying water vapor during heating of hydroxyapatite does not change phase composition, however slight changes in the sample structure are detected with FTIR analysis. After thermal treatment in the water vapor, absorption intensity of [OH] group increases in ratio to [PO_4], comparing with a sample heated in the Air, which could improve HAp properties and make this phase more stable.

5. CONCLUSIONS

This summarizing work can help to evaluate both synthesed CaP and structure, phase composition and properties of the CaP bioceramic products. FTIR spectrometry along with XRD is one of the most important, quickest and most available methods for studying CaP materials and in many cases more sensitive than XRD in order to detect forming of new phases. Using these methods, it was possible to detect synthesis parameters in order to obtain a pure and thermally stable HAp and b-TCP, as well as biphasic mixtures with controlable ratio. From the results of above mentioned studies it can be concluded that a thorough selection of environment is required for processing HAp powder and it is also required to monitor behavior of the material during the heating, in order to obtain the desirable product. By variating conditions of thermal treatment, it is possible to improve structure of the synthesized HAp, for example, eliminate carbonate groups included in the structure during synthesis, increase number of [OH] groups, as well as slow (by thermal treatment with water vapor presemce) or quicken (by thermal treatment in vacuum) HAp decomposition.

6. REFERENCES

1. Barinov S.M., Rau J.V., Cesaro S.N. (2006). Carbonate release from carbonated hydroxyapatite in the wide temperature range// J. Mater. Sci. Mater. Med., Vol.17., pp. 597.–604
2. Destainville A., Champion E., Bernache-Assollante D., et al. (2003). Synthesis, characterization and thermal behaviour of apatite tricalcium phosphate // Materials Chemistry and Physics, No. 80, pp. 269 – 277 Dorozhkin S.V. (2009). Calcium Orthophosphates in Nature, Biology and Medicine. Materials, 2: pp. 399-498
3. Dorozhkin S.V. (2009). Calcium orthophosphate-based biocomposites and hybrid biomaterials. J. Mater. Sci., 44(9): pp. 2343-2387
4. Dorozhkin S.V. (2009). Calcium Orthophosphate Cements and Concretes. Materials, 2: pp. 221-291
5. El Kady A.M., K.R.M., El Bassyouni G.T. (2009). Fabrication, characterization and bioactivity evaluation of calcium pyrophosphate/polymeric biocomposites. Ceram. Int.
6. Ji G., Zhu H., Jiang X., et. al. (2009). Mechanical Strenght of Epoxy Resin Composites Reinforced by Calcined Pearl Shell Powders. J. Appl. Polym. Sci., Vol.114, pp. 3168.- 3176
7. Han J-K., Song H-Y., et al. (2006). Synthesis of height purity nano-sized hydroxyapatite powder by microwave-hydrothermal method. Materials Chemistry and Physics, No. 99, pp 235 – 239
8. Kothapalli C., Wei M., Vasiliev A., et. al. (2004). Influence of temperature and concentration on the sintering behavior and mechanical properties of hydroxyapatite// Acta Mater, Vol.52, pp. 5655-5663
9. Kwon S-H., Jun Y-K., Hong S-H., el al. (2003). Synthesis and dissolution behaviour of b- TCP and HA/b-TCP composite powders. Journal of European Ceramic Society, No. 23, pp. 1039–1045
10. Landi E., Tampieri A., Celotti G., et. al. (2000). Densification behaviour and mechanisms of synthetic hydroxyapatites. J. Eur. Ceram. Soc., Vol. 20, pp. 2377–2387
11. Lioua S.-C., Chena S.-Y., Lee H.-Y., et. al. (2004). Structural characterization of nano-sized calcium deficient apatite powders. Biomaterials, Vol. 25, pp. 189-196.
12. Meejoo S., Maneeprakorn W., Winotai P. (2006). Phase and thermal stability of nanocrystalline hydroxyapatite prepared via microwave heating. Thermochimica Acta, No. 447, pp. 115–120

13. Mobasherpour I., Heshajin M. (2007). Synthesis of nanocrystalline hydroxyapatite by using precipitation method. Journal of Alloys and Compounds, N 430, pp. 330 – 333
14. Nilen R.W.N., Richter P.W. (2008). The thermal stability of hydroxyapatite in biphasic calcium phosphate ceramics. J. Mater. Sci. Mater. Med., Vol. 19(4), pp. 1693–1702
15. Ratner B., Hoffman A., Schoen F. et. al. (2004). Biomaterials Scienc. An Introduction to Materials in Medicine, Second Edition // Academic Press, pp. 851
16. Raynaud S., Champion E., Bernache-Assollant D. Et al. (2002). Calcium phosphate apatite with variable Ca/P atomic ratio I. Synthesis, characterisation and thermal stability of powders. Biomaterials No 23, pp. 1065–1072 Shi D., (2006). Introduction to biomaterials. World Scientific Publishing, p. 253.
17. Siva Rama Krishna D., Siddharthan A., Seshadri S. K., et. al. (2007). A novel route for synthesis of nanocrystalline hydroxyapatite from eggshell waste. J. Mater. Sci. - Mater. Med., Vol.18., pp. 1735-1743
18. Ślósarczyk A., Paszkiewicz Z., Zima A. (2010). The effect of phosphate source on the sintering of carbonate substituted hydroxyapatite. Ceram Int., Vol. 36, pp. 577-582

7 Chapter

FTIR SPECTROSCOPY OF ADSORBED PROBE MOLECULES FOR ANALYZING THE SURFACE PROPERTIES OF SUPPORTED PT (PD) CATALYSTS

1. INTRODUCTION

Supported metal catalysts are important for many fields of applied chemistry, including chemical synthesis, petrochemistry, environmental technology, and energy generation/storage. For prediction of catalyst performance in a chosen reaction and optimization of its functions, it is necessary to know the composition of the surface active sites and have methods for estimating their amount and strength. One of the most available and well-developed methods for studying the composition and structure of the surface functional groups of supported metal catalysts is vibrational spectroscopy, in particular with the use of adsorbed probe molecules.

Although Fourier transform infrared (FTIR) spectroscopy is widely employed for characterization of the catalyst surface (Paukshtis, 1992; Ryczkowski, 2001), it is still unclear whether the regularities obtained under conditions of spectral pretreatments and measurements (evacuation, temperature) can be used for interpreting and predicting the surface properties during adsorption of a precursor or in a catalytic reaction. Thus, Aim of the present work is not only to demonstrate the possibilities of FTIR spectroscopy of adsorbed molecules for investigation of the surface functional groups in the chosen catalytic systems, but also to compare FTIR spectroscopy data with the data obtained for supported metal catalysts by other physicochemical methods and with the catalyst properties in model and commercially important reactions. Main emphasis will be made on quantitative determination of various surface groups and elucidation of the effect of their ratio on the acid-base, adsorption and catalytic properties of the surface.

The study was performed with model and commercially important supports and catalysts: gamma alumina, which is among the most popular supports in the synthesis of supported metal catalysts for oil refining, petrochemistry, and gas emissions neutralization; supported platinum and palladium catalysts containing sulfated zirconia (Pt/SZ, Pd/SZ) or alumina-promoted SZ (Pt/SZA), which are suitable for low-temperature isomerization of n-alkanes and hydroisomerization of benzene-containing fractions of gasoline.

2. VIBRATIONAL SPECTROSCOPY OF ADSORBED PROBE MOLECULES FOR INVESTIGATION OF SUPPORTED CATALYSTS – ESTIMATION OF THE STRENGTH AND CONCENTRATION OF VARIOUS SURFACE SITES

2.1 FTIR spectra of adsorbed probe molecules

FTIR spectroscopy of adsorbed probe molecules is one of the most available and well- developed methods for studying the composition and structure of the surface functional groups of supported metal catalysts. As the vibrational spectrum reflects both the properties of the molecule as a whole and the characteristic features of separate chemical bonds, FTIR spectroscopy offers the fullest possible information on the perturbation experienced by a molecule on contact with the solid surface, and often determines the structure of adsorption complexes and of surface compounds. Examination of supported metal catalysts deals with two types of surfaces strongly differing in their properties: surface of a support and surface of a metal-containing particle. Various species can reside on the support surface: hydroxyl groups of different nature; Lewis acid sites (coordinatively unsaturated surface cations); base sites (bridging oxygen atoms or oxygen atoms of OH groups); structures formed by impurity anions that remain after the synthesis (sulfate, nitrate and ammonia groups) or form upon contacting with Air (carbonate-carboxylate structures).

Various spectroscopic probe molecules are widely used for characterization of Lewis and Brønsted acid sites on the surfaces of oxide catalysts. Among such probes are strong bases: amines, ammonia and pyridine, and weak bases: carbon oxide, carbon dioxide and hydrogen (Knözinger, 1976a; Kubelková et al., 1989; Kustov, 1997; Morterra & Magnacca, 1996; Paukshtis, 1992). Being a weaker base than ammonia, pyridine interacts with the sites widely varying in acidity. However, within each type of Lewis acid site, which is determined with pyridine as a probe molecule, there are distinctions in acidity that cannot be revealed with the use of strong bases. In this connection, very advantageous is the adsorption of weak bases like CO. The application of such probe molecules as CO or pyridine makes it possible to estimate both the concentration and the acid strength of OH groups and Lewis acid sites in zeolites, oxide and other systems (Knözinger, 1976b; Paukshtis, 1992). Concentration of the surface groups accessible for identification by FTIR spectroscopy is above 0.1 mmol/g.

In the case of base surface sites, the concentration and strength can be characterized with deuterochloroform (Paukshtis, 1992). Surface of a metal-containing particle may consist of metal atoms with various oxidation states or different charge states caused by the metal- support interaction. Metal cations and atoms on the surface can be detected only from changes in the spectra of adsorbed molecules, Since vibrations of the metal-oxygen bonds on the surface belong to the same spectral region as lattice vibrations and thus are not observed in the measurable spectra, whereas vibrational frequencies of the metal-metal bonds are beyond the measuring range of conventional FTIR spectrometers. Surface atoms and nanoparticles of metals and metal ions are usually identified by the method of spectroscopic probe molecules such as CO (Little, 1966; Sheppard & Nguyen, 1978). Examination of the nature of the binding in Men+CO complexes suggests that the frequency of adsorbed CO (ν_{CO}) should depend on the valence and coordination states of the cations, that is, on their abilities to accept π-donation (increasing ν_{CO}) and to donate π-orbitals of the CO (decreasing ν_{CO}, as for carbonyl complexes) (Davydov, 2003; Hadjiivanov & Vayssilov, 2002; Little, 1966). Carbonyls involving p-donation can have different structures – linear or bridged – and the number of metal atoms bonded to the CO molecules can also be different. Complexes involving cations only be linear (terminal) because in an M-O-M situation the distance between cations is to great to form a bond between a CO molecules and two cation sites simultaneously.

2.2 Experimental

The FTIR spectra were measured on a Shimadzu FTIR-8300 spectrometer over a range of 700-6000 cm^{-1} with a resolution of 4 cm^{-1} and 100 scans for signal accumulation. Before spectra recording, powder samples were pressed into thin self-supporting wafers (8-30 mg/cm^2) and activated in a special IR cell under chosen conditions and further in vacuum (p < 10–3 mbar). FTIR spectra are presented in the optical density units referred to a catalyst sample weight (g) in 1 cm^2 cross-section of the light flux.

Quantitative measurements in FTIR spectroscopy are based on the empirical Beer–Lambert– Bouguer law interrelating the intensity of light absorption and the concentration of a substance being analyzed. For FTIR spectroscopy of adsorbed molecules, this law is applied in the integral form:

$$A \approx \int \log(T_0/T)_v dv, \qquad (1)$$

where A is the integral absorbance (cm^{-1}), T0 and T are the Transmittance along the base line, and the band contour, respectively.

The concentration of active sites on the catalyst surface was estimated by the formula

$$N[\mu mol/g] = \frac{A \cdot S}{p \cdot A_0}, \qquad (2)$$

where A_0 is the integral absorption coefficient (integral intensity of absorption band (a.b.) for 1 μmol of the adsorbate per 1 cm^2 cross-section of the light flux), p is the weight of a sample wafer (g), and S is the surface area of a sample wafer (cm^2).

The concentration of surface OH groups in g-Al_2O_3 was determined from the integral intensities of absorption bands νOH in the region of 3650-3800 cm^{-1} using the integral absorption coefficient A_0 = 5.3 cm/mmol (Baumgarten et al., 1989).

In the present study, acidic properties of the samples were examined by FTIR spectroscopy using CO adsorption at -196 °C and CO pressure 0.1-10 mbar. An increase in the νCO band frequency of adsorbed CO relative to the value of free CO molecules (2143 cm^{-1}) is caused by the formation of complexes with Lewis or Brønsted acid sites. Complexes with Lewis acid sites are characterized by the bands with a frequency above 2175 cm^{-1}, whereas the frequency range from 2150 through 2175 cm^{-1} is typical of CO complexes with OH groups.

The concentration of Lewis acid sites was measured by the integral intensity of CO band in the range of 2170-2245 cm^{-1}. For alumina and compositions with prevailing fraction of alumina (Al_2O_3 and SO_4^{2-}-ZrO_2-Al_2O_3), the following A_0 values (cm/mmol) were used: 1.25 (2245-2220 cm^{-1}), 1.0 (2200 cm^{-1}), 0.9 (2190 and 2178-2180 cm^{-1}); for zirconia and sulfated zirconia, A_0 was equal to 0.8 cm/mmol (Paukshtis, 1992). A value of the upward νCO frequency shift determines the strength of Lewis acid sites, as it is related to the heat of complex formation by the following formula (Paukshtis, 1992):

$$Q_{CO} = 10.5 + 0.5 \cdot (\nu_{CO} - 2143) \quad (3)$$

The concentration of Brønsted acid sites in the modified aluminum oxides was measured by the integral intensity of CO band in the range of 2170-2175 cm^{-1}, A_0 = 2.6 cm/mmol. The concentration of Brønsted acid sites in sulfated samples was determined from the integral intensity of absorption band due to the pyridinium ion with a maximum at 1544 cm^{-1} (A_0 = 3.5 cm/μmol) (Paukshtis, 1992). Adsorption of carbon monoxide on Brønsted acid sites at -196 °C results in the shift of OH bands to the lower frequency region due to perturbation of OH stretch by hydrogen bonding with CO molecule. The higher the shift of OH stretching vibration, the stronger the acidity of this OH group (Maache et al., 1993; Paze et al., 1997).

Basic properties of the samples were studied by FTIR spectroscopy using $CDCl_3$ adsorption at 20 °C. At the formation of H bonds, deuterochloroform behaves as a typical acid. A decrease in the frequency of ν_{CD} band of $CDCl_3$ adsorbed relative to the value of physisorbed molecules (2265 cm^{-1}) is caused by the formation of complexes with base sites. The strength of base site was determined by the band shift of the CD stretching vibrations that occurred under $CDCl_3$ adsorption. The strength can be recalculated into the proton affinity (PA) scale using the formula (Paukshtis, 1992):

$$\log(\Delta\nu_{CD}) = 0.0066PA - 4.36 \quad (4)$$

The concentration of base sites was measured by the integral intensity of CD band in the range of 2190-2255 cm^{-1}. The integral absorption coefficient was calculated from the correlation equations (Paukshtis, 1992):

$$A_0 = 0.375 + 0.0158\Delta\nu_{CD} \text{ for } \Delta\nu_{CD} > 13\ cm^{-1} \quad (5)$$

$$A_0 = 0.125 + 0.0034\Delta\nu_{CD} \text{ for } \Delta\nu_{CD} < 13\ cm^{-1} \quad (6)$$

3. INVESTIGATION OF SUPPORTS AND CATALYSTS BY FTIR SPECTROSCOPY

3.1 Gamma alumina. The role of studying the surface functional groups for understanding the processes of adsorption and catalysis

Owing to its unique acid-base and structural properties, aluminum oxide, first of all γ-Al_2O_3, remains the most popular catalyst and catalyst support. The analysis of catalytic reactions usually deals with Lewis acid and base sites of Al_2O_3. However, in the catalyst synthesis, adsorption properties of the surface during its interaction with aqueous solutions strongly determine the composition of surface hydroxyl cover of alumina. It should be noted that modern concepts of the surface structure of aluminum oxides, which were developed in recent 50 years, are based mainly on the vibrational spectroscopy data. Various structural models of the aluminum oxide surface were suggested to explain the experimental data (Egorov, 1961; Peri, 1965; Tsyganenko & Filimonov, 1973; Zamora & Córdoba, 1978; Knözinger & Ratnasamy, 1978). These models are based on the spinel structure of transitional alumina modifications. Recent attempts to develop advanced models of the surface structure or refine the existing models were made by Tsyganenko and Mardilovich (Tsyganenko & Mardilovich, 1996) as well as Liu and Truitt (Liu & Truitt, 1997). Such advanced models admit the existence of fragments on the alumina surface, which comprise pentacoordinated aluminum atom.

At present, 7 absorption bands characterizing the isolated OH groups are commonly distinguished in FTIR spectrum of γ-Al_2O_3. Low-frequency bands are assigned to the bridging OH groups located between aluminum atoms with different coordination: 3665-3675 ($Al^{V}(OH)Al^{IV}$), 3685-3690 cm^{-1} ($Al^{VI}(OH)Al^{IV}$), 3700-3710 cm^{-1} ($Al^{VI}(OH)Al^{V}$), and 3730-3740 ($Al^{VI}(OH)Al^{VI}$)1. High-frequency bands correspond to the terminal OH groups bound to one aluminum atom with different coordination: 3745-3758 ($Al^{VI}OH$), 3765-3776 ($A^{V}OH$), and 3785-3792 ($Al^{IV}OH$) cm^{-1}. In addition, there is a broad a.b. at 3600 cm^{-1} attributed to hydrogen-bonded OH groups (Paukshtis, 1992).

The surface Lewis acidity is formed by electron-acceptor sites represented by coordinatively unsaturated aluminum cations on the Al2O3 surface. The use of CO as a probe molecule makes it possible to estimate both the strength and the amount of Lewis acid sites, which is essential when alumina is employed as a catalyst or catalyst support. CO is adsorbed on the γ-Al_2O_3 surface to form three types of surface complexes (Della Gatta et al., 1976; Zaki & Knözinger, 1987; Zecchina et al., 1987). FTIR spectra of adsorbed CO show the following a.b. corresponding to stretching vibrations of CO molecule: 2180-2205 cm^{-1} (weak Lewis acid sites), 2205-2220 cm^{-1} (medium strength Lewis acid sites), and 2220-2245 cm^{-1} (strong Lewis acid sites). The high-frequency bands are related with two types of Lewis acid sites including Al^{IV} ions located in configurations with crystallographic defects. The low- frequency bands correspond to Al^{VI} ions in the regular defects of low-index faces of crystallites.

3.1.1 FTIR spectroscopy for determining the sites of precursor anchoring during synthesis of Pt/Al2O3 catalysts

In the synthesis of supported platinum catalysts, chloride complexes of platinum (IV) are commonly used as precursors. Their sorption on the alumina surface occurs from aqueous solutions and implies the involvement of OH groups of the support surface. Therewith, two main mechanisms of the interaction between metal complex and support are considered, implementation of each mechanism depending both on the chemical composition of a complex (degree of hydrolysis) and the ratio of various OH groups (Belskaya et al., 2008, 2011; Bourikas, 2006; Lycourghiotis, 2009). The first mechanism consists in the formation of outer sphere complexes; it implies electrostatic interaction between chloroplatinate and alumina surface, which is protonated and positively charged at low pH of the solution:

$$\sim Al-OH + H^+ + [PtCl_6]^{2-} \leftrightarrow \sim Al-OH_2^+-[PtCl_6]^{2-} \leftrightarrow \sim Al-[PtCl_6]^{2-} + H_2O \quad (7)$$

The formation of inner sphere complexes is accompanied by a deeper interaction of a complex with the oxide surface via ligand exchange with the surface OH groups:

$$\sim Al-OH + [PtCl_6]^{2-} \leftrightarrow \sim Al-[(OH)-PtCl_5]^- + Cl^- \quad (8)$$

FTIR spectroscopy is widely used to investigate state of the surface at different stages of catalyst synthesis. When studying the precursor-support interaction, this method allows identification of OH groups involved in chemisorption of the metal complex. Analysis of experimental data presented in Fig. 1 and Table 1 shows that adsorption of platinum complexes followed by anchoring of oxide platinum species forming on the g-Al_2O_3 surface decreases the intensity only of high-frequency bands. As the amount of supported platinum species increases (platinum content of 0.5 and 1.0 wt%), the concentration of all types of terminal groups ($A^{IV}OH$, $Al^{VI}OH$, $Al^{IV}OH$)

decreases; so does the concentration of bridging OH groups $Al^{VI}(OH)Al^{VI}$ bonded to octahedral aluminum. Exactly these types of OH groups seem to be involved in anchoring the anionic complexes of platinum (IV). Having more strong base properties, they are capable of interacting with chloroplatinate by mechanism (8), acting as the attacking ligand. However, the bridging groups with a.b. 3665-3710 cm^{-1} ($Al^{VI}(OH)Al^{IV}$, AlV(OH)Al^{IV}, Al^{VI}(OH)AlV) are virtually not involved in platinum anchoring (Table 1).

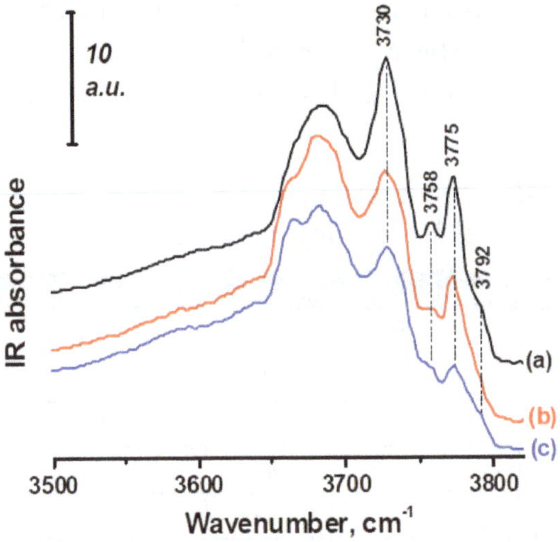

Fig. 1. FTIR spectra of surface hydroxyl groups of g-Al2O3 (a), 0.5% Pt/g-Al2O3 (b) and 1% Pt/g-Al2O3 (c). The samples were calcined and outgassed at 500 °C

Table 1. Types and concentrations of hydroxyl groups in calcined alumina and Pt/alumina samples as determined by FTIR spectroscopy data

Sample Type of OH group		γ-Al_2O_3	0.5%Pt/γ-Al_2O_3	1%Pt/γ-Al_2O_3	γ-Al_2O_3 Hydrothermal treatment 180 °C, 3 h
Al^{IV}OH (3790-3795 cm^{-1})	Concentration, µmol/g	12	9	8	6
A^{IV}OH (3775 cm^{-1})		35	29	16	33
Al^{VI}OH (3758 cm^{-1})		34	27	20	28
$Al^{VI}(OH)Al^{VI}$ (3730-3740 cm^{-1})		107	96	78	89
AlV(OH)Al^{VI} (3705-3710 cm^{-1})		50	54	52	52
$Al^{VI}(OH)Al^{IV}$ (3690 cm^{-1})		62	75	67	62
AlV(OH)Al^{IV} (3665-3670 cm^{-1})		36	35	34	43
Σ OH		336	325	275	313

According to analysis of the spectra of adsorbed CO, in the region of CO stretching vibrations all the samples have a.b. at 2245 and 2238 cm^{-1} characterizing CO complexes with strong Lewis acid sites, absorption bands at 2220 and 2205 cm^{-1} characterizing CO complexes with medium strength Lewis acid sites, and absorption bands at 2189-2191 cm^{-1} characterizing CO complex with weak Lewis acid sites. Deposition of platinum raises the concentration of nearly all types of Lewis acid sites, which is related with introduction of Cl– ion of the complex. Along with this, a substantial decrease in the concentration of weak Lewis acid sites is observed (Table 2). These electron-deficient sites may also take part in the anchoring of anionic platinum complexes via electrostatic interaction (Kwak et al., 2009; Mei et al., 2010).

Table 2. Types and concentrations of Lewis acid sites according to FTIR spectroscopy of adsorbed CO

Type of Lewis acid site	Super strong	Strong	Medium I	Medium II	Weak
n_{CO}, cm^{-1}	2245	2238	2220	2205	2189-2191
QCO, kJ/mol	61.5	58	48.5	41.5	34
Samples	Concentration, mmol/g				
g-Al_2O_3	0.9	2.5	5	9	460
1% Pt/g-Al_2O_3	1.2	2.9	6	27	380
g-Al_2O_3 Hydrothermal treatment 180 oC, 3 h	1.1	4.2	5	24	470

Thus, the analysis of FTIR spectra of alumina provides data on the nature and amount of various surface sites; moreover, it allows identification of the sites where active component precursor is anchored during catalyst synthesis, and makes it possible to hypothesize about mechanism and strength of the metal complex-support interaction.

3.1.2 Novel approaches to varying the composition of surface functional groups of alumina

Variation of the acid-base properties of alumina surface is commonly performed by chemical modifying via the introduction of additional anions (halogens, sulfates, phosphates) or cations (alkaline or alkaline earth metals) (Bocanegra et al., 2006; Ghosh & Kydd, 1985; Lisboa et al., 2005; López Cordero et al., 1989; Marceau et al., 1996; Requies et al., 2006; Rombi et al., 2003; Scokart et al., 1979; Wang et al., 1994). These methods complicate the process of catalyst synthesis and can lead to non-reproducible results.

This Section presents some unconventional approaches to alumina modifying for controlling the state of its surface functional cover. One of approaches consists in altering the relative content of hydroxyl groups and Lewis acid sites on the γ-Al_2O_3 surface without changes in the chemical composition of support (Mironenko et al., 2009, 2011). For this purpose, two techniques are employed: chemisorption of aluminum oxalate complexes followed by their thermal decomposition, and hydrothermal treatment of γ-Al_2O_3. Besides, there is an approach leading to considerable enhancement of acidic properties of the surface. Such effect is provided by γ-Al_2O_3 promotion with silica. The formation of SiO_2 takes place in the pore space of alumina during thermal decomposition of preliminarily introduced silicon- containing precursor.

3.1.2.1 Modifying the functional **cover of the γ-Al2O3 surface using aluminum oxalate complexes**

The proposed method of modifying implies the chemisorption (in distinction to conventional methods of incipient wetness impregnation) of anionic aluminum oxalate complexes [Al(C_2O_4)2(H_2O)2]- and [Al(C_2O_4)3]3- on the g-Al_2O_3 surface. It is essential that this approach excludes both the formation of a bulk alumina phase in the porous space after decomposition of supported complexes, and considerable changes in the texture parameters. Analysis of FTIR spectra of the modified alumina surface hydroxyl cover revealed (Fig. 2) that chemisorption of the oxalate complexes and subsequent formation of aluminum oxide compounds supported on g-Al_2O_3 (calcination at 550 °C) decreased mainly the intensity of two a.b. at 3670 and 3775 cm^{-1}. These bands characterize the bridging and terminal OH groups bound to pentacoordinated aluminum atom. Probably these are exactly the groups that are involved in anchoring of aluminum oxalate complexes.

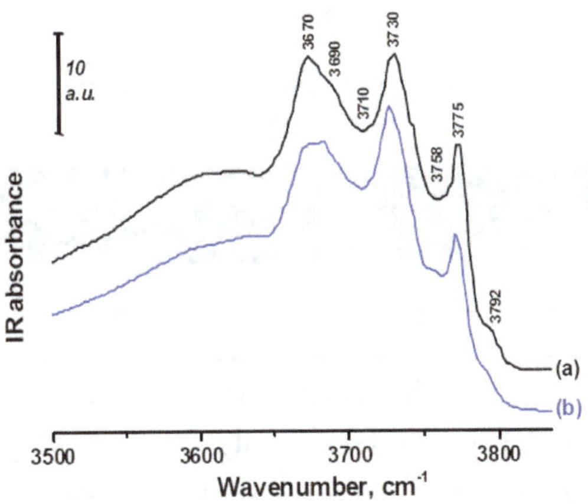

Fig. 2. FTIR spectra of surface hydroxyl groups of g-Al2O3 (a) and 3%Al2O3/γ-Al2O3 (b). The samples were calcined and outgassed at 500 °C

However, the formation of aluminum oxide compounds having their own surface OH groups (Al_2O_3/γ-Al_2O_3) resulted in substantial changes in the adsorption and acidic properties of the γ-Al_2O_3 surface. Thus, investigation of the sorption of chloride complexes of platinum (IV) on the modified γ-Al_2O_3 showed a 1.5-fold increase in the sorption capacity and an increased strength of the metal complex-support interaction (Mironenko et al., 2009).

According to FTIR spectroscopy of adsorbed CO (Fig. 3, Table 3), the anchoring of aluminum oxide compounds on the surface of initial γ-Al_2O_3 support decreased the concentration of weak Lewis acid sites with a.b. νCO = 2191 cm^{-1} (regular defects of the alumina surface including the octahedral aluminum ion) without changes in the concentration of other types of Lewis acid sites.

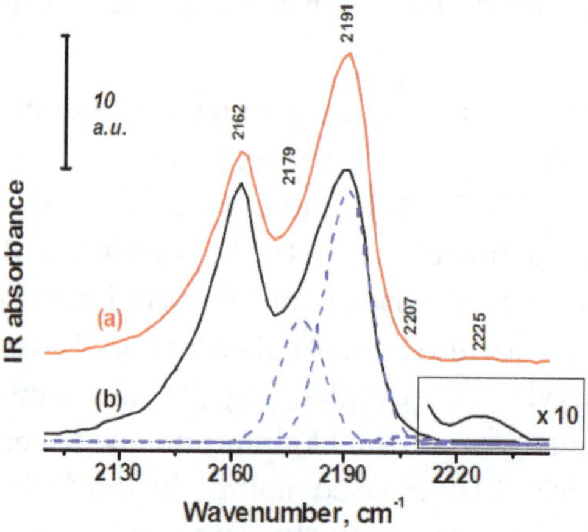

Fig. 3. FTIR spectra of CO adsorbed at -196 °C and a CO pressure of 4 mbar: (a) γ-Al2O3, (b) 3%Al2O3/γ-Al2O3. The dashed line shows the deconvolution of spectrum (b) into its components. Inset: portion of spectrum (b) magnified 10 times

Table 3. Concentrations of Lewis acid sites characterized by different absorption bands in the FTIR spectrum of adsorbed CO

Sample	Lewis acid site concentration, mmol/g				
	2179 cm^{-1}	2191 cm^{-1}	2207 cm^{-1}	2225 cm^{-1}	Σ
γ-Al_2O_3	155	360	8	3	526
3%Al_2O_3/γ-Al_2O_3	155	330	7	2	494

A decrease in the surface acidity revealed by FTIR spectroscopy of adsorbed CO is in good agreement with the results of catalytic testing (Fig. 4). Alumina samples before and after modifying were compared in a model reaction of 1-hexene double-bond isomerization, which is sensitive to amount and strength of Lewis acid sites. Although the reaction conditions radically differ from conditions of spectral measurements, the observed decrease in 1-hexene conversion can be predicted and interpreted using FTIR spectroscopy data.

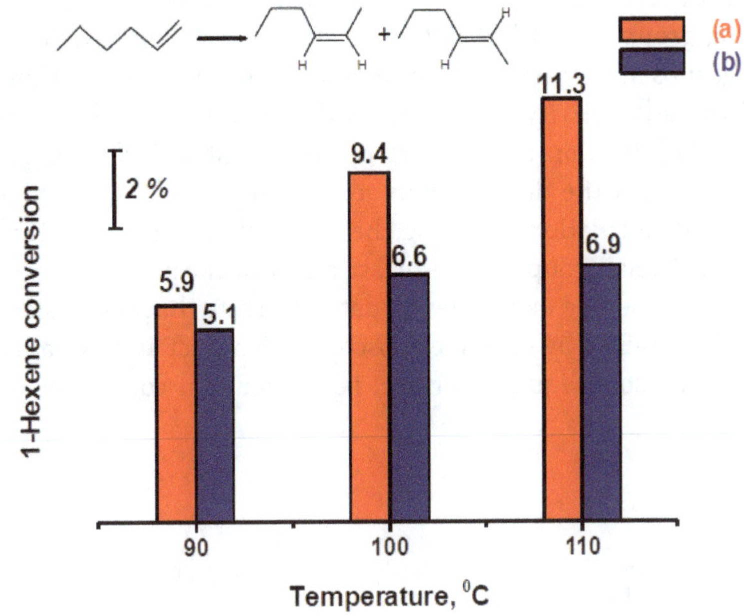

Fig. 4. Temperature dependence of the 1-hexene conversion: (a) γ-Al2O3, (b) 3%Al2O3/γ-Al2O3. Reaction conditions: atmospheric pressure, T = 90-110 °C, He : C6H[1]2 molar ratio 3.3

3.1.2.2 Modifying the functional cover of the γ-Al2O3 surface at hydrothermal treatment

Hydrothermal treatment of γ-Al_2O_3 is commonly used for alteration of the porous structure parameters (Chertov et al., 1982). Our study demonstrated that this technique is efficient for controlling the state of the oxide surface. Hydrothermal treatment of γ-Al_2O_3 was carried out in a temperature range of 50-200 °C with the treatment time varying from 0.5 to 12 h. This produced a hydroxide phase of boehmite AlO(OH) on the γ-Al_2O_3 surface, which amount can be readily controlled by the treatment conditions. After hydrothermal treatment, the samples were calcined at 550 °C to reduce the oxide phase.

The FTIR spectroscopic examination revealed the hydrothermal treatment effect on the concentration and ratio of functional groups on the γ-Al_2O_3 surface. Figure 5 shows FTIR spectra of the surface hydroxyl cover of initial γ-Al_2O_3 and γ-Al_2O_3 subjected to hydrothermal treatment at various temperatures with subsequent calcination at 550 °C. The quantitative analysis of FTIR spectroscopy data (Fig. 5 and Table 1) showed changes in the relative content of different surface OH groups of γ-Al_2O_3 with elevation of hydrothermal treatment temperature. This modifying technique was found to increase the fraction of low- frequency bridging hydroxyl groups (vOH = 3710-3670 cm^{-1}) from 50 to 70% of all OH groups and decrease the content of terminal hydroxyl groups (vOH = 3790-3760 cm^{-1}) and especially the bridging group $Al^{VI}(OH)Al^{VI}$ (vOH = 3730 cm^{-1}). A decrease in the content of basic OH groups necessary for the anchoring of chloride platinum complexes decreased the adsorptivity of support with respect to $[PtCl_6]^{2-}$. The adsorption isotherms of $H_2[PtCl_6]$ on the support pretreated at different temperatures (Fig. 6) demonstrate that the difference in adsorptivity can be quite high (more than a twofold), and calcination restoring the oxide phase cannot restore the relative content of functional groups of the surface and its adsorption properties in aqueous solutions. The presented experimental data illustrate that conclusions on the state of the surface obtained by FTIR spectroscopy can reflect and explain the processes occurring at the solid-liquid interface, i.e. under real conditions of the catalyst synthesis.

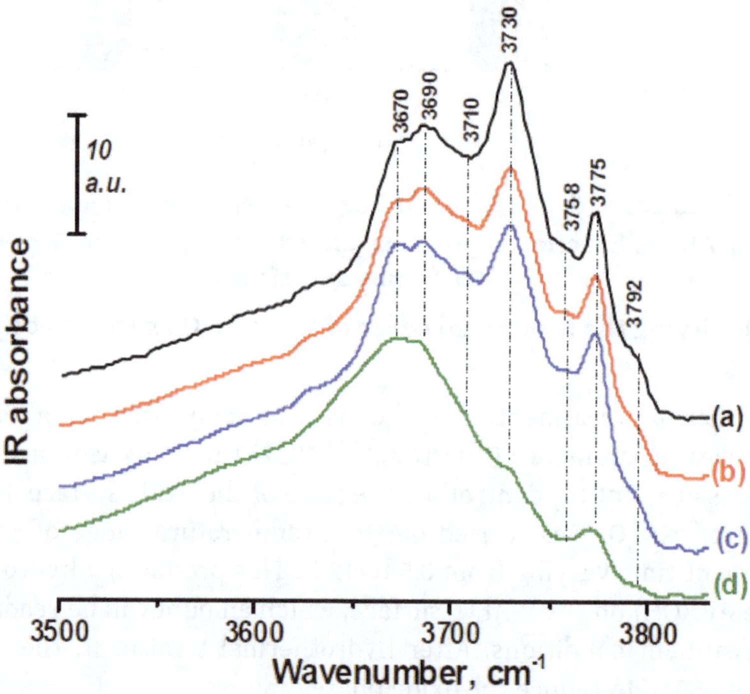

Fig. 5. FTIR spectra of hydroxyl cover of γ-Al2O3 (a) and γ-Al2O3 after hydrothermal treatment for 3 h at 150 (b), 180 (c) and 200 °C (d). The samples were calcined and outgassed at 500 °C

Analysis of the FTIR spectra of adsorbed CO shows (Table 2) that introduction of the hydrothermal treatment step increases the concentration of all Lewis acid site types: weak (2180 and 2190 cm^{-1}), medium strength (2205 cm^{-1}), and strong (2225 cm^{-1}). The obtained result is important for application of this modifying method in the synthesis of catalytic compositions for the reactions requiring the presemce of acid sites.

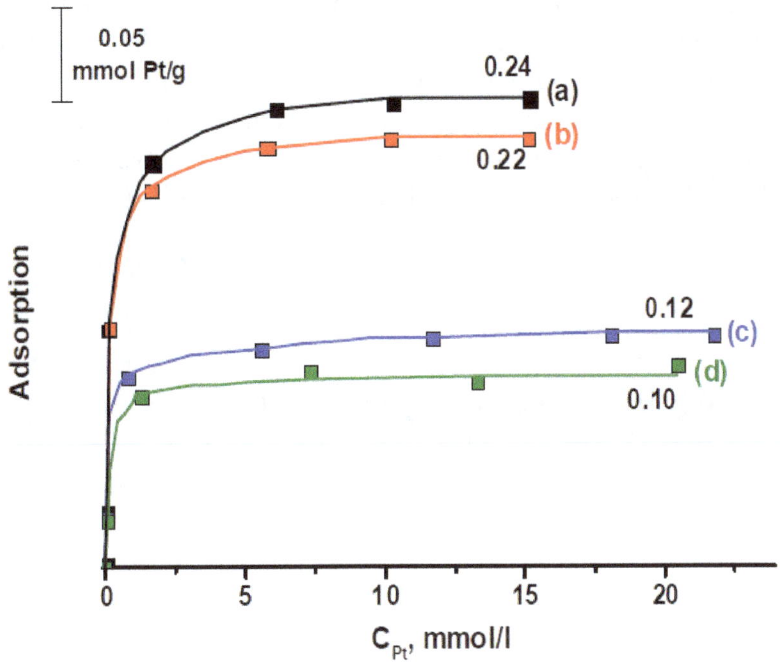

Fig. 6. Isotherms of $H_2[PtCl_6]$ adsorption from aqueous solutions on γ-Al2O3 (a) and γ-Al2O3 after hydrothermal treatment for 3 h at 150 (b), 180 (c) and 200 °C (d) followed by heat treatment at 550°C

Thus, the FTIR spectroscopy study demonstrated that the main effect of the proposed method for modifying of the alumina surface consists in changing the ratio of different type sites capable of anchoring the active component precursor, in particular, in diminishing the fraction of more basic OH groups capable of coordination binding of the complexes (the formation of inner sphere complexes). The conclusions based on FTIR spectroscopy data and concerning changes in the surface state able to affect the mechanisms of precursor-support interaction and strength of such interaction were supported by independent 195Pt NMR study. 195Pt MAS NMR can be used to acquire data on the composition of adsorbed complexes and their interaction with the surface (Shelimov et al., 1999, 2000). Among advantages of this method for investigation of platinum complexes is a wide overall range of chemical shift (ca. 15000 ppm). This allows a relatively simple identification of Pt (IV) complexes with different structure from their 195Pt chemical shift, which is very sensitive to the ligand environment. Thus, substitution of a Cl– ligand in $[PtCl_6]^{2-}$ by H_2O or OH– produces chemical shifts by 500 and 660 ppm, respectively.

However, in the study of complexes adsorbed on the support surface, the 195Pt NMR signals are observed only if octahedral symmetry of the complexes is retained or slightly distorted during the adsorption. Coordination anchoring of a complex on alumina, when one or several chloride ligands of $[PtCl_6]^{2-}$ are substituted by hydroxyl groups of the support (see equation (8)), is accompanied by a substantial decrease in intensity and broadening of the peaks; sometimes NMR signals are not detected. Such situation is observed at the adsorption of complexes on the unmodified g-Al_2O_3. At a maximum possible concentration of a metal complex for the chemisorption (platinum content of 4.5 wt%), the spectrum has only a broad peak with low intensity in the region characterizing $[PtCl_6]^{2-}$ (Fig. 7(a)).

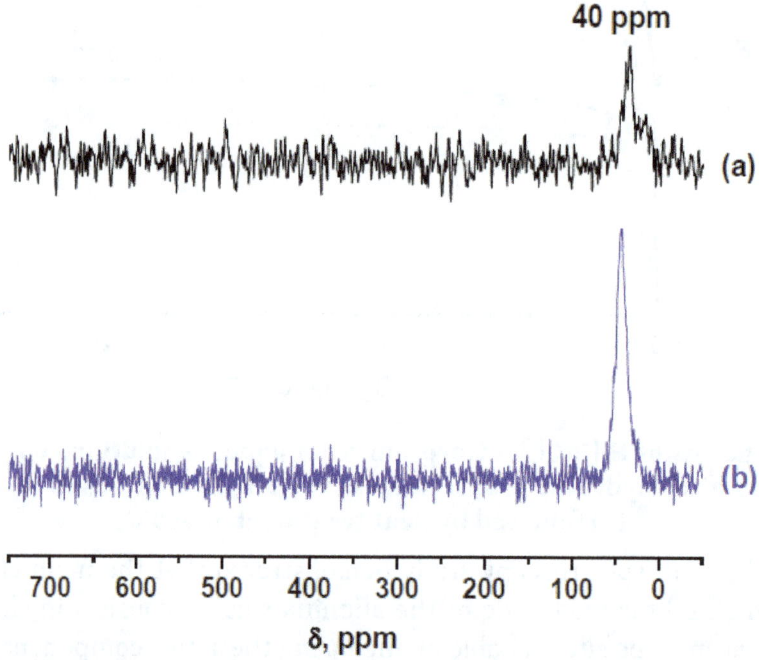

Fig. 7. 195Pt MAS NMR spectra of 4.5%Pt/γ-Al2O3 (unmodified support) (a) and 2.0%Pt/Al2O3 with modified support (hydrothermal treatment at 180 °C, 3 h) (b)

This fact agrees with the diffuse Reflectance electron spectroscopy and EXAFS examination of adsorbed complexes (Belskaya et al., 2008, 2011) showing that platinum on the g-Al_2O_3 surface is mainly a component of the hydrolyzed coordinatively anchored complexes.

Meanwhile, in the case of modified support (Fig. 7(b)), there is an intense peak in the spectrum even at a two times lower platinum content corresponding to chloride complex $[PtCl_6]^{2-}$. Hence, anchoring of the complex does not produce noticeable changes in its chemical composition; moreover, there is a considerable decrease in the contribution of coordination binding with the surface involving OH groups of the support. Electrostatic interaction of a metal complex with the modified support with respect to equation (7) seems to prevail here. In addition, temperature-programmed reduction followed by chemisorption of H_2 and CO probe molecules was used to confirm that a decrease in the bond strength between precursor and support decreases the reduction temperature of adsorbed platinum species and diminishes the dispersion of supported particles by a factor of more than 3.

3.1.2.3 The effect of g-Al2O3 modifying with silica on acid-base properties of the surface

Due to its high thermal stability, alumina modified with silica is a promising support for exhaust neutralization catalysts. Acid-base properties of the developed composition can be optimized by means of FTIR spectroscopy of adsorbed probe molecules. Composition of the support was varied by changing the silica concentration from 1.5 to 10 wt%. The modifying SiO_2 compound was formed in the pore space of γ-Al_2O_3 support during thermal decomposition of preliminarily introduced tetraethoxysilane $Si(OC_2H_5)_4$.

Figure 8 shows FTIR spectra in the region of OH group stretching vibrations of the aluminum oxide samples modified with silica in comparison with the spectrum of initial γ-Al_2O_3. The spectra of all modified oxides show a decrease in intensity of absorption bands corresponding to OH groups of the bridging and terminal types as compared to initial γ-Al_2O_3. The additional absorption bands appear at 3740-3745 cm^{-1}, their intensity growing with silicon content of the sample. The band at 3745 cm^{-1} can be assigned to the terminal Si(OH) groups.

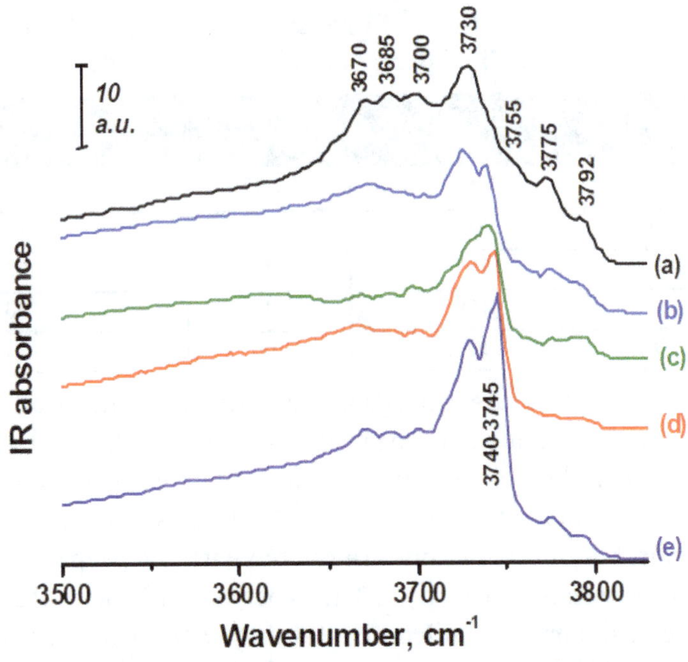

Fig. 8. FTIR spectra of hydroxyl groups for g-Al2O3 calcined at 550 °C (a) and of 1.5% SiO_2/ γ-Al2O3 (b), 5% SiO_2/γ-Al2O3 (c), 7% SiO_2/γ-Al2O3 (d), 10% SiO_2/γ-Al2O3 (e) calcined at 450 °C. The samples were outgassed at 450 °C

Adsorption of CO on γ-Al_2O_3 at -196 °C results in the shift of OH bands with nOH 3700 and 3775 cm^{-1} to the lower frequency region due to perturbation of OH stretch by hydrogen bonding. Aluminum oxides have no Brønsted acidity according to the minor shift of OH stretching vibration (ΔvOH/CO) by 120-130 cm^{-1} (Maache et al., 1993; Paze et al., 1997). A large shift of OH stretching vibrations for the silicon-containing samples with ΔvOH/CO equal to 240-290 cm^{-1} indicates the formation of strong Brønsted acid sites on the sample surface. As demonstrated by Crépeau et al. (Crépeau et al., 2006), high acidity for amorphous silica- alumina can be shown by free silanol groups located nearby an Al atom. Acidity of OH groups at 3740 cm^{-1} of the 10% SiO_2/Al_2O_3 sample approaches the acidity of bridged Si(OH)Al groups in zeolites with the band at 3610 cm^{-1} according to the value of the low frequency shift of OH vibrations with adsorbed CO (ΔvOH/CO= 300 cm^{-1}).

For all tested systems, the spectrum of adsorbed CO has a.b. with nC_O = 2158-2163 cm^{-1} attributed to CO hydrogen-bonded with hydroxyl groups, and a.b. with nC_O = 2130-2135 cm^{-1} related with absorption of physisorbed CO. Besides, for all silicon-containing systems, there is a.b. with $nC_O \approx$ 2170 cm^{-1}, which was assigned to CO hydrogen-bonded with Brønsted acid sites. Concentrations of these sites estimated from the intensity of a.b. with $nC_O \approx$ 2170 cm^{-1} are listed in Table 4. One may see that concentration of Brønsted sites increases with the silica content in the sample.

Table 4. Surface concentrations of Brønsted acid sites and specific surface area of silica promoted aluminum oxides

Sample	Specific surface area, m^2/g	DnOH, cm^{-1}	Concentration of Brønsted acid sites, mmol/g
g-Al_2O_3	162	120-130	-
1.5% SiO_2/Al_2O_3	120	240	14
5% SiO_2/Al_2O_3	131	240	28
7% SiO_2/Al_2O_3	179	240	30
10% SiO_2/Al_2O_3	191	290	33

By using low-temperature CO adsorption, the concentration and strength of coordinatively unsaturated surface sites of the synthesized oxide systems were also estimated. The absorption bands and their intensities observed in the spectrum of γ-Al_2O_3 calcined at 550 °C (Table 5) are close to those reported in the previous Sections. Minor distinctions are related with the preparation procedures and texture characteristics of γ-Al_2O_3. The types of Lewis acid sites identified by CO adsorption on the surface of silicon-containing systems are close to the types of Lewis acid sites for g-Al_2O_3. The concentration of super strong Lewis acid sites (QCO = 59.5 kJ/mol) decreases, whereas the concentration of strong Lewis acid sites (QCO = 54 kJ/mol) and medium strength Lewis acid sites (QCO = 42.5 kJ/mol) increases. The type of Lewis acid sites with $nC_0 \approx 2230\ cm^{-1}$ is typical of aluminosilicate structures and can be assigned to aluminum in a defect octahedral coordination, which is bonded to silicon atom in the second coordination sphere (Paukshtis, 1992).

Four types of base sites were identified on the alumina surface using deuterochloroform as a probe molecule. The spectra are shown in Fig. 9; site strengths and concentrations are listed in Table 6. Strong base sites of alumina are characterized by a.b. with νCD 2193 and 2217 cm^{-1}, which correspond to the calculated PA values of 942 and 915 kJ/mol. The band at 2235 cm^{-1} corresponds to medium-strength base sites, whereas the band at 2250 cm^{-1} is attributed to weak base sites. Paukshtis (Paukshtis, 1992) hypothesized that the strong and medium-strength base sites are bridging oxygen atoms (Al–O–Al), whereas the weak sites are oxygen atoms of OH groups (Al–OH). The introduction of 1.5-5% SiO_2 results in disappearance of super strong base sites and an abrupt decrease in the concentration of other types of base sites, which can be related with sequential blocking of the surface by silica. The introduction of 7-10% SiO_2 increases the concentration of strong and medium strength base sites, which evidences the formation of a new surface phase, probably aluminosilicate one. This phase blocks only partially the surface of initial alumina. All the samples modified with silica have high-frequency a.b. of adsorbed deuterochloroform at 2255-2258 cm^{-1} characterizing the sites which basicity is close to that of silica gel OH groups.

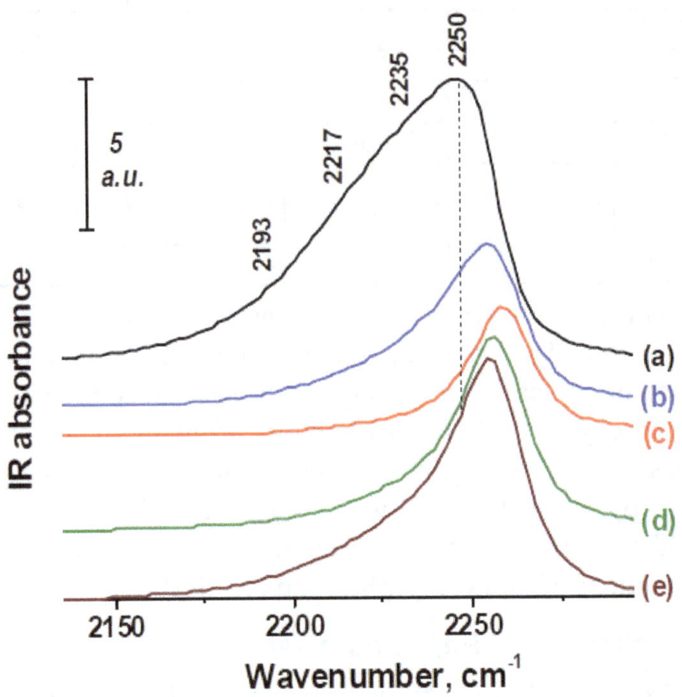

Fig. 9. FTIR spectra of adsorbed deuterochloroform on γ-Al_2O_3 (a) and aluminum

oxides modified with 1.5% (b), 5% (c), 7% (d), and 10% (e) of SiO2

Table 5. Types and concentrations of Lewis acid sites according to FTIR spectroscopy of adsorbed CO

Type of Lewis acid site	Super strong	Strong	Medium I	Medium II	Weak
ν_{CO}, cm^{-1}	2240	2230	2220	2205	2190
QCO, kJ/mol	59.5	54	48.5	42.5	33
Samples	Concentration, μmol/g				
γ-Al_2O_3	0.4	1	8	6	300
1.5% SiO_2/Al_2O_3	0.3	5	4	22	330
5% SiO_2/Al_2O_3	-	6	4	15	290
7% SiO_2/Al_2O_3	-	7	4	18	300
10% SiO_2/Al_2O_3	-	12	4	17	320

Table 6. Types and concentrations of base sites according to FTIR spectroscopy of adsorbed CDCl3. * The concentration may be overrated due to close proximity of the band of adsorbed and physisorbed deuterochloroform

Type of base site	Super strong	Strong	Medium	Weak I	Weak II
ΔνCD, cm^{-1}	72	48-43	30-23	15	10-7
PA, kJ/mol	942	915-908	885-857	839	
Samples	Concentration, μmol/g				
γ-Al_2O_3	48	130	163	210	
1.5% SiO_2/Al_2O_3	-	38	74	-	570*
5% SiO_2/Al_2O_3	-	-	21	-	560*
7% SiO_2/Al_2O_3	-	26	56	-	730*
10% SiO_2/Al_2O_3	-	80	124	-	700*

Thus, Section 3.1 demonstrated the conventional approaches employed in FTIR spectroscopy for investigation of functional groups on the alumina surface. Original methods for modifying of this most popular support were reported. Applicability of FTIR spectroscopy for estimating the effect of modifying on the surface acid-base properties and optimizing the composition of surface sites was shown. Thus, FTIR spectroscopy in combination with other methods can explain changes in adsorption and catalytic characteristics and predict the behavior of oxide surface under real conditions of catalyst synthesis and testing.

3.2 State of palladium in Pd/SZ catalysts

Sulfated zirconia (SZ) promoted with noble metals is a very effective catalyst for isomerization of alkanes due to its high activity at low temperatures and high selectivity of isomers formation (Song & Sayari, 1996). Information on the state of metal in isomerization catalyst is quite topical. For example, metallic platinum or palladium improve the dehydrogenating capacity of the catalyst, which affects the formation of isoalkanes and enhance the production of atomic hydrogen which is necessary for the removal of coke precursors (Vera et al., 2002, 2003). It should be noted that state of the metal is determined to a great extent by the conditions of oxidative and reductive treatment of catalysts before the reaction. Thus, the challenge is to find the optimal pretreatment temperatures allowing the formation of acid sites and retaining the metallic function.

A convenient method for solving this problem is FTIR spectroscopy of adsorbed CO molecules. In our earlier work (Belskaya et al., 2010), this method was used for studying the state of supported palladium particles in Pd/SZ under different conditions of catalyst pretreatment. In the experiment, the catalyst treatment in various gas media (Air, hydrogen) and at different temperatures (100-400 °C) was performed directly in a spectrometer cell. CO adsorption was carried out over a pressure range of 0.1 to 10 mbar at room temperature.

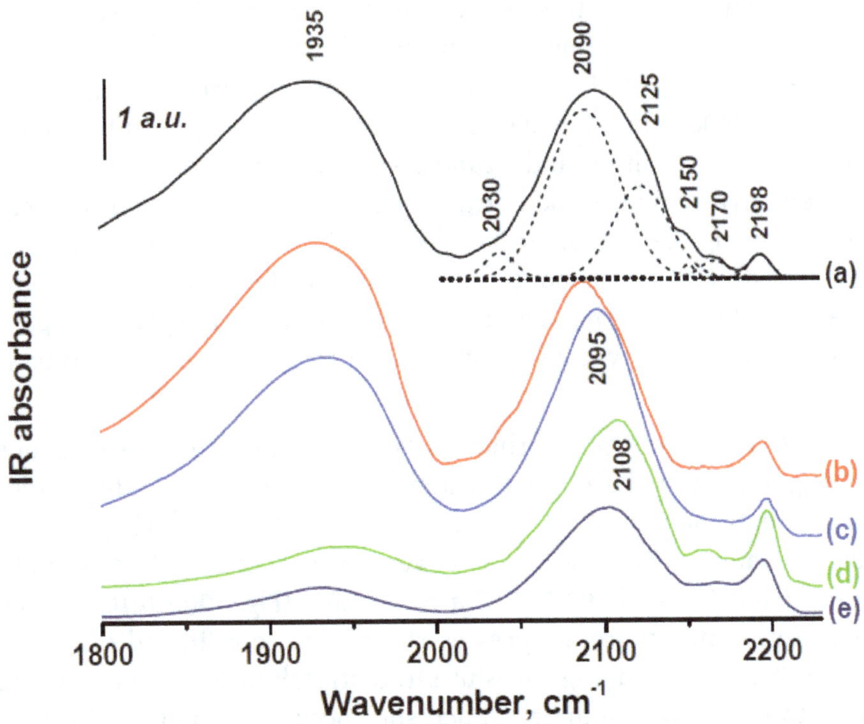

Fig. 10. FTIR spectra of CO adsorbed on Pd/SZ (25 °C; 10 mbar) after oxidation in Air at 400 °C (a), after reduction in hydrogen at 150 °C (b), 200 °C (c), 300 °C (d) and 350 °C (e).

IR spectra of CO adsorbed on the surface of Pd/SZ pretreated under different conditions are shown in Fig. 10. The spectrum of CO adsorbed on the sample that was activated in Air shows several a.b. located at 1935, 2030, 2090, 2125, 2150, 2170 and 2198 cm^{-1} (Fig. 10(a)). The sharp band at 2090 cm^{-1} and the broad band around 1935 cm^{-1} are ascribed to terminal and bridge-coordinated CO on Pd0, respectively (Sheppard & Nguyen, 1978). The bands at 2030, 2125 and 2150 cm^{-1} can be attributed, respectively, to bridged CO complex with Pd+ and linear CO complexes with Pd+ and Pd2+ isolated ions; the band at 2170 cm^{-1} can be assigned to CO linearly adsorbed on Pd2+ in PdO species (Hadjiivanov & Vayssilov, 2002). Besides, the band at 2196–2198 cm^{-1} is present in all spectra and may be assigned to CO adsorption on Zr4+ ions (Morterra et al., 1993). The highest intensity of this a.b. is observed for the sample reduced in H_2 at 300 °C. This fact confirms studies concerning the necessity of high-temperature reduction of SZ to obtain the greatest Lewis acidity of the catalyst after metal incorporation.

All spectra are background subtracted. Spectra were offset for clarity

According to FTIR spectroscopy data, in oxidized Pd/SZ sample a part of palladium is presented as Pd0 (high intensities of a.b. at 1935 and 2090 cm^{-1}). Supposedly, the formation of metallic palladium can be caused by evacuation at high temperatures during the pretreatment in IR cell. This assumption was confirmed in a special experiment by means of UV-vis spectroscopy (Belskaya et al., 2010). Pd/SZ catalyst after the oxidation is characterized by a.b. at 20500, 34000, 39500 and 46000 cm^{-1}. The a.b. at 20500 and 39500 cm^{-1} can be attributed, respectively, to d–d transition and ligand-to-metal charge transfer of Pd2+ ions in D4h oxygen environment (Rakai et al., 1992). Evacuation at 300 °C decreases the concentration of Pd2+ ions in PdO (a decrease in the intensity of a.b. at 20500 cm^{-1} was observed). This experiment clearly demonstrates that a possible effect of pretreatment conditions on the state of catalyst surface in IR spectroscopy study (in this case, the effect of evacuation at elevated temperatures) should be taken into account.

In the FTIR spectra of CO adsorbed on Pd/SZ samples that were reduced at 150–200 °C (Fig. 10(b), (c)), the a.b. at 1930 and 2090–2095 cm^{-1} attributed to metallic palladium dominate. Bands at 2125–2170 cm^{-1}, assigned to CO adsorbed on oxidized Pd ions, almost vanish. The differences between spectra of adsorbed CO on Pd/SZ samples reduced at 200 and 300-350 °C are dramatic (Fig. 10(c)-(e)). An increase in the reduction temperature suppresses the bridge coordinated CO IR bands, decreases the intensity and slightly shifts the stretching frequency of linearly adsorbed CO to higher wavenumbers. Such shift occurs when CO is chemisorbed on the sulfur-saturated Pd surface (Guerra, 1969; Jorgensen & Madix, 1985). So, appearance of the band at 2105–2108 cm^{-1} can be attributed to linear CO complexes with palladium in an electron-deficient state with high S-coverage. According to (Jorgensen & Madix, 1985; Ivanov & Kustov, 1998) a possible reason for the decrease in concentration of bridging CO is the partial covering of Pd surface by sulfur species. Thus, we suppose that low CO chemisorption capacity of Pd0 atoms in Pd/SZ samples reduced at high temperatures is due to partial sulfur coating of the metal surface. Therewith, new a.b. at 2135 and 2160 cm^{-1} are observed in the spectra of CO adsorbed on Pd/SZ samples that were reduced at 300–350 °C. These bands appear on admission of 0.1 mbar CO and grow in intensity with increasing CO pressure (Belskaya et al., 2010); they are attributed to CO complexes with oxidized Pd+ and Pd2+ species, probably in Pdn+–S2- sites (Vazquez-Zavala et al., 1994).

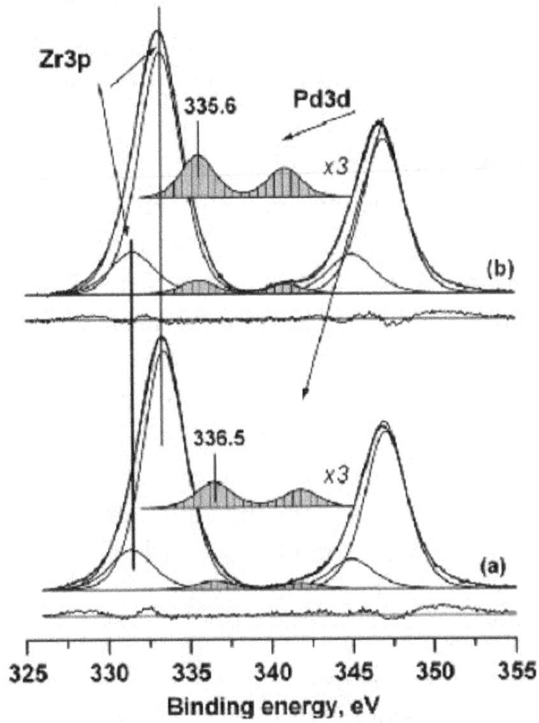

Fig. 11. Zr3p and Pd3d core-level spectra of the samples after oxidation in Air at 400 °C (a) and after reduction in hydrogen at 300 °C (b). For clear identification, Pd3d spectra are multiplied by 3

IR spectroscopic data on the state of supported palladium in Pd/SZ after oxidation and reduction were compared with X-ray photoelectron spectroscopy (XPS) data for the same samples. Figure 11 shows the X-ray photoelectron spectra in the spectral region of Zr3p and Pd3d core-level lines. The difference curves between experimental spectrum and the envelope of the fit are presented under each spectrum. After calcination in Air at 400 °C (Fig. 11(a)), palladium spectrum can be described by one doublet line with binding energy (E_b) of Pd3d$_{5/2}$ 336.5 eV. Such value of E_b is close to that for the oxidized palladium species in palladium oxide PdO (Brun et al., 1999; Pillo et al., 1997). After the action of H_2 (Fig. 11(b)), Pd3d spectrum gives a peak with E_b (Pd3d$_{5/2}$) 335.6 eV, which is assigned to small metal particles. Bulk metal palladium is known to have E_b of Pd3d$_{5/2}$ at 335.2 eV (Brun et al., 1999; Otto et al., 1992). In our case, a small increase of E_b is likely to originate from the size effect (Mason, 1983). The formation of bulk palladium sulfide was not observed because the corresponding value of E_b of Pd3$d_{5/2}$, which is about 337.2 eV (Chaplin et al., 2007), was not detected. Thus, changes of the metal surface after high temperature reduction in the H_2 atmosphere, which are demonstrated by the FTIR spectra, have no significant effect on the electronic state of metal particles: according to XPS data, Pd remains in the metallic state. H_2S resulting from sulfate reduction blocks the active metal surface, most likely without formation of PdS in large amounts. However, it cannot be unambiguously concluded from Pd3d and S2p core-level spectra whether the sulfide film forms on the palladium surface after reduction or not.

For assessment of the state of palladium, we also used a model reaction which is commonly employed to test the metallic function of catalysts -low-temperature (50-90 °C) hydrogenation of benzene. There was a clear effect of pretreatment conditions on the conversion of benzene to cyclohexane. Pd/SZ catalyst reduced at the temperature corresponding to metal formation (according to TPR data) demonstrated the highest conversion. In the case of reduction temperature above 120 °C, before the catalytic test H_2S was detected in the exhaust gases, and poisoning of the metal function was observed. Thus, in Pd/SZ catalysts, after the reduction treatment at 300 °C, hydrogenation activity of palladium is strongly inhibited. However, the constant activation energy and complete recovery of hydrogenation activity under mild regeneration conditions (Belskaya et al., 2010) indicate that the metal surface is only blocked by sulfate decomposition products without their chemical interaction with palladium.

Thus, FTIR spectroscopy of adsorbed CO used to examine the state of supported palladium in Pd/SZ catalysts provided data that agree well with the data obtained by independent methods – XPS and a model reaction for testing the metal function. Analysis of changes in the state of surface revealed by FTIR spectroscopy can be useful for explaining the adsorption and catalytic properties as well as for optimizing the conditions of thermal stages during catalyst synthesis.

3.3 Alumina promoted Pt/SO42−-ZrO2

The approaches for controlling the SO_4^{2-}-ZrO_2 acidity are of great practical importance, as they can change the catalyst activity and selectivity in various acid-catalyzed reactions (Hua et al., 2000; Lavrenov et al., 2007; Zalewski et al., 1999). In our works (Kazakov et al., 2010, 2011, 2012), we optimized the acidic and hydrogenation properties of bifunctional Pt/SO_4^{2-}- ZrO_2 catalyst for the one-step hydroisomerization of benzene-containing fractions, which is intended for elimination of benzene in gasoline while minimizing the octane loss. The introduction of alumina into the catalyst was suggested as the main modifying procedure. IR spectroscopy allowed us to elucidate the effect of catalyst composition on the properties of surface functional groups, to reveal the role of alumina in the formation of metal and acid sites, and provided a Detailed characterization of the surface properties of optimal hydroisomerization catalyst. The work was performed with catalysts Pt/SO_4^{2-}-ZrO_2 (4.5 wt% SO_4^{2-}), Pt/SO_4^{2-}-ZrO_2-Al_2O_3, (3.1 wt% SO_4^{2-} and 67.8 wt% Al_2O_3) and Pt/Al_2O_3. Samples were denoted as Pt/SZ, Pt/SZA and Pt/A, respectively. Platinum concentration was 0.3 wt%. We studied also the supports used for the catalyst synthesis (SZ, SZA and A, respectively). The preparation procedure for supports and catalysts is reported in (Kazakov et al., 2010, 2011).

3.3.1 The effect of alumina introduction on the state of supported platinum

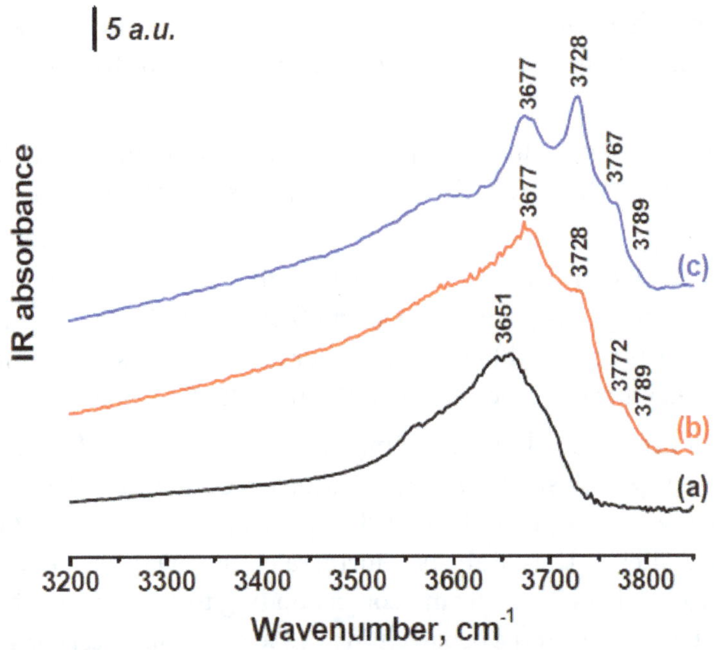

Fig. 12. FTIR spectra of the catalyst supports in the OH stretching region: SZ (a), SZA (b), A (c). Spectra were offset for clarity. Prior to recording, the samples were evacuated at 400 °C

The formation of platinum sites and the state of metal in a finished catalyst strongly depend on the interaction of precursor with the surface groups of support. As it was shown earlier, the composition and amount of hydroxyl groups on the support surface play a significant role in the platinum compounds anchoring from a solution of $H_2[PtCl_6]$. Infrared spectra of the OH stretching region for SZ, SZA and A supports are shown in Fig.12. The spectrum of sample SZ is represented by a.b. at 3651 cm^{-1}, which corresponds to bridging OH groups with acidic properties

(Kustov et al., 1994; Manoilova et al., 2007). The FTIR spectrum of SZA sample has a.b. at 3772 and 3789 cm^{-1} assigned to terminal OH groups, and a.b. 3677 and 3728 cm^{-1} corresponding to bridging OH groups (Knözinger & Ratnasamy, 1978). The presemce of these types of OH groups is typical for the γ-Al_2O_3 surface (sample A in Fig. 12).

Significant distinctions in the hydroxyl cover suggest different mechanisms of the interaction between metal complex and support. Indeed, the presemce of only the hydroxyl groups with acidic properties on the SZ surface explains the absence of chemisorption anchoring of the anionic $[PtCl_6]^{2-}$ complex. After the introduction of 67.8 wt% alumina into sulfated zirconia sites for chloroplatinate ions sorption appear on the surface of mixed SO_4^{2-}–ZrO_2-Al_2O_3 support. However, the concentration of bridging $Al^{VI}(OH)Al^{VI}$ groups (3728 cm^{-1}) that are most active in chloroplatinate anchoring, and terminal groups (3772 and 3789 cm^{-1}) on SZA support is lower than their concentration on the alumina surface, which causes a smaller fraction of complexes anchored by chemisorption (81% for SZA and 100% for A).

The reduction temperature of platinum species anchored on SZA support, which characterizes the strength of precursor - support interaction, also has a medium value.

According to TPR data, platinum reduction on SZ surface (sample Pt/SZ) in the absence of chemisorption interaction starts from 90 °C. However, in the samples with alumina this process shifts toward higher temperatures, and a maximum rate of hydrogen consumption is observed at 210 and 225 °C for Pt/SZA and Pt/A.

The state of supported platinum in finished catalysts after reduction in hydrogen was investigated by FTIR spectroscopy of adsorbed CO (Fig. 13). The band with νCO 2200-2208 cm^{-1}, which is present in all the spectra, corresponds to CO complexes with Lewis acid sites of the catalysts (Morterra et al., 1993). The spectrum of Pt/A sample shows a.b. with νCO 2065 cm^{-1} corresponding to stretching vibrations of CO linearly adsorbed on Pt0, and a broad band at 1830 cm^{-1} characterizing the bridging CO species on Pt0 (Apesteguia et al., 1984; Kooh et al., 1991). After CO adsorption on Pt/SZ, there appear bands at 2100 and 2150 cm^{-1}, which have close intensities and correspond to linear CO complexes with Pt0 and Ptδ+, respectively (Grau et al., 2004; Morterra et al., 1997). In the case of Pt/SZA, the frequencies of CO (a.b. 2085 cm^{-1}) adsorbed on metal platinum particles are intermediate in comparison with frequencies for samples Pt/A and Pt/SZ.

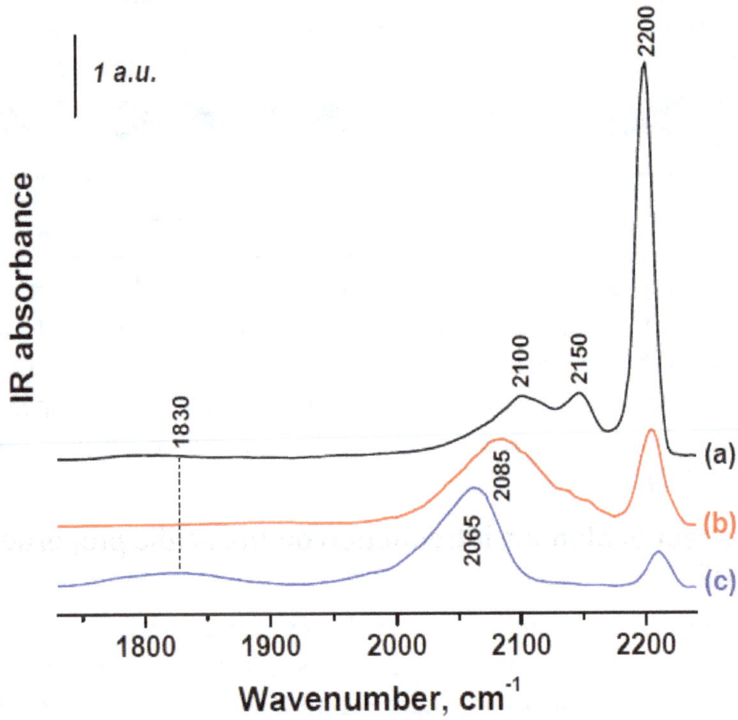

Fig. 13. FTIR spectra of CO (25 °C; 10 mbar) adsorbed on the catalysts: Pt/SZ (a), Pt/SZA (b), Pt/A (c). Prior to recording, the samples were reduced in hydrogen flow at 300 °C and then evacuated at 500 °C

In comparison with Pt/SZ, the band corresponding to CO – Pt0 complexes for samples Pt/SZA and Pt/SZ is shifted toward higher frequencies by 20 and 35 cm^{-1}, respectively. Such upward shift of vCO was also observed for Pd/SZ samples (Section 3.2) and can be related to the presemce of sulfur species on the metal surface (Apesteguia et al., 1984, 1987; J.R. Chang & S.L. Chang, 1998). These sulfur species are formed both at the stage of oxidative treatment and during the reduction; they can poison the metal partially or completely (Dicko et al., 1994; Iglesia et al., 1993). As a result, Pt/SZ demonstrates very poor hydrogenation activity and does not chemisorb hydrogen (Table 7). Pt/SZA sample has an enhanced hydrogenation activity in comparison with Pt/SZ; nevertheless, it is lower than that observed for Pt/A.

Table 7. Hydrogen chemisorption and benzene hydrogenation over Pt/SZ, Pt/SZA and Pt/A catalysts. Benzene hydrogenation conditions: 200 °C, 0.1 MPa, weight hourly space velocity 4.0 h-1, H2 : C6H6 molar ratio 8

Catalyst	H/Pt	Benzene conversion, %
Pt/SZ	0.00	1.8
Pt/SZA	0.44	40.7
Pt/A	0.85	97.1

The increasing accessibility of platinum sites for adsorption and catalytic reaction in a series Pt/SZ < Pt/SZA < Pt/A can be attributed both to a decrease in the content of sulfur compounds in the catalyst upon dilution of sulfated zirconia with alumina, and to a higher resistance to poisoning of more disperse supported platinum crystallites produced by chemisorption anchoring of a precursor (J.R. Chang et al., 1997).

3.3.2 The effect of alumina introduction on the acidic properties of Pt/SO42−-ZrO2

The strength and concentration of acid sites of catalysts Pt/SZ, Pt/SZA and Pt/A reduced at 300 °C were estimated from FTIR spectra of adsorbed CO and pyridine molecules. After CO adsorption, spectra of all the samples had the a.b. 2180-2208 cm^{-1} corresponding to CO complexes with Lewis acid sites of different strength, a.b. 2168-2171 cm^{-1} assigned to CO complexes with Brønsted acid sites, a.b. 2160-2162 cm^{-1} characterizing CO complexes with hydroxyl groups having weak acidic properties, and a.b. 2133-2148 cm^{-1} corresponding to the adsorption of physisorbed CO molecules. In all cases, only the medium strength (a.b. with vCO 2204-2208 cm^{-1} for Pt/SZ, 2202-2208 cm^{-1} for Pt/SZA, and 2204 cm^{-1} for Pt/A) and weak Lewis acid sites (a.b. with vCO 2194, 2188, 2180 cm^{-1} for Pt/SZ, and 2192 cm^{-1} for Pt/SZA and Pt/A) were detected. After the pyridine adsorption, we observed a.b. corresponding to complexes of pyridine molecules with Brønsted acid sites (1544 cm^{-1}) and Lewis acid sites (1445 cm^{-1}).

Data on the concentration of Lewis acid sites (calculated from the integral intensities of adsorbed CO a.b.) and Brønsted acid sites (calculated from the integral intensities of adsorbed pyridine a.b.) for the tested catalysts are listed in Table 8. The Pt/SZ catalyst has the highest content both of Lewis and Brønsted acid sites. For sample Pt/A, Brønsted acid sites able to protonate pyridine were not observed. Pt/SZA has intermediate position with respect to its acidic properties. The Brønsted acid sites content in this sample is 3.5 times lower as compared to Pt/SZ, which virtually corresponds to a decrease of ZrO_2 amount in its composition (29.1 against 95.5 wt%, respectively). The total amount of Lewis acid sites in comparison with Pt/SZ decreases twofold. However, the ratio of medium strength and weak Lewis acid sites for Pt/SZA sample corresponds to the ratio revealed for Pt/SZ. Thus, the introduction of alumina into Pt/SZ system decreases the amount of Lewis and Brønsted acid sites, which is related to the effect of its dilution with a component having a lower intrinsic acidity. The observed nonadditive change in the concentration of acid sites, in particular Lewis acid sites, may be caused by interaction of the system components, which was noted earlier (Kazakov et al., 2010).

Table 8. Acidic properties of Pt/SZ, Pt/SZA and Pt/A catalysts according to FTIR spectroscopy of adsorbed CO and pyridine. Prior to recording, samples were reduced in hydrogen flow at 300 °C and then evacuated at 500 °C

Catalyst	νCO, cm^{-1}	Lewis acid sites, μmol/g	Lewis acid sites, μmol/g		Total Lewis acid sites, μmol/g	Total Brønsted acid sites, μmol/g
			medium	weak		
	2208	25				
	2204	90				
Pt/SZ	2194	260	115	530	645	32
	2188	240				
	2180	30				
	2208	15				
Pt/SZA	2202	45	60	250	310	9
	2192	250				
Pt/A	2204	29	29	350	379	0
	2192	350				

Results of FTIR spectroscopic study are in good agreement with the model acid-catalyzed reactions of n-heptane and cyclohexane isomerization. The introduction of alumina into Pt/SZ decreases the total catalyst activity in n-C7H16 isomerization, which shows up as increase of the temperature of 50% n-heptane conversion from 112 to 266 °C (Table 9). For isomerization of cyclohexane to methylcyclopentane, higher operating temperatures are thermodynamically more favorable (Tsai et al., 2011). As a result, Pt/SZA catalyst is more efficient for cyclohexane isomerization due to higher selectivity at higher temperatures (Table 10).

Table 9. Isomerization of n-heptane over Pt/SZ, Pt/SZA and Pt/A catalysts. Reaction conditions: 1.5 MPa, weight hourly space velocity 4.0 h–1, H2 : n-C7H16 molar ratio 5. X – conversion

Catalyst	X n-C7, %	t, °C	Iso-C7 yield, %	Selectivity to iso-C7, %
Pt/SZ	50.0	112	43.3	86.5
Pt/SZA	50.0	266	47.1	94.2
Pt/A	12.9	300	6.3	47.0

Table 10. Isomerization of cyclohexane over Pt/SZ, Pt/SZA and Pt/A catalysts. Reaction conditions: 1.5 MPa, weight hourly space velocity 4.0 h–1, H2 : C6H¹2 molar ratio 5. X – conversion; CH – cyclohexane; MCP – methylcyclopentane

Catalyst	200 °C			275 °C		
	X CH, %	MCP yield, %	Selectivity to MCP, %	X CH, %	MCP yield, %	Selectivity to MCP, %
Pt/SZ	70.0	57.1	81.7	91.8	26.2	29.6
Pt/SZA	4.4	4.4	99.3	74.4	69.6	93.7
Pt/A	-	-	-	0.2	0.0	-

Thus, FTIR spectroscopy applied to investigation of Pt/SZA system proved to be a highly informative method, which allowed us to elucidate the role of alumina both in the formation of platinum sites and in the catalyst behavior in the acid-catalyzed reactions. Although state of the surface under conditions of FTIR spectroscopic examination strongly differ from its state upon contacting with aqueous solutions of metal complexes or in catalytic reactions, FTIR spectroscopy data on the state of supported platinum as well as on the nature and strength of acid sites can be used to optimize the composition of bifunctional catalyst Pt/SZA designed for hydroisomerization of benzene-containing fractions.

4. CONCLUSION

The possibilities of FTIR spectroscopy, in particular with the use of adsorbed CO, pyridine or deuterochloroform probe molecules, for investigation of some model and industrially important supports and catalysts were demonstrated. The effect of chemical composition of a support (Al_2O_3, Al_2O_3-SiO_2, SO_4^{2-}-ZrO_2, SO_4^{2-}-ZrO_2-Al_2O_3) and modification technique on the concentration and ratio of different types of OH groups and coordinatively unsaturated surface sites was shown.

Concentrations of the surface sites on supports before and after anchoring of the active metal component were compared to demonstrate a relation between composition of the functional surface groups, adsorption capacity of the support and strength of the interaction between metal complex precursor and support, and to identify the sites involved in anchoring of the active component. The impact of support nature and composition, conditions of oxidation and reduction treatments on the metal-support interaction and ratio of oxidized and reduced forms of supported metal (platinum or palladium) was revealed.

FTIR spectroscopy data for the examined catalytic systems were compared with the data of XPS, diffuse Reflectance electron spectroscopy, H_2 and CO chemisorption for determination of supported metal dispersion, and temperature-programmed reduction as well as with the results of testing in the following catalytic reactions: double-bond isomerization of 1-hexene, hydrogenation of benzene and isomerization of n-heptane and cyclohexane.

5. REFERENCES

1. Apesteguia, C.R., Brema, C.E., Garetto, T.F., Borgna, A., & Parera, J.M. (1984). Sulfurization of Pt/Al_2O_3-Cl Catalysts: VI. Sulfur-Platinum Interaction Studied by Infrared Spectroscopy. Journal of Catalysis, Vol.89, No.1, (September 1984), pp. 52-59, ISSn 0021-9517

2. Apesteguia, C.R., Garetto, T.F., & Borgna, A. (1987). On the Sulfur-Aided Metal-Support Interaction in Pt/Al_2O_3-Cl Catalysts. Journal of Catalysis, Vol.106, No.1, (July 1987), pp. 73-84, ISSn 0021-9517

3. Baumgarten, E., Wagner, R., & Lentes-Wagner, C. (1989). Quantitative Determination of Hydroxyl Groups on Alumina by IR Spectroscopy. Fresenius Zeitschrift für Analytische Chemie, Vol.334, No.3, (January 1989), pp. 246-251, ISSn 0016-1152

4. Bel'skaya, O.B., Karymova, R.Kh., Kochubey, D.I., & Duplyakin, V.K. (2008). Genesis of the Active-Component Precursor in the Synthesis of Pt/Al_2O_3 Catalysts: I. Transformation of the $[PtCl_6]^{2-}$ Complex in the Interaction between Chloroplatinic Acid and the γ-Al_2O_3 Surface. Kinetics and Catalysis, Vol.49, No.5, (September 2008), pp. 720-728, ISSn 0023-1584

5. Belskaya, O.B., Danilova, I.G., Kazakov, M.O., Gulyaeva, T.I., Kibis, L.S., Boronin, A.I., Lavrenov, A.V., & Likholobov, V.A. (2010). Investigation of Active Metal Species Formation in Pd-Promoted Sulfated Zirconia Isomerization Catalyst. Applied Catalysis A: General, Vol.387, No.1-2, (October 2010), pp. 5-12, ISSn 0926-860X

6. Belskaya, O.B., Duplyakin, V.K., & Likholobov, V.A. (2011). Molecular Design of Precursor in the Synthesis of Catalytic Nanocomposite System Pt-Al_2O_3. Smart Nanocomposites, Vol.1, No.2, pp. 99-133, ISSn 1949-4823

7. Bocanegra, S.A., Castro, A.A., Guerrero-Ruíz, A., Scelza, O.A., & de Miguel, S.R. (2006). Characteristics of the Metallic Phase of Pt/Al_2O_3 and Na-Doped Pt/Al_2O_3 Catalysts for Light Paraffins Dehydrogenation. Chemical Engineering Journal, Vol.118, No.3, (May 2006), pp. 161-166, ISSn 1385-8947

8. Bourikas, K., Kordulis, C., & Lycourghiotis, A. (2006). The Role of the Liquid-Solid Interface in the Preparation of Supported Catalysts. Catalysis Reviews - Science and Engineering, Vol.48, No.4, (December 2006), pp.363-444, ISSn 0161-4940

9. Brun, M., Berthet, A., & Bertolini, J.C. (1999). XPS, AES and Auger Parameter of Pd and PdO. Journal of Electron Spectroscopy and Related Phenomena, Vol.104, No.1-3, (July 1999), pp. 55-60, ISSn 0368-2048

10. Chang, J.R., Chang, S.L., & Lin, T.B. (1997). γ-Alumina-Supported Pt Catalysts for Aromatics Reduction: A Structural Investigation of Sulfur Poisoning Catalyst Deactivation. Journal of Catalysis, Vol.169, No.1, (July 1997), pp. 338-346, ISSn 0021-9517

11. Chang, J.R., & Chang, S.L. (1998). Catalytic Properties of γ-Alumina-Supported Pt Catalysts for Tetralin Hydrogenation: Effects of Sulfur-Poisoning and Hydrogen Reactivation. Journal of Catalysis, Vol.176, No.1, (May 1998), pp. 42-51, ISSn 0021- 9517

12. Chaplin, B.P., Shapley, J.R., & Werth, C.J. (2007). Regeneration of Sulfur-Fouled Bimetallic Pd-Based Catalysts. Environmental Science & Technology, Vol.41, No.15, (June 2007), pp. 5491-5497, ISSn 0013-936X

13. Chertov, V.M., Zelentsov, V.I., & Lyashkevich, B.N. (1982). Production of Finely Divided Boehmite Powder. Journal of Applied Chemistry of the USSR, Vol.55, pp. 2120-2122, ISSn 0021-888X

14. Crépeau, G., Montouillout, V., Vimont, A., Mariey, L., Cseri, T., & Maugé F. (2006). Nature, Structure and Strength of the Acidic Sites of Amorphous Silica Alumina: An IR and NMR Study. Journal of Physical Chemistry B, Vol.110, No.31, (August 2006), pp. 15172-15185, ISSn 1520-6106

15. Davydov, A. (2003). Molecular Spectroscopy of Oxide Catalyst Surfaces, John Wiley & Sons Ltd, ISBN 0-471-98731-X, Chichester, England

16. Della Gatta, G., Fubini, B., Ghiotti, G., & Morterra, C. (1976). The Chemisorption of Carbon Monoxide on Various Transition Aluminas. Journal of Catalysis, Vol.43, Nos.1-3, (June 1976), pp. 90-98, ISSn 0021-9517

17. Dicko, A., Song, X.M., Adnot, A., & Sayari, A. (1994). Characterization of Platinum on Sulfated Zirconia Catalysts by Temperature Programmed Reduction. Journal of Catalysis, Vol.150, No.2, (December 1994), pp. 254-261, ISSn 0021-9517

18. Egorov, M.M. (1961). The Nature of the Surface of Catalytically Active Aluminum Oxide. Proceedings of the Academy of Sciences of the USSR, Physical Chemistry Section, Vol.140, Nos.1-6, (September-October 1961), pp. 697-700, ISSn 0271-5007

19. Ghosh, A.K., & Kydd, R.A. (1985). Fluorine-Promoted Catalysts. Catalysis Reviews – Science and Engineering, Vol.27, No.4, (December 1985), pp. 539-589, ISSn 0161-4940

20. Grau, J.M., Yori, J.C., Vera, C.R., Lovey, F.C., Condo, A.M., & Parera, J.M. (2004). Crystal Phase Dependent Metal-Support Interactions in Pt/SO_4^{2-}-ZrO_2 Catalysts for Hydroconversion of n-Alkanes. Applied Catalysis A: General, Vol.265, No.2, (July 2004), pp. 141-152, ISSn 0926-860X

21. Guerra, C.R. (1969). Infrared Spectroscopic Studies of CO Adsorption by Metals. The Effect of Other Gases in Adsorption. Journal of Colloid and Interface Science, Vol.29, No.2, (February 1969), pp. 229-234, ISSn 0021-9797

22. Hadjiivanov, K.I., & Vayssilov, G.N. (2002). Characterization of Oxide Surfaces and Zeolites by Carbon Monoxide as an IR Probe Molecule, In: Advances in Catalysis, B.C. Gates,

23. H. Knözinger, (Ed.), pp. 307-511, Academic Press, Inc, ISBN 0-12-007847-3, New York, The United States of America

24. Hua, W., Xia, Y., Yue, Y., & Gao, Z. (2000). Promoting Effect of Al on SO_4^{2-}/MxOy (M=Zr, Ti, Fe) Catalysts. Journal of Catalysis, Vol.196, No.1, (November 2000), pp. 104-114, ISSn 0021-9517

25. Iglesia, E., Soled, S.L., & Kramer, G.M. (1993). Isomerization of Alkanes on Sulfated Zirconia: Promotion by Pt and by Adamantyl Hydride Transfer Species. Journal of Catalysis, Vol.144, No.1, (November 1993), pp. 238-253, ISSn 0021-9517

26. Ivanov, A.V., & Kustov, L.M. (1998). Investigation of the State of Palladium in the Pd/SO4/ZrO_2 System by Diffuse-Reflectance IR Spectroscopy. Russian Chemical Bulletin, Vol.47, No.1, (January 1998), pp. 55-59, ISSn 1066-5285

27. Jorgensen, S.W., & Madix, R.J. (1985). Steric and Electronic Effects of Sulfur on CO Adsorbed on Pd(100). Surface Science, Vol.163, No.1, (November 1985), pp. 19-38, ISSn 0039- 6028

28. Kazakov, M.O., Lavrenov, A.V., Mikhailova, M.S., Allert, N.A., Gulyaeva, T.I., Muromtsev, I.V., Drozdov, V.A., & Duplyakin, V.K. (2010). Hydroisomerization of Benzene- containing Gasoline Fractions on a Pt/SO_4^{2-}-ZrO_2-Al_2O_3 Catalyst: I. Effect of Chemical Composition on the Phase State and Texture Characteristics of SO_4^{2-}- ZrO_2-Al_2O_3 Supports. Kinetics and Catalysis, Vol.51, No.3, (June 2010), pp. 438-443, ISSn 0023-1584

29. Kazakov, M.O., Lavrenov, A.V., Danilova, I.G., Belskaya, O.B., & Duplyakin, V.K. (2011). Hydroisomerization of Benzene-containing Gasoline Fractions on a Pt/SO_4^{2-}-ZrO_2- Al_2O_3 Catalyst: II. Effect of Chemical Composition on Acidic and Hydrogenating and the Occurrence of Model Isomerization Reactions. Kinetics and Catalysis, Vol.52, No.4, (July 2011), pp. 573-578, ISSn 0023-1584

30. Kazakov, M.O., Lavrenov, A.V., Belskaya, O.B., Danilova, I.G., Arbuzov, A.B., Gulyaeva, T.I., Drozdov, V.A., Duplyakin, V.K. (2012). Hydroisomerization of Benzene- containing Gasoline Fractions on a Pt/SO_4^{2-}-ZrO_2-Al_2O_3 Catalyst: III. Hydrogenating Properties of the Catalyst. Kinetics and Catalysis, Vol.53, No.1, (January-February 2012), Accepted, In Press, ISSn 0023-1584

31. Knözinger, H. (1976a). Hydrogen Bonds in Systems of Adsorbed Molecules, In: The Hydrogen Bond: Resent Developments in Theory and Experiments, P. Schuster, G. Zundel, & C. Sandorfy, (Ed.), pp. 1263-1364, North-Holland Pub. Co., ISBN 0720403154, Amsterdam, The Netherlands

32. Knözinger, H. (1976b). Specific Poisoning and Characterization of Catalytically Active Oxide Surfaces, In: Advances in Catalysis, D.D. Eley, H. Pines, & P.B. Weisz, (Ed.), pp. 184- 271, Academic Press, Inc, ISBN 0-12-007825-2, New York, The United States of America

33. Knözinger, H. & Ratnasamy, P. (1978). Catalytic Aluminas: Surface Models and Characterization of Surface Sites. Catalysis Reviews – Science and Engineering, Vol.17, No.1, (January 1978), pp. 31-70, ISSn 0161-4940

34. Kooh, A.B., Han, W.-J., Lee, R.G., & Hicks, R.F. (1991). Effect of Catalyst Structure and Carbon Deposition on Heptane Oxidation over Supported Platinum and Palladium. Journal of Catalysis, Vol.130, No.2, (August 1991), pp. 374-391, ISSn 0021- 9517

35. Kubelková, L., Beran, S., & Lercher, J.A. (1989). Determination of Proton Affinity of Zeolites and Zeolite-Like Solids by Low-Temperature Adsorption of Carbon Monoxide. Zeolites, Vol.9, No.6, (November 1989), pp. 539-543, ISSn 0144-2449

36. Kustov, L.M., Kazansky, V.B., Figueras, F., & Tichit, D. (1994). Investigation of the Acidic Properties of ZrO_2 Modified by SO_4^{2-} Anions. Journal of Catalysis, Vol.150, No.1, (November 1994), pp. 143-149, ISSn 0021-9517

37. Kustov, L.M. (1997). New Trends in IR-Spectroscopic Characterization of Acid and Basic Sites in Zeolites and Oxide Catalysts. TopiCs in Catalysis, Vol.4, Nos.1-2, (November 1997), pp. 131-144, ISSn 1022-5528

38. Kwak, J.H., Hu, J., Mei, D., Yi, C.-W., Kim, D.H., Peden, C.H.F., Allard, L.F., & Szanyi, J. (2009). Coordinatively Unsaturated Al^{3+} Centers as Binding Sites for Active Catalyst Phases of Platinum on γ-Al_2O_3. Science, Vol.325, No.5948 (September 2009), pp. 1670-1673, ISSn 0036-8075

39. Lavrenov, A.V., Basova, I.A., Kazakov, M.O., Finevich, V.P., Belskaya, O.B., Buluchevskii, E.A., & Duplyakin, V.K. (2007). Catalysts on the Basis of Anion-Modified Metal Oxides for Production of Ecologically Pure Components of Motor Fuels. Russian Journal of General Chemistry, Vol.77, No.12, (December 2007), pp. 2272-2283, ISSn 1070-3632

40. Lisboa, J. da S., Santos, D.C.R.M., Passos, F.B., & Noronha, F.B. (2005). Influence of the Addition of Promoters to Steam Reforming Catalysts. Catalysis Today, Vol.101, No.1, (March 2005), pp. 15-21, ISSn 0920-5861

41. Little, L.H. (1966). Infrared Spectra of Adsorbed Species, Academic Press, Inc, ISBN 0124521509, London, England

42. Liu, X., & Truitt, R.E. (1997). DRFT-IR Studies of the Surface of γ-Alumina. Journal of the American Chemical Society, Vol.119, No.41, (October 1997), pp. 9856-9860, ISSn 0002- 7863

43. López Cordero, R., Gil Llambías, F.J., Palacios, J.M., Fierro, J.L.G., & López Agudo, A. (1989). Surface Changes of Alumina Induced by Phosphoric Acid Impregnation. Applied Catalysis, Vol.56, No.2, (August 1989), pp. 197-206, ISSn 0166-9834

44. Lycourghiotis, A. (2009). Interfacial Chemistry, In: Synthesis of Solid Catalysts, K.P. de Jong, (Ed.), pp. 13-31, WILEY-VCH Verlag GmbH & Co. KGaA, ISBN 978-3-527-32040-0, Weinheim, Germany

45. Maache, M., Janin, A., Lavalley, J.C., Joly, J.F., & Benazzi E. (1993). Acidity of Zeolites Beta Dealuminated by Acid Leaching: An FTi.r. Study Using Different Probe Molecules (Pyridine, Carbon Monoxide). Zeolites, Vol.13, No.6, (July-August 1993), pp. 419- 426, ISSn 0144-2449

46. Manoilova, O.V., Olindo, R., Otero Areán, C., & Lercher, J.A. (2007). Variable Temperature FTIR Study on the Surface Acidity of Variously Treated Sulfated Zirconias. Catalysis Communications, Vol.8, No.6, (June 2007), pp. 865-870, ISSn 1566-7367

47. Marceau, E., Che, M., Saint-Just, J., & Tatibou't, J.M. (1996). Influence of Chlorine Ions in Pt/Al_2O_3 Catalysts for Methane Total Oxidation. Catalysis Today, Vol.29, Nos.1-4, (May 1996), pp. 415-419, ISSn 0920-5861

48. Mason, M.G. (1983). Electronic Structure of Supported Small Metal Clusters. Physical Review B, Vol.27, No.2, (January 1983), pp. 748-762, ISSn 0163-1829

49. Mei, D., Kwak, J.H., Hu, J., Cho, S.J., Szanyi, J., Allard, L.F., & Peden, C.H.F. (2010). Unique Role of Anchoring Penta-Coordinated Al^{3+} Sites in the Sintering of γ-Al_2O_3- Supported Pt Catalysts. Journal of Physical Chemistry Letters, Vol.1, No.18, (September 2010), pp. 2688-2691, ISSn 1948-7185

50. Mironenko, R.M., Belskaya, O.B., & Likholobov, V.A. (2009). Investigation of the Interaction of Chloride Complexes of Platinum (IV) with Aluminum Oxide in a Structural Modification of the Surface (in Russian), Proceedings of 24th International Chugaev Conference on Coordination Chemistry, pp. 116-117, ISBN 5-85263-026-8, Saint Petersburg, Russia, June 15-19, 2009

51. Mironenko, R.M., Belskaya, O.B., Danilova, I.G., Talsi, V.P., & Likholobov, V.A. (2011). Modifying the functional Cover of the γ-Al_2O_3 Surface Using Organic Salts of Aluminum. Kinetics and Catalysis, Vol.52, No.4, (August 2011), pp. 629-636, ISSn 0023-1584

52. Morterra, C., Cerrato, G., Emanuel, C., & Bolis, V. (1993). On the Surface Acidity of Some Sulfate-Doped ZrO_2 Catalysts. Journal of Catalysis, Vol.142, No.2, (August 1993), pp. 349-367, ISSn 0021-9517

53. Morterra, C., & Magnacca, G. (1996). A Case Study: Surface Chemistry and Surface Structure of Catalytic Aluminas, as Studied by Vibrational Spectroscopy of Adsorbed Species. Catalysis Today, Vol.27, Nos.3-4, (February 1996), pp. 497-532, ISSn 0920- 5861

54. Morterra, C., Cerrato, G., Di Ciero, S., Signoretto, M., Pinna, F., & Strukul, G. (1997). Platinum-Promoted and Unpromoted Sulfated Zirconia Catalysts Prepared by a One-Step Aerogel Procedure: 1. Physico-Chemical and Morphological Characterization. Journal of Catalysis, Vol.165, No.2, (January 1997), pp. 172-183, ISSn 0021-9517

55. Otto, K., Haack, L.P., & de Vries, J.E. (1992). Identification of Two Types of Oxidized Palladium on γ-Alumina by X-ray Photoelectron Spectroscopy. Applied Catalysis B: Environmental, Vol.1, No.1, (February 1992), pp. 1-12, ISSn 0926-3373

56. Paukshtis, E.A., & Yurchenko, E.N. (1983). Study of the Acid-Base Properties of Heterogeneous Catalysts by Infrared Spectroscopy. Russian Chemical Reviews, Vol.52, No.3, (March 1983), pp. 242-258, ISSn 0036-021X

57. Paukshtis, E.A. (1992). Infrared Spectroscopy of Heterogeneous Acid-Base Catalysis (in Russian), Nauka, ISBN 5-02-029281-8, Novosibirsk

58. Paze, C., Bordiga, S., Lamberti, C., Salvalaggio, M., Zecchina, A., & Bellussi, G. (1997). Acidic Properties of H-β Zeolite as Probed by Bases with Proton Affinity in the 118-204 kcal mol-1 Range: A FTIR Investigation. Journal of Physical Chemistry B, Vol.101, No.24, (June 1997), pp. 4740-4751, ISSn 1520-6106

59. Peri, J.B. (1965). A Model for the Surface of γ-Alumina. Journal of Physical Chemistry, Vol.69, No.1, (January 1965), pp. 220-230, ISSn 0022-3654

60. Pillo, T., Zimmermann, R., Steiner, P., & Hufner, S. (1997). The Electronic Structure of PdO Found by Photoemission (UPS and XPS) and Inverse Photoemission (BIS). Journal of Physics: Condensed Matter, Vol.9, No.19, (May 1997), pp. 3987-3999, ISSn 0953-8984

61. Rakai, A., Tessier, D., & Bozon-Verduraz, F. (1992). Palladium-Alumina Catalysts – a Diffuse Reflectance Study. New Journal of Chemistry, Vol.16, No.8-9, (August-September 1992), pp. 869-875, ISSn 1144-0546
62. Requies, J., Cabrero, M.A., Barrio, V.L., Cambra, J.F., Güemez, M.B., Arias, P.L., La Parola, V., Peña, M.A., & Fierro, J.L.G. (2006). Nickel/Alumina Catalysts Modified by Basic Oxides for the Production of Synthesis Gas by Methane Partial Oxidation. Catalysis Today, Vol.116, No.3, (August 2006), pp. 304-312, ISSn 0920-5861
63. Rombi, E., Cutrufello, M.G., Solinas, V., De Rossi, S., Ferraris, G., & Pistone, A. (2003). Effects of Potassium Addition on the Acidity and Reducibility of Chromia/Alumina Dehydrogenation Catalysts. Applied Catalysis A: General, Vol.251, No.2, (September 2003), pp. 255-266, ISSn 0926-860X
64. Ryczkowski, J. (2001). IR Spectroscopy in Catalysis. Catalysis Today, Vol.68, No.4, (July 2001), pp. 263-381, ISSn 0920-5861
65. Scokart, P.O., Selim, S.A., Damon, J.P., & Rouxhet, P.G. (1979). The Chemistry and Surface Chemistry of Fluorinated Alumina. Journal of Colloid and Interface Science, Vol.70, No.2, (June 1979), pp. 209-222, ISSn 0021-9797
66. Shelimov, B.N., Lambert, J.-F., Che, M., & Didillon, B. (1999). Initial Steps of the Alumina- Supported Platinum Catalyst Preparation: a Molecular Study by 195Pt NMR, UV- Visible, EXAFS and Raman Spectroscopy. Journal of Catalysis, Vol.185, No.2, (July 1999), pp. 462-478, ISSn 0021-9517
67. Shelimov, B.N., Lambert, J.-F., Che, M., & Didillon, B. (2000). Molecular-Level Studies of Transition Metal – Support Interactions During the First Steps of Catalysts Preparation: Platinum Speciation in the Hexachloroplatinate/Alumina System. Journal of Molecular Catalysis A: Chemical, Vol.158, No.1, (September 2000), pp. 91-99, ISSn 1381-1169
68. Sheppard, N., & Nguyen, T.T. (1978). The Vibrational Spectra of Carbon Monoxide Chemisorbed on the Surface of Metal Catalysts – a Suggested Scheme of Interpretation, In: Advances in Infrared and Raman Spectroscopy, R.J.H. Clark, R.E. Hester, (Ed.), pp. 67-147, Heyden & Son Inc, ISBN 0-85501-185-8, Philadelphia, The United States of America
69. Song, X., & Sayari, A. (1996). Sulfated Zirconia-Based Strong Solid-Acid Catalysts: Recent Progress. Catalysis Reviews - Science and Engineering, Vol.38, No.3, (August 1996), pp. 329-412, ISSn 0161-4940
70. Tsai, K.-Y., Wang, I., & Tsai, T.-C. (2011). Zeolite Supported Platinum Catalysts for Benzene Hydrogenation and Naphthene Isomerization. Catalysis Today, Vol.166, No.1, (May 2011), pp. 73-78, ISSn 0920-5861

71. Tsyganenko, A.A., & Filimonov, V.N. (1973). Infrared Spectra of Surface Hydroxyl Groups and Crystalline Structure of Oxides. Journal of Molecular Structure, Vol.19, (December 1973), pp. 579-589, ISSn 0022-2860

72. Tsyganenko, A.A., & Mardilovich, P.P. (1996). Structure of Alumina Surfaces. Journal of the Chemical Society, Faraday Transactions, Vol.92, No.23, (December 1996), pp. 4843- 4852, ISSn 0956-5000

73. Vazquez-Zavala, A., Fuentes, S., & Pedraza, F. (1994). The Influence of Sulfidation on the Crystalline Structure of Palladium, Rhodium and Ruthenium Catalysts Supported on Silica. Applied Surface Science, Vol.78, No.2, (June 1994), pp. 211-218, ISSn 0169- 4332

74. Vera, C.R., Pieck, C.L., Shimizu, K., Yori, J.C., & Parera, J.M. (2002). Pt/SO_4^{2-}-ZrO_2 Catalysts Prepared from Pt Organometallic Compounds. Applied Catalysis A: General, Vol.232, No.1-2, (June 2002), pp. 169-180, ISSn 0926-860X

75. Vera, C.R., Yori, J.C., Pieck, C.L., Irusta, S., & Parera, J.M. (2003). Opposite Activation Conditions of Acid and Metal functions of Pt/SO_4^{2-}-ZrO_2 Catalysts. Applied Catalysis A: General, Vol.240, No.1-2, (February 2003), pp. 161-176, ISSn 0926-860X

76. Wang, H., Tan, S., & Zhi, F. (1994). IR Characterization of Base Heterogeneity of Solid Catalysts, In: Acid-Base Catalysis II, H. Hattori, M. Misono, & Y. Ono, (Ed.), pp. 213- 216, Elsevier, Inc, ISBN 978-0-444-98655-9, Amsterdam, The Netherlands

77. Zaki, M.I., & Knözinger, H. (1987). Carbon Monoxide – a Low Temperature Infrared Probe for the Characterization of Hydroxyl Group Properties on Metal Oxide Surfaces. Materials Chemistry and Physics, Vol.17, Nos.1-2, (April-May 1987), pp. 201-215, ISSn 0254-0584

78. Zalewski, D.J., Alerasool, S., & Doolin, P.K. (1999). Characterization of Catalytically Active Sulfated Zirconia. Catalysis Today, Vol.53, No.3, (November 1999), pp. 419-432, ISSn 0920-5861

79. Zamora, M., & Córdoba, A. (1978). A Study of Surface Hydroxyl Groups on γ-Alumina. Journal of Physical Chemistry, Vol.82, No.5, (March 1978), pp. 584-588, ISSn 0022- 3654

80. Zecchina, A., Escalona Platero, E., & Otero Areán, C. (1987). Low Temperature CO Adsorption on Alum-Derived Active Alumina: An Infrared Investigation. Journal of Catalysis, Vol.107, No.1, (September 1987), pp. 244-247, ISSn 0021-9517

8 Chapter

HYDROTHERMAL TREATMENT OF HOKKAIDO PEAT

AN APPLICATION OF FTIR AND 13C NMR SPECTROSCOPY ON EXAMINING OF ARTIFICIAL COALIFICATION PROCESS AND DEVELOPMENT

1. INTRODUCTION

There have been great changes in attitude toward the use of peat as an energy source Since World War II (WEC, 2001). In Japan, peatland covers over 2500 km^2 and accounts for a total energy resource of approximately 1.99 GJ.1010 (Spedding, 1988). Peatland in Japan is widely distributed throughout Hokkaido, which is the northernmost area of the country's four main islands. Although peatland also exists in other regions, its distribution is extremely localized. Peatland is distributed over an area of approximately 2000 km^2 in Hokkaido (Noto, 1991), which is equivalent to approximately 6% of the flat area on this island. Peatland is also widespread in the northeastern part of Sapporo, which is the largest city in Hokkaido. Peatland in Japan is often lacustrine peat, which is formed when lakes and marshes become filled with dead plants from their surrounding areas and are then transformed into land. This type of peat is characterized by the spongy formation of plant fiber. In the peatland of Hokkaido, peat usually accumulates to a thickness of three to five meters on the ground surface, while the soft clay layer underlying is the peat is often over 20 meters thick. In some areas, a sand layer exists between the peat and the clay layers.

One approach to study artificial coalification process is dewatering and conversion by hydrothermal treatment. Hydrothermal treatment of peat has been studied recently by the authors (Mursito et al., 2010; Mursito et al., 2010). In this method, raw peat is directly transformed without pretreatment or drying, which leads to greatly reduced costs. However, few studies have been conducted to evaluate the hydrothermal treatment of raw peat by means of all coalification process. Despite this lack of study, experiments imitating coalification by subjecting materials to heating with high pressure water were reported first by Bergius in 1913, who termed the method hydrothermal carbonization.

The conversion of cold climate peat into liquid fuel has been studied and conducted in Germany, Canada, Sweden, Finland, Iceland and Israel (Björnbom et al., 1986). Although there has been much less interest in peat liquefaction than coal liquefaction, a large number of batch autoclave studies have evaluated the use of cold climate peat as the raw material for the formation of liquid fuel. The conversion of peat to liquid fuel in Sweden produced an organic product similar to very heavy oil after raw peat was treated with CO under high pressures and temperatures (Björnbom et al., 1981). In Canada, the conversion of peat to gas and liquid employed CO and/or H_2 and water (Cavalier & Chornet, 1977).

Hydrothermal treatment of Hokkaido cold climate peat has also been investigated. However, it is still necessary to evaluate the liquid and gas products formed during the process to facilitate its energy and chemical utilization. The aim of this chapter is to characterize and determine the effectiveness of hydrothermal treatment for upgrading and dewatering processes on the solid products of Hokkaido cold climate peat as well as to determine its artificial coalification process by applying of FTIR and 13C NMR spectroscopy. A fundamental study of the effects of processed temperature on the products of hydrothermal treatment of cold climate peat is also described in this Chapter.

2. EXPERIMENTAL

2.1 Materials

Raw cold climate peat samples were obtained from peatland areas owned by the Takahashi Peat Moss Company, Hokkaido, Japan. The site is located in a peat mining area that consists of about 40 ha that already contained an open and systematic drainage system. The peat mining method used at the site is the cut and block and dry method, and the peat moss products are primarily used for agriculture and gardening. The peat in the study area is approximately 5–10 m thick and the water level is about 50 cm. Prior to World War II, about 1200 ha of peatland in this area were owned by the Japan Oil-Petroleum Company, which converted the harvested peat into oil. The typical properties of the peat from this mining site are shown in Table 1.

2.2 Apparatus and experimental procedure

All experiments were conducted in a 0.5 L batch-type reactor (Taiatsu Techno MA22) that was equipped with an automatic temperature controller and had a maximum pressure of 30 MPa and a maximum temperature of 400°C (Fig. 1) (Mursito et al., 2010). The raw peat samples were introduced to the reactor without any pretreatment except for milling. The amount of the raw peat added to the reactor was 300 g, which corresponded to 40 g of moisture-free peat. The reactor was pressurized with n_2 to 2.0 MPa at ambient temperature, after which the raw peat was agitated at 200 rpm while the reaction temperature was automatically adjusted from 150°C to 380°C at an average heating rate of 6.6°C/min. Under supercritical conditions (380°C), the charge was 230 g and the initial pressure was 0.1 MPa. After the desired reaction time of 30 min, the reactor was cooled immediately.

After cooling, the gas products were released through a gasometer (Shinagawa DC-1) and their volume was determined by collection into a gas micro syringe (ITO MS-GANX00). The evolved gas composition was then determined by gas chromatography (GC) using a GC equipped with a thermal conductivity detector (Shimadzu GC-4C) using Molecular Sieve 5A and Porapak Q columns. The column temperature was set at 60° C and argon was applied as the carrier gas at a rate of 30 mL/min. The results were recorded using a Shimadzu C-R8A Chromatopac data processor. The results of GC analysis are discussed elsewhere. The solid and liquid phases were then collected from the reactor and separated by filtration (ADVANTEC 5C) using a water aspirator. The total moisture content of the filtered solid products was then determined using a moisture content analyzer (Sartorius MA 150).

Table 1. Proximate and ultimate analysis, yield of solid products and calorific value of Takahashi peat moss and hydrothermally upgraded peat.

	150oC	200oC	250oC	270oC	300oC	330oC	350oC	380oC
Proximate analysis (wt%) Moisture (a.r) 86.9	-	-	-	-	-	-	-	-
Equilibrium Moisture (X) (a.r) 13.3	12.3	12.8	5.6	5.3	4.8	3.2	3.0	2.4
Volatile Matter (d.a.f) 68.1	67.5	65.0	56.4	52.8	48.4	45.4	44.0	37.2
Fixed Carbon (d.a.f) 31.9	32.5	35.0	43.6	47.2	51.6	54.6	56.0	62.8
Ash (d.b) 4.4	12.8	9.1	8.3	7.1	6.0	7.4	7.5	8.2
Ultimate analysis (wt%) (d.a.f)								
C 54.7	57.6	59.6	67.0	68.0	72.2	74.3	75.2	79.1
H 5.7	5.8	5.7	5.5	5.4	5.5	5.5	5.4	5.4
N 1.1	2.2	1.7	1.7	1.6	1.6	1.7	1.7	1.7
O (diff.) 38.0	33.8	32.3	25.1	24.3	20.0	17.9	17.2	13.3
S 0.5 Yield of solid products	0.7	0.6	0.6	0.6	0.6	0.5	0.5	0.5
(Y) (wt%) (d.b) - Calorific value (CV)	77.7	77.6	65.3	65.2	61.2	53.9	50.4	48.3
(kJ.kg^{-1}) (d.b) 21,527	20508	22427	25008	25836	27985	28299	29296	30047
(ECV) (kJ.kg^{-1}) 17307	16727	18263	22436	23268	25457	26242	27264	28233

d.b = dry basis; *a.r* = as received basis; *d.a.f* = dry ash free basis; *diff.* = differences

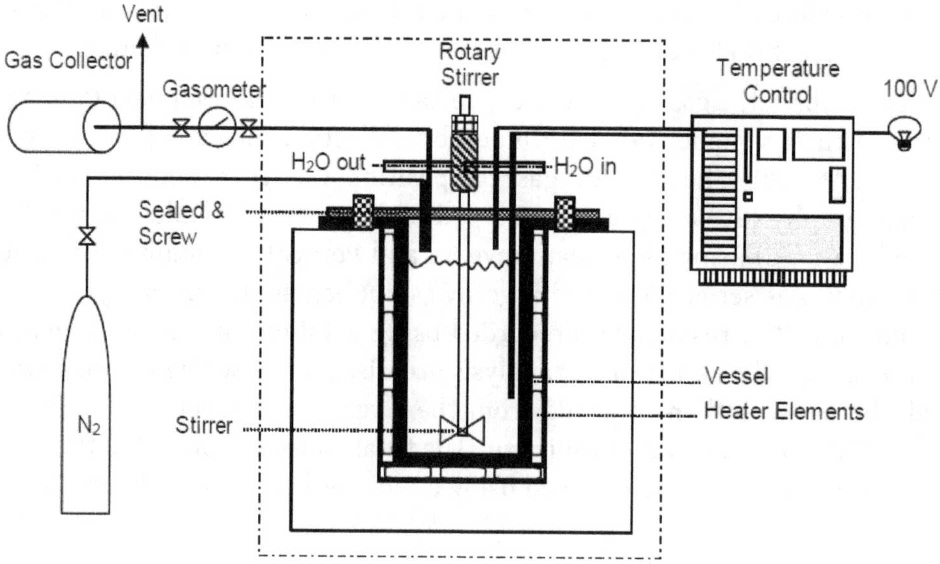

Fig. 1. Schematic figures of hydrothermal batch type reactors.

2.3 Analysis

The liquid product was filtered through a sheet of Advantec 0.45 μm pore size membrane filter prior to analysis. After dilution by a factor of 1000 using ultra-pure water, the total organic carbon (TOC) content and total inorganic carbon (TIC) content of the liquid product was determined using a Shimadzu TOC-5000A VCsH TOC analyzer. The organic compounds in the liquid product were identified and quantified by GC-MS (Agilent 6890N and JEOL Jms- Q1000GC (A)) using a J&W Scientific methyl silicon capillary column measuring 0.32 mm x 60 m. The split ratio was 99 and the column temperature was maintained at 40° C for 3 minutes, followed by an increase of 15 °C/min. to 250° C, which was maintained for 10 minutes. The composition of the sugar compounds in the liquid products was determined by high- performance liquid chromatography (HPLC) using a JASCO RI-2031 refractive index detector and Shodex KS-811, with 2 mM HClO4 applied at 0.7 mL/min. as the eluent. The results of liquid products content analysis are discussed elsewhere.

The elemental composition of the raw peat and solid product was determined using an elemental analyzer (Yanaco CHN Corder MT-5 and MT-6). Additionally, proximate analysis (based on JIS M 8812) total sulfur analysis (based on JIS M 8819) and calorific analysis (based on JIS M 8814) were conducted separately. The gross calorific value (CV) was measured using the bomb calorimetric method and the effective calorific value (ECV) of the sample at a constant pressure was determined based on JIS M 8814, which is followed by ISO 1928. The equilibrium moisture content of the dried solid product was further analyzed while maintaining their moisture contents according to JIS M 8811. Briefly, an aliquot of the sample was placed inside a desiccator containing saturated salt solution and then measured rapidly using a moisture content analyzer (Sartorius MA 150).

The primary components and the chemical structure of the raw peat and the solid product were further analyzed by Fourier transform infrared spectroscopy (FTIR) (JASCO 670 Plus) using the Diffuse Reflectance Infrared Fourier Transform Spectroscopy (DRIFTS) technique and the JASCO IR Mentor Pro 6.5 software for spectral analysis. The cross polarization/magic angle spinning (CP/MAS) 13C NMR spectrum of raw peat and the solid product was measured using a solid state spectrometer (JEOL CMX-300). The measurement conditions were as follows: spinning speed in excess of 12 kHz, contact time of 2 ms, pulse repetition time of 7 s and scan number of 10,000. Chemical shifts are in ppm referenced to hexamethylbenzene. The curve fitting analysis of the spectrum was conducted using the Grams/Ai 32 Ver. 8.0 software (Galactic Industries Corp., USA).

2.4 Sequential extraction of peat bitumen, plant constituents and humic substances

After drying at room temperature, the raw peat was milled and sieved through an 80-mesh screen, after which the plant constituents (hemicelluloses, cellulose and lignin), peat bitumen, humic substances (humic acid (HA), fulvic acid (FA) and humin (Hm)) and other insoluble contents were fractionated using methods that have been previously described. Briefly, the peat bitumen (benzene/ethanol-soluble) was extracted in a Soxhlet apparatus for 8 hours with a mixture of benzene and ethanol (4:1 vol./vol.). The bitumen-free peat was then extracted in ultra-pure water at 100° C for 5 hours to obtain the water-soluble materials. Next, the samples were dried at room temperature, after which the residual peat was sequentially extracted with 2% HCl at 100 °C for 5 hours to obtain the hemicelluloses. The samples were then incubated in 72% H_2SO4 at room temperature for 4 hours, after which the concentration of H_2SO4 was adjusted to 4% by the addition of ultra-pure water and the samples were heated at 100 °C for 3 hours to obtain the cellulose.

The residual peat was then correlated with the humic substances, lignin and other insoluble compounds. To accomplish this, the humic substances were removed by acid and alkali extraction. Briefly, the Air-dried peat was washed twice with 0.01 M HCl and then treated with NaOH at pH 13.5 for 48 hours, which gave a supernatant fraction (humic and fulvic acids), and a fraction that contained the humin and other insolubles. All fractions were separated by filtration and centrifugation.

3. RESULTS AND DISCUSSION

3.1 Sequential extraction of raw peat and solid products

The liquid content of the raw peat was 86.9 wt.%, which corresponded to the moisture content, and the products increased from 87.3 wt.% to 89.4 wt.% as the temperature increased. As the temperature increased, the gas products content increased from 1.6 wt.% to 4.8 wt.%. The increase in the liquid and gas products in response to increasing temperature suggest that dewatering and decomposition occurred during the process. Takahashi peat that contained most of the organic constituents of the original plant materials were least decomposed and peatification occurred shortly. The amounts and concentration of plant constituents, peat bitumen (benzene/ethanol soluble) and humic substances in Takahashi moss peat are described in Table 2.

Table 2. Humic substances, plant constituents and bitumen of Takahashi moss peat.

Humic substances, insoluble fractions and lignin	(wt.%) dry-base
Humic and fulvic acids (HA and FA)	0.9
Humin (Hm), insoluble fractions and lignin	4.2
Carbohydrates	
Hemicelluloses	30.7
Cellulose	40.7
Extracted peat bitumen (benzene/ethanol soluble)	9.0
Water soluble compounds	14.5

Extracted peat bitumen can be formed during natural decomposition and accounted for 9.0 wt.% of the raw peat, suggesting that the extracted materials correspond to the wax like hydrophobic formations. The hemicellulose and cellulose content was 30.7 wt.% and 40.7 wt.%, respectively. The high levels of carbohydrate indicate that the plant constituents still remained and were decomposed during peatification. The materials from which Takahashi moss peat is formed consist of cattails and reed grass (in Japanese, gama and ashi, which differ greatly from the raw materials that lead to the formation of tropical peat. This difference may explain why Takahashi peat differs from Pontianak peat. The water soluble content of the raw peat was 14.5 wt.%. The total humic substances, lignin and other insoluble fractions comprised 5.1 wt.% of the raw peat, of which HA and FA comprised 0.9 wt.%.

Figure 2 shows the effect of the processing temperature on the contents of peat bitumen, water soluble compounds and carbohydrates in the solid products. All other compounds consisted of char and other insoluble materials. The peat bitumen content in the solid products ranged from 4.7 to 26.0 wt.%, which suggests that peat bitumen formed in solid products formed in response to hydrothermal treatment. The peat bitumen decreased at low temperatures (150°C to 250°C), then increased slightly at 250°C and continued to increase with increasing temperature. Under supercritical conditions, the peat bitumen decreased slightly when compared to the peat bitumen content at 350°C, which may have been due to the intensive decomposition of organics in raw peat during the formation of gaseous and liquid products as a result of the supercritical reaction. The water soluble content accounted for 14.5 wt.% of the raw peat and 6.2 wt.% of the product formed at 380°C, indicating that this fraction decreased with increasing temperature. Both carbohydrates (hemicellulose and cellulose) decreased with increasing temperature and were no longer present in products formed at 300°C and above. Hydrothermal dewatering was highly affected by the decomposition of carbohydrates at 150°C to 270°C, suggesting that organics containing polysaccharides will be obtained in the liquid products.

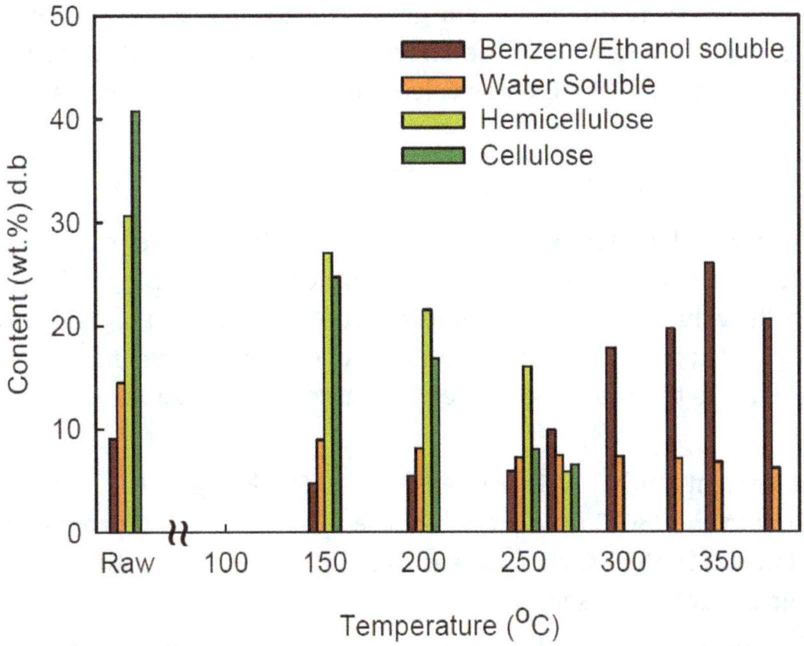

Fig. 2. The effect of temperature on the concentration of peat bitumen (benzene/ ethanol soluble), water soluble compounds and carbohydrates in solid products

3.2 Properties of solid products

Table 1 show the effects of temperature on the yield and moisture contents of solid products, respectively. The yield decreased as the temperature increased. Specifically, the maximum solid product yield was 77.7 wt.% at the lower temperature of 150°C, while the minimum yield was 48.3 wt.% at the supercritical temperature (380°C). The decrease in the yield of the solid product was due to the extensive thermal decomposition of the raw material into liquid and gaseous products. Dewatering also occurred during the decomposition reaction. Indeed, the minimum moisture content of the filtered solid products was 20.9 wt.% when the reaction was conducted at 380°C, while the maximum moisture content of 47.9 wt.% was obtained at 150°C. Moreover, the equilibrium moisture content of the solid products obtained at 380°C was 2.4 wt.%, while it was 13.3 wt.% at 150°C. As the temperature increased, the equilibrium moisture content of the solid products decreased when the products were maintained at a constant humidity (77–79%), indicating that the product may be hydrophobic when subjected to high temperatures. In general, the use of higher temperatures for the hydrothermal treatment of raw peat improves the dewaterability of the solid product; therefore, solid products produced at higher temperatures have better resistance against moisture adsorption when they are maintained at high humidity.

Lower equilibrium moisture contents resulted in the solid products having higher calorific values. In addition, the effective calorific value of peat fuel decreased as the level of moisture increased. The gross calorific value (CV) determined by the bomb calorimetric method and the effective calorific value (ECV) of the samples are shown in Table 1. The ECV was calculated using the equation as described on (JIS M 8814):

$$ECV = [CV - 212.2 \times H - 0.8 \times (O + N)] \times (1 - 0.01\,X) - 24.43\,X \quad (1)$$

where ECV is the effective calorific value, which is the net calorific value at a constant pressure of the equilibrium moisture content sample in kJ/kg, CV is the gross calorific value at a constant volume of the dry sample in kJ/kg, H, O and N are the hydrogen, oxygen and nitrogen contents of the dry sample in mass percentage, respectively, X is the moisture content in mass percentage for which the ECV is desired.

In the present study, CV increased from 20,508 kJ/kg in the 150°C product to 30,047 kJ/kg in the 380°C product, while the ECV of the 150°C product was 16,727 kJ/kg and that of the 380°C product was 28,233 kJ/kg. However, the yield decreased as the temperature increased.

In addition, the fixed carbon content increased from 31.9 wt.% to 62.8 wt.%, and the volatile matter decreased from 68.1 wt.% to 37.2 wt.% as the temperature increased. The ash content of the raw peat and solid products also increased from 4.4 wt.% and 8.2 wt.%. All of the solid products had lower levels of volatile material than the raw peat. Moreover, the chemical variations in the C, H, N and O contents of the solid products following the hydrothermal reaction of peat at different temperatures were also very interesting. Hydrothermal treatment decomposed the raw peat, which resulted in the oxygen content decreasing from 38.0 wt.% to 13.3 wt.% as the temperature increased. These findings suggest that oxygen loss corresponds to dewatering and the decreased yield of solid product.

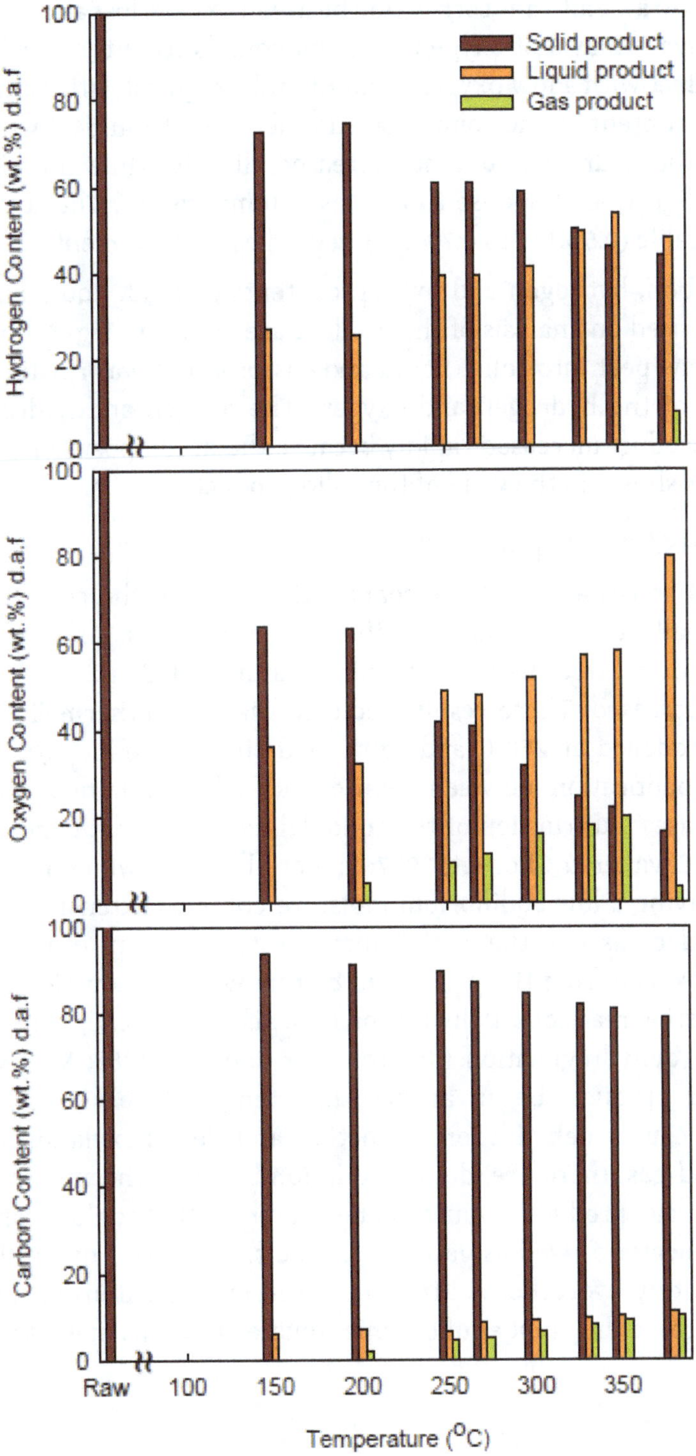

Fig. 3. The effect of temperature on the conversion of C, H and O of raw peat into solid, liquid and gas products.

There was also a significant correlation between oxygen loss and the calorific value. The extensive removal of oxygen-rich compounds from the raw peat resulted in a solid product with a low oxygen content and a high calorific value. Additionally, the carbon content of the solid products increased from 54.7 wt.% to 79.1 wt.%. Moreover, the hydrogen content decreased slightly while the nitrogen content increased slightly in response to increased temperature. The sulfur content was relatively stable (0.5 wt.% to 0.7 wt.%), regardless of treatment.

The carbon, hydrogen and oxygen contents of solid, liquid and gas products calculated based on analysis of the product are shown in Fig. 3. Conversion of the carbon in raw peat into solid, liquid and gas products was relatively slower than conversion of the hydrogen and oxygen. The oxygen and hydrogen content in the liquid product increased rapidly because the decomposition of peat increased suddenly, as shown in the yield of the solid products.

3.3 Coalification properties of solid products

Figure 4 shows a plot of the coalification band of the cold climate peat and hydrothermally upgraded and coalified solid products. As the temperature of the hydrothermal treatment increased, the atomic H/C and O/C ratio of the solid products decreased. These results indicate that hydrothermally upgraded solid products produced at 250°C and 380°C had similar atomic H/C and O/C ratios following coalification between lignite and sub- bituminous coals. Heat and pressure causes a disruption of the colloidal nature of peat during hydrothermal treatment (Cavalier & Chornet, 1977; Lau et al., 1987), which results in the solid products having a low equilibrium moisture content. Extensive losses of oxygen also led to decreases in the equilibrium moisture content of the solid products. Moreover, oxygen from the peat could be removed by reduction (loss of oxygen) and dehydration reactions. Dehydration followed decarboxylation, while reduction followed by dehydrogenation (Kalkreuth & Chornet, 1982; Van Krevelen, 1950) of the solid products began at the same temperature (150°C). Hydrothermal dewatering causes dehydration, reduction and decarboxylation of the product to liquid and gas; therefore, decarboxylation by hydrothermal treatment of raw tropical peat can produce organic soluble materials that contain carboxylic groups in the wastewater as well as gaseous products. Moreover, disruption of colloidal forms of peat by hydrothermal treatment can lead to extensive dehydration and possibly increase the number of organic soluble materials in wastewater.

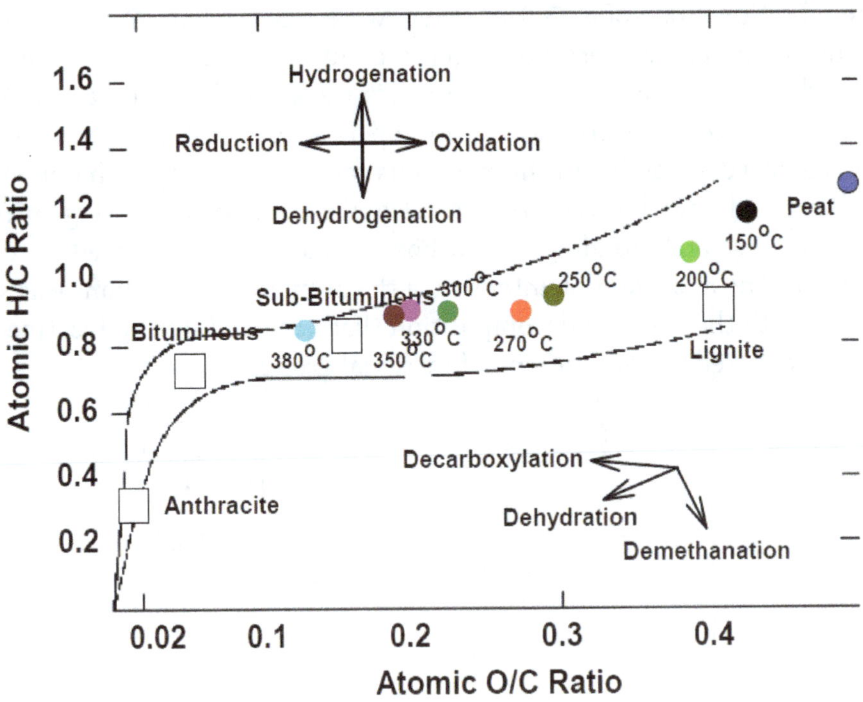

Fig. 4. Coalification band representing raw moss peat and hydrothermally coalified solid products.

3.4 FTIR results of solid products

Figure 5 shows the FTIR spectra of the raw peat and solid products. Assignments of the peaks in each spectrum of the main functional groups were conducted using the JASCO IR Mentor Pro 6.5 software and several publications (Kalkreuth & Chornet, 1982; Van Krevelen, 1950; Orem et al., 1996; Painter et al., 1981; Ibarra & Juan, 1985; Ibarra et al., 1996; Xuguang, 2005). Examination in a range of 3500–3300 cm^{-1} zone revealed a progressive lowering in relative peak intensity in –OH stretching mode of the solid products at 380oC. This peak is somewhat diminished in relative intensity, probably due to the dewatering of raw peat during hydrothermal treatment and the loss of hydroxyl-functionalized carbohydrates (Kalkreuth & Chornet, 1982). The spectrum in a range of 3000–2800 cm^{-1} showed existence of –CHx stretching mode in an aliphatic carbon. In addition, significant changes can also be observed in a range of 1800–1100 cm^{-1}. The carbonyl C=O stretching vibration mode of carboxylic acid at 1770 cm^{-1} was observed initially but the signal was almost completely disappeared with treatment at 350-380°C. The relative intensity of the ketone carboxyl band C=O groups was clearly observed at 1680 cm^{-1} and shifted slightly at higher temperatures. A peak at 1470–1511 cm^{-1} assigned to the stretching vibration mode of C=C in aromatic ring carbons (Kalkreuth & Chornet, 1982), was gradually sharpened in relative intensity with increase in temperature. Peak assigned to the bending vibration mode of C–O–R in ethers were observed at 1080 cm^{-1}, and with increasing temperature, were no longer present at 270°C. Peak assigned to the bending vibration mode of -C-H in aromatics was also observed at 900–700 cm^{-1} during the hydrothermal process at all temperatures. Peak assigned to aromatic nuclei CH at 880 cm^{-1} tended to sharpen in its relative intensity with temperature. Dehydration and decarboxylation were also affected, with lowering in relative intensity of OH stretch bonding and carboxyl groups being observed in response to treatment. The relative peak intensity of the aromatic ring carbons was sharply observed due to the thermal decomposition that occurred during treatment, and this greatly affected the other carbon functional groups.

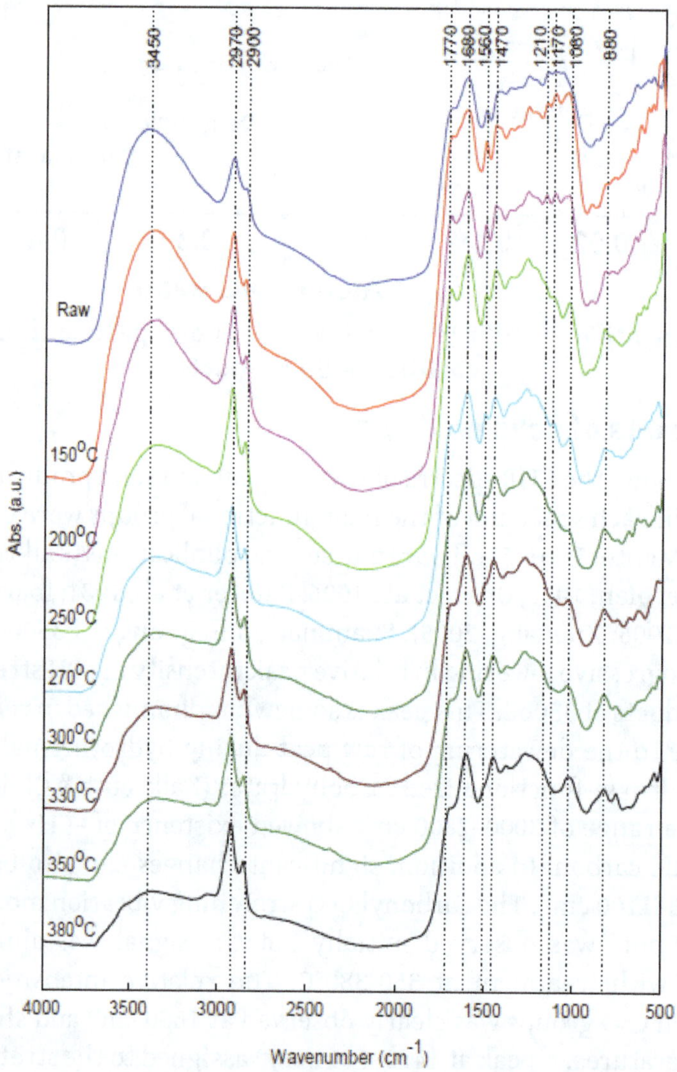

Fig. 5. FTIR spectra of raw peat and solid products produced at all processing temperatures.

3.5 13C NMR results of solid products

The 13C NMR spectra of the carbon-functional groups of raw peat and hydrothermally treated solid products produced at different treated temperature are shown in Fig. 6. Determination and assignment of the peak area distribution of carbon-functional groups was based on several publications (Orem et al., 1996; Hammond et al., 1985; Freitas et al., 1999; Yoshida et al., 1987; Yoshida et al., 2002). The strongest peak in the 13C NMR spectrum was at 74–76 ppm, which corresponds to methoxyl carbons (OCH3) may have been related to the presemce of carbohydrate carbons (i.e., hemicelluloses, cellulose). These were confirmed on Section 3.1 that the major components of Hokkaido peat are hemicelluloses and cellulose for about 71.4 wt.%. Secondly, the peaks at 30–32 ppm in raw peat containing aliphatic carbons (CHx (CH_2 and CH3)), which was likely due to the occurrence of humic acids and related substances (i.e., humic substances). Peat contains the most important organic fraction in nature, humic substances, which are composed of humic acid (HA), fulfic acid (FA) and humin (Hm) (Cavalier & Chornet, 1977). Lignin, cellulose and hemicellulose decreased as the humification of peat increased. In addition, the peak areas of the aliphatic carbons (CHx) decreased progressively as the temperature increased. The spectrum of raw peat contained a peak area at 56–59 ppm and 64–65 ppm, which may have been due to ether carbons in lignin and cellulose in the 13C NMR spectrum. These results suggest that most organic constituents in the original plant material were least biodegraded, decomposed during peatification. These peaks decreased with increasing temperature and the ether carbon peak was nearly completely gone at 330°C. The area at 74□76 ppm, which corresponds to methoxyl carbons (OCH3) may have been related to the presemce of carbohydrate carbons. This peak area decreased with increasing temperature, eventually decreasing to almost undetectable limits. These findings indicate that carbohydrate carbons were decomposed easily by hydrothermal treatment. The peak representative of aromatic carbons bound to the hydrogen (Ar–H), aromatic non-oxygenated carbon (Ar–C) and aromatic oxygenated carbon (Ar–O) were observed at the area of 100–106 ppm, 127–130 ppm and 151–155 ppm respectively. Furthermore, increasing the temperature of the hydrothermal treatment resulted in an increase in relative area intensity of the aromatic carbons. Finally, the peak area of carboxyl carbons (COOH) in the region of 171–180 ppm and carbonyl carbon (C=O) peaks at 195–200 ppm were observed in the spectrum of raw peat.

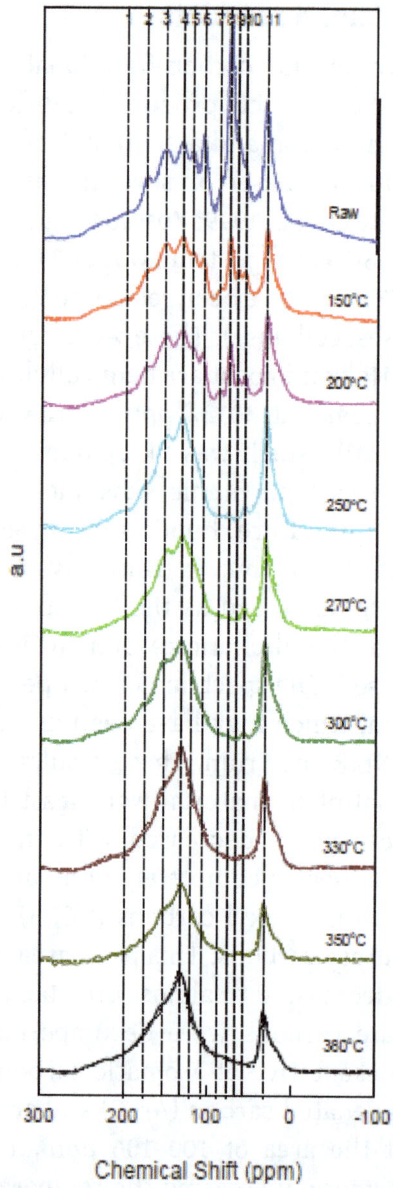

1: C=O; 2: COOH; 3: Ar–O; 4: Ar–C; 5: Ar–H; 6, 7, 8, 9 and 10: methoxyl carbons OCH3; 11: aliphatic carbons CHx (CH_2 and CH3)

Fig. 6. 13C NMR spectra of raw peat and solid products produced at all processing temperatures.

4. CONCLUSION

In this Chapter, the effectiveness of hydrothermal upgrading and dewatering of Hokkaido cold climate peat was evaluated at temperatures ranging from 150°C to 380°C, a maximum final pressure of 25.1 MPa and a residence time of 30 minutes.

The hydrothermally dewatered peat fuel product had a significantly higher ECV than raw peat, with the raw peat having an ECV of 17,307 kJ/kg and the products having ECV values ranging from 16,727 kJ/kg to 28,233 kJ/kg. Hydrothermal dewatering may also have impacted the extensive dehydration process by causing a significant loss in the oxygen content. Additionally, the carbon content of the solid products increased from 54.7 wt.% to 79.1 wt.% as the temperature increased. The hydrothermally upgraded peat fuel also had an equilibrium moisture content that ranged from 2.4 wt.% to 13.3 wt.%. A significant loss of oxygen could result in the formation of solid products with low equilibrium moisture.

An application of FTIR and 13C NMR spectroscopy on hydrothermal coalification could determine the decomposition of organic compounds in peat at different treated temperature. increasing the temperature of the hydrothermal treatment resulted in an increase in relative area intensity of the aromatic carbons bound to the hydrogen (Ar–H), aromatic non- oxygenated carbon (Ar–C) and aromatic oxygenated carbon (Ar–O). These mean and also correspond to the increasing of aromaticity as well as coalification degree.

5. REFERENCES

1. Björnbom, E.; Olsson, B. & Karlsson, O. (1986). Thermochemical refining of raw peat prior to liquefaction. Fuel, Vol.65, pp. 1051-1056.
2. Björnbom, P.; Granath, L.; Kannel, A.; Karlsson, G.; Lindstrijm, L. & Björnbom, EP. (1981).
3. Liquefaction of Swedish peats. Fuel, Vol.60, pp. 7–13.
4. Cavalier, JC. & Chornet, E. (1977). Conversion of peat with carbon monoxide and water.
5. Fuel, Vol.56, pp. 57–64.
6. Freitas, JCC.; Bonagamba, TJ. & Emmerich, FG. (1999). 13C High-resolution solid-state NMR study of peat carbonization. Energ Fuel, Vol.13, pp. 53–59.
7. Hammond, TE.; Cory, DG.; Ritchey, M. & Morita, H. High resolution solid state 13C n.m.r. of Canadian peats. Fuel 1985;64:1687–1695.
8. Ibarra, JV. & Juan, R. (1985). Structural changes in humic acids during the coalification process. Fuel, Vol.64, pp. 650–656.
9. Ibarra, JV.; Muñoz, E. & Moliner, R. (1996). FTIR study of the evolution of coal structure during the coalification process. Org Geochem, Vol.24, pp. 725-735.
10. Japanese Industrial Standards Committee. JIS M 8814. (2003). Coal and coke. Determination of gross calorific value by the bomb calorimetric method, and calculation of net calorific value. Japanese Standards Association. Tokyo

11. Kalkreuth, W. & Chornet, E. (1982). Peat hydrogenolysis using H_2/CO mixtures: Micropetrological and chemical studies of original material and reaction residues. Fuel Process Technol, Vol.6, pp. 93–122.

12. Lau, FS.; Roberts, MJ.; Rue, DM.; Punwani, DV; Wen, WW. & Johnson, PB. (1987). Peat beneficiation by wet carbonization. Int J Coal Geol, Vol.8, pp. 111–121.

13. Mursito, AT.; Hirajima, T. & Sasaki, K. (2010). Upgrading and dewatering of raw tropical peat by hydrothermal treatment. Fuel, Vol.89, pp. 635–41.

14. Mursito, AT.; Hirajima, T.; Sasaki, K. & Kumagai S. (2010). The effect of hydrothermal dewatering of Pontianak tropical peat on organics in wastewater and gaseous products. Fuel, Vol.89, pp. 3934–3942.

15. Noto, S. (1991). Peat engineering handbook. Civil Engineering Research Institute of Hokkaido Development Bureau

16. Orem, WH.; Neuzil, SG.; Lerch, EL. & Cecil, CB. (1996). Experimental early-stage coalification of a peat sample and a peatified wood sample from Indonesia. Org Geochem, Vol.24, pp. 111–125.

17. Painter, PC.; Snyder, RW.; Starsinic, M.; Coleman, MM.; Kuehn, DW. & Davis, A. (1981). Concerning the application of FT-IR to the study of coal: a critical assessment of band assignments and the application of spectral analysis programs. Appl Spectrosc, Vol.35, pp. 475–485.

18. Spedding, PJ. (1988). Peat. Fuel, Vol.67, pp. 883–900.

19. Van Krevelen, DW. (1950). Graphical-statistical method for the study of structure and reaction processes of coal. Fuel, Vol.29, pp. 269–284.

20. World Energy Council (WEC). (2001). Survey of Energy Resources. Peat.

21. Xuguang, S. (2005). The investigation of chemical structure of coal macerals via transmitted- light FT-IR microspectroscopy. Spectrochim Acta, Vol.62, pp. 557–564.

22. Yoshida, T. & Maekawa, Y. (1987). Characterization of coal structure by CP/MAS carbon-13 NMR spectrometry. Fuel Process Technol, Vol.15, pp. 385–395.

23. Yoshida, T.; Sasaki, M.; Ikeda, K.; Mochizuki, M.; Nogami, Y. & Inokuchi, K. (2002) Prediction of coal liquefaction reactivity by solid state 13C NMR spectral data. Fuel, Vol.81, pp. 1533–1539.

SECTION-2
POLYMERS AND BIOPOLYMERS

9 Chapter

FTIR – AN ESSENTIAL CHARACTERIZATION TECHNIQUE FOR POLYMERIC MATERIALS

1. INTRODUCTION

Infrared spectroscopy is an essential and crucial characterization technique to elucidate the structure of matter at the molecular scale. The chemical composition and the bonding arrangement of constituents in a homopolymer, copolymer, polymer composite and polymeric materials in general can be obtained using Infrared (IR) spectroscopy (Bhargava et al. 2003)

The FTIR spectrometers obtain the IR spectrum by Fourier transformation of the signal from an interferometer with a moving mirror to produce an optical transform of the infrared signal. Numerical Fourier analysis gives the relation of intensity and frequency, that is, the IR spectrum. The FTIR technique can be used to analyze gases, liquids, and solids with minimal preparation. (Lee, 1977).

The spectrometer may operate in transmission or reflection, but also in attenuated total reflection (ATR) mode, which have been widely used during the last two decades. The transmission mode is very suitable for quantitative analysis, Since the main parameters to apply on the Beer-Lambert law are known or easily estimated. The reflection mode is used when polymer is not well dissolved at room temperature, and a film or pellets of the sample is characterized. Such is the case of polyolefins and some engineering polymers like PET.

The ATR objective has a polished face of diamond, germanium or zinc selenide (ZnSe) that is pressed into contact with the sample (Sawyer et al. 2008). Infrared reflection is attenuated by absorption within a surface layer a few micrometers deep. In this case, a good contact is required, but that is easy for most polymers, and the advantage is that no sample preparation is required. Therefore, powders, films, gels and even polymer solutions can be characterized.

The absorption versus frequency characteristics of light transmitted through a specimen irradiated with a beam of infrared radiation provides a fingerprint of molecular structure. Infrared radiation is absorbed when a dipole vibrates naturally at the same frequency in the absorber. The pattern of vibrations is unique for a given molecule, and the intensity of absorption is related to the quantity of absorber. In the IR region, each group has several and different patterns of vibration such as: stretching, bending, rocking, etc. The absorbance of these bands is proportional to their content based on the Lambert–Beer law. Because of the complexities of the structures (for example, polybutadiene unit has 1,4-cis, 1,4-trans and 1,2-vinyl microstructure while polyisoprene unit has other additional 3,4-microstructure), the general methods for quantitative analysis require external standards, which are usually characterized by primary technique like Nuclear Magnetic Resonance (Zhang et al, 2007). Nevertheless, it is possible to characterize polymers, quantitatively, using FTIR as it will be discussed later.

Thus, infrared spectroscopy permits the determination of components or groups of atoms that absorb in the infrared at specific frequencies, permitting identification of the molecular structure (Bower, 1989; Koenig, 2001). These techniques are not limited to chemical analysis. In addition, the tacticity, crystallinity, and molecular strain can also be measured. Copolymer compositions can be determined as block copolymers absorb additively, and alternating copolymers deviate from this additivity due to interaction of neighboring groups.

The Aim of this chapter is to present some examples in the context of different areas, in order to remark the importance of FTIR to determine different aspects as identification of functional groups, to elucidate mechanisms of reaction or even to determine the time of reaction.

The track of a reaction in real time when xerogels are synthesized, interaction of different materials like chitin with polyurethane during synthesis of biocomposite, identification of functional groups attached to nanotubes during their modification and establishment of microstructure of polybutadiene during its hydrogenation are the main examples of the use of FTIR, even being semi-quantitative, quantitative or qualitative analysis, that will be presented and analyzed.

The synthesis of xerogels involves basically two steps, hydrolysis and condensation. The follow of such steps is important to establish the time needed to change process conditions and optimize the xerogel. Even when FTIR analysis is a qualitative technique, this analysis could be used as a semi-quantitative indirect measure of the variation of functional groups, during hydrolysis or condensation reactions.

Biocomposites are very fascinating materials Since they offer characteristics of two or more different materials, in order to have very specific features that would be practically impossible to obtain by every single material of biocomposite. Chitin is an abundant biopolymer obtained from shrimp, insects and some vegetal species. This material is capable to remove some contaminants like fluoride from water. Nevertheless, in order to improve the mechanical characteristics of chitin, in order to be applied in water treatment in real conditions, it must be supported. Polyurethane is a very versatile polymer due to its chemical structure. During its synthesis, interactions between functional groups take place in order to create the urethane group. The synthesis of biocomposite must bear in mind that interaction between compounds is essential to create a mechanical and chemical resistant material. FTIR with ATR analysis was carried out to characterize a biocomposite based on chitin and polyurethane, demonstrating that interaction between them occurs.

The characterization of carbon nanotubes usually is carried out using Raman spectroscopy. Nonetheless, the use of FTIR allows the determination of functional groups when carbon nanotubes are chemically treated in order to improve its interaction with other entities like polymeric matrix, for example. In this sense, oxidized nitrogen-doped nanotubes were analyzed by FTIR, corroborating that functional groups are present in these carbon nanotubes.

2. XEROGELS

The xerogels are materials that have precise microstructure and they are synthesized by sol- gel synthesis. Two steps are implied in the sol-gel process, hydrolysis and condensation, which are very sensitive to reaction conditions, being faster or slower and giving xerogels with different characteristics, in terms of structure, surface area, and porosity (Fidalgo and Ilharco, 2005).

In this sense, the trace of each reaction with time, hydrolysis and condensation, is important in order to establish the end and beginning of each of these reactions. FTIR is a very useful characterization technique to follow reactions during sol-gel synthesis, Since there are specific functional groups that participate in the sol-gel process.

In the case of hydrolysis, silanol is produced from reaction between silane and water molecules. The silanol exhibits a strong absorbance around 945 cm^{-1}, which can be followed during the reaction.

The condensation reaction, where the silanol groups react between them, creating siloxane groups, gives an intense absorbance band at 1080 cm^{-1}. Therefore, it is also possible to see variation of such band with time (Andrade et al., 2010).

2.1 Kinetics by FTIR

The main functional group that is produced during hydrolysis of silanes is silanol, which has a characteristic absorbance band at 945 cm^{-1}. The hydrolysis reaction, usually, is carried out in presemce of some alcohol, depending of the silane. For example, if the hydrolysis is carried out with tetra methoxi silane, then the chosen alcohol is methanol, or with tetra ethoxy silane, the preferred alcohol is ethanol.

This is because the by-product of hydrolysis reaction is an alcohol. Therefore, the intense signals in the infrared bands located at 945 and 880 cm^{-1}, corresponding to silanol and ethanol (for example) can be followed during hydrolysis reaction.

In the Figure 1 it is shown the spectra for hydrolysis of tetra ethoxy silane (TEOS) in presemce of ethanol at different reaction time. The absorbance band at 945 cm^{-1} can be assigned to the vibration of the Si-O from the silanol residual groups [Si-(OH)4], whereas the band at 880 cm^{-1} can be assigned to OH groups from ethanol. The time at which no more increment of the absorbance band of silanol and OH groups occurs was established as the time at which the hydrolysis reaction was completed.

From Figure 1, it can be seen that the absorbance bands corresponding to silanol and OH, from ethanol, increased up to 30 min. After this time, there was no increase of these bands (spectra not shown), which can indicate that the hydrolysis reaction has been completed.

Determination of the time needed to carry out the hydrolysis reaction is critical Since the knowledge of the complete conversion to silanol groups is required to promote a more efficient condensation reaction.

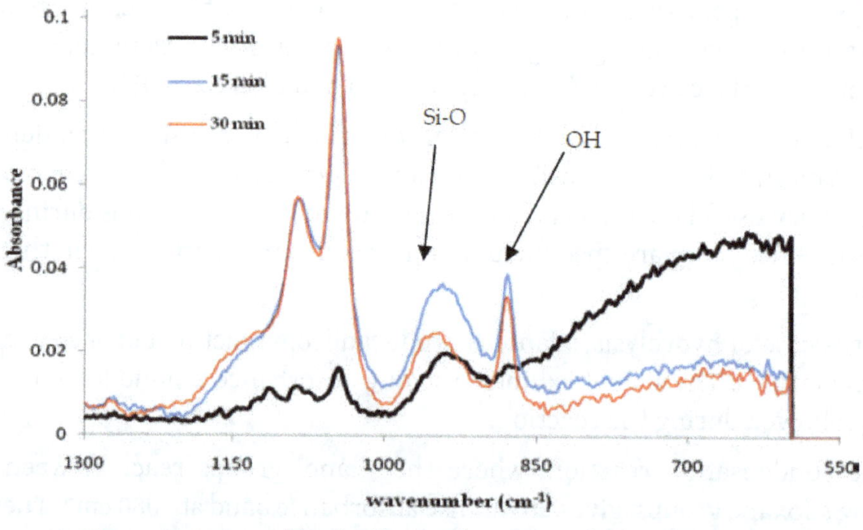

Fig. 1. Spectra of hydrolysis of TEOS at different times.

In the case of the condensation reaction, the silanols are reacting, between them, to produce siloxane groups. The siloxane group has a characteristic absorbance band, corresponding to the vibration mode (Si-O-Si) at 1080 cm^{-1}, that can be followed with time. In fact, from Figure 1, it is noted that after 15 minutes of reaction, there is an important increment of the band at 1080 cm^{-1} compared to the spectrum at 5 minutes of reaction. It is important to have in mind that hydrolysis and condensation can take place at same time, but condensation is catalyzed with alkaline chemicals, as amines. The fact that the band at 1080 cm^{-1} had not an increment, from 15 to 30 minutes, could be attributed to the absence of alkaline catalyst in the media reaction. It was shown (Andrade et al., 2010) that the absorbance was increased when an alkaline catalyst was added to the media reaction, and the band corresponding to silanol (945 cm^{-1}) was reduced with time.

3. BIOCOMPOSITES

Biocomposites i.e. composite materials comprising one or more phase(s) derived from a biological origin (Fowler et al., 2006), are very interesting materials Since they offer characteristics of two or more different materials, in order to have very specific features that would be practically impossible to obtain by every single material of biocomposite. Many types of biocomposites have been proposed, depending of the application objective of each material. Water treatment researchers have recently proposed the use of biocomposites as adsorbent materials, in order to remove many different contaminants from water. Synthesis of biocomposites to use them as adsorbents has been necessary, for example, because biosorbents (adsorbents from a biological origin) usually have poor physical/chemical resistance. The roll of the matrix in biocomposite is to reinforce the biosorbent, by establishing physical and/or chemical links between both phases.

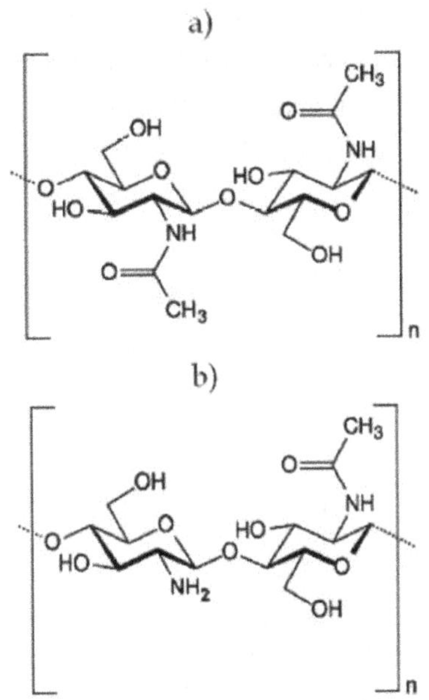

Fig. 2. Chemical structures of a) chitin and b) chitosan (Elnashar, 2010).

One of the most studied biosorbent is chitin, which is an abundant biopolymer found in crustaceans, insects and fungus. This biopolymer is commercially purified by alkaline deproteinization, acid demineralization and decoloration by organic solvents of crustaceans wastes (Pastor, 2004). An additional stronger alkaline treatment of chitin produces deacetylated chitin. If the acetylation degree (DA) decreases at 39% or less, the biopolymer is named chitosan. Hence, the DA of chitin is variable and depends on the process conditions (alkali concentration, contact time, temperature, etc.), which produces DA values from 100 to 0%. Because of this, chitin is known as the biopolymer which has a DA from 100 to 40%; likewise, when the chitinous biopolymer has DA lower than 40%, the biopolymer is named chitosan. Chitosan is, therefore, a biopolymer with structure very similar to that of chitin (see Figure 2); however, chitosan solubility is much greater, especially in acid mediums.

Chitin is capable to remove many contaminants from water. A wide variety of studies have demonstrated the ability of chitin to uptake substances as metals, anions and organics.

Nevertheless, in order to improve the mechanical characteristics of chitin to be applied for water treatment in real conditions, it must be supported. The use of polymers as supporting matrix of biosorbents has given encouraging results. In this sense, polyurethane has been one of the most used polymers.

Polyurethane includes a group of polymers derived from the isocyanates, organic compounds which have the isocyanate group (-N=C=O) in their structures and, owing to this, are very reactive. Polyurethane shows advantages as easy handling and the possibility to obtain malleable and resistant biocomposites. The Figure 3 shows a general polyurethane prepolymer structure and how isocyanate group reaction gives rise to different type of chemical linkages.

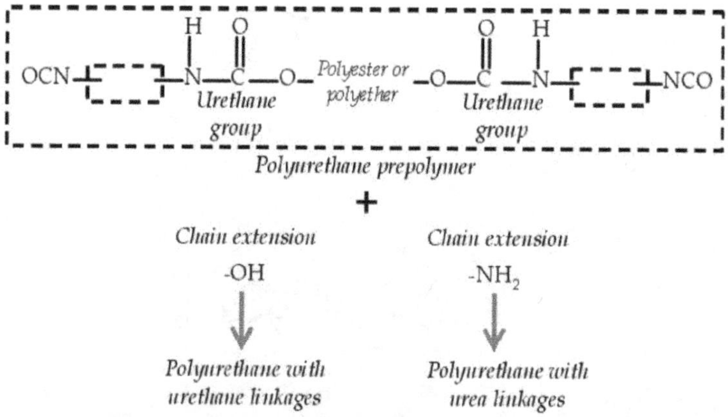

Fig. 3. General chemical structure of polyurethane prepolymer (modified from Hepburn, 1982).

Chemical reaction between the isocyanate and a primary amine group produces the urea group, whereas the reaction between isocyanate and hydroxyl group forms the urethane group. Both urea and urethane are very stable chemical groups, which make the polyurethane a polymer very useful in applications that must resist extreme conditions of temperature, friction and UV radiation (Hepburn, 1982). Due to the high reactivity of the isocyanate with the primary amine and hydroxyl groups, it is valid to assume that the polyurethane establish strong unions with biosorbents to form biocomposites. Hence, such unions make polyurethane a good support matrix for biosorbents. In fact, more and more studies have been recently published about the use of polyurethane as support of biosorbents as seaweed (Alhakawati & Banks, 2004; Zhang & Banks, 2006), moss, sunflower waste and maize plant (Zhang & Banks, 2006), bacteria (Vullo et al., 2008; Mao et al., 2010), fungi (Sudha & Abraham, 2003; Li et al., 2008) and chitin (Davila-Rodriguez et al., 2009). Most of these studies have focused on the biocomposite biosorption capacity; however, the interaction polyurethane-biosorbent has been practically not studied. Infrared spectroscopy is a good methodology to characterize such interaction.

3.1 Methods

One research has recently reported characterization of the polyurethane-chitin interaction in a novel fluoride biosorbent biocomposite, using attenuated total reflection Fourier transform infrared spectroscopy (FTIR-ATR) (Davila-Rodriguez et al., 2009). Such a biocomposite was produced by mixing chitin flakes with polyurethane (60:40 w/w) during the polymerization reaction, which was carried out under intense stirring. The biocomposite consisted, therefore, of chitin flakes covered by a thin film of polyurethane of approximately 1 mm thickness, according to different observations performed by scanning electron microscopy. The novel composite material showed greater chemical resistance compared to pure chitin, measured as a decrease of around ten times in the loss of mass when the material was submerged in an acid aqueous medium at pH 5 (from 19.6 to 1.5%). These results demonstrated the strong interaction between chitin and polyurethane on their contact surface. Nevertheless, FTIR-ATR methodology was used in order to get a better understanding of the biopolymer-matrix links formed during the biocomposite synthesis. For the FTIR-ATR analysis, a sample of the chitin-polyurethane biocomposite was used as produced. The FTIR-ATR instrument used was a Nicolet 6700 (Thermo Scientific) operating in the wavenumber range of 650 to 4000 cm^{-1}, with ZnSe crystal, 32 scans and 4 cm^{-1} resolution.

3.2 Results and discussion

Analysis of FTIR-ATR spectra showed that the biocomposite spectrum was very similar to that of polyurethane. In addition, it is important to point out that main bands of chitin spectrum were at the same wavenumber intervals that those of polyurethane (see Figure 4).

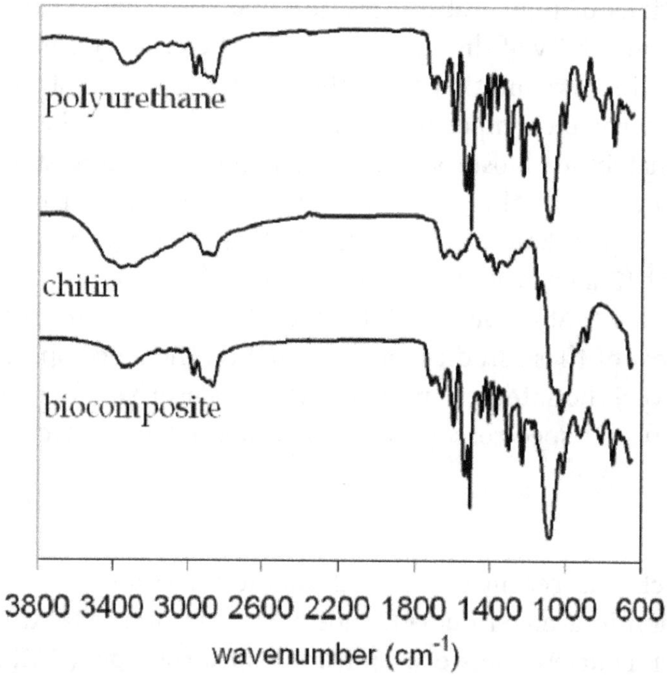

Fig. 4. FTIR-ATR spectra of chitin, polyurethane and chitin-polyurethane biocomposite (modified from Davila-Rodriguez et al., 2009).

Such an aspect contributed to produce the similarity between biocomposite and polyurethane spectra. This fact was predictable because chitin and polyurethane have similar chemical groups on their structures, including amine, amide and carbonyl. Nevertheless, some of the bands observed in the biocomposite spectrum were different in intensity compared to those of polyurethane and chitin. Bands located at 660, 1590 and 3300 cm^{-1}, corresponding to amine/amide groups, and the band located at 3350 cm^{-1}, corresponding to hydroxyl group, experimented appreciable changes in intensity. This phenomenon could be due to the chitin-polyurethane chemical interaction. As previously mentioned, primary amine and hydroxyl groups (both present in chitin structure) are able to react chemically with the polymer functional groups i.e. isocyanate, to form urea and urethane groups, respectively.

On the other hand, it is important to say that chitin-polyurethane interaction has an important physical component, Since attraction forces as Van der Waals interactions and hydrogen bonds are present at the interface chitin-polyurethane. This aspect could have caused that chitin-polyurethane chemical interaction was not strong enough to be more clearly observed by means of the FTIR-ATR analysis. In addition, chitin was added to the biocomposite in solid form, which limits the achievability of amine and hydroxyl groups of chitin surface due to a steric impediment. Other important aspect to point out is that polyurethane film thickness over the interface chitin-polyurethane could be irregular, which could difficult to reach the chitin-polyurethane interface by infrared light.

FTIR-ATR characterization of interface chitin-polyurethane could be improved by controlling the uniformity of polyurethane film thickness. This would be achieved by polymerizing the polyurethane in the necessary quantity to cover a high-sized chitin flake (e.g. 1 cm diameter), trying to produce a very thin film. As reported in literature, infrared light penetration depth can range from some hundred nanometers to several micrometers (Kane et al., 2009). Therefore, 1 μm thickness for polyurethane film is sufficient to infrared light reaches the chitin-polyurethane interface; this is part of future work in this research.

4. CARBON NANOTUBES: SYNTHESIS, PROPERTIES AND MODIFICATION

Carbon nanotubes (CNT) are key elements in nanotechnology. The structure of carbon nanotubes is depicted as a rolled segment of a graphene sheet, formed by linking each carbon atom to three equivalent neighbors by covalent bonds resulting in a hexagonal network. Carbon nanotubes can be structured by only one graphene sheet (single-wall carbon nanotubes, SWCNT) or by several coaxial graphene sheets (multiwall carbon nanotubes, MWCNT) (Dresselhaus et al., 2001). Synthesis of carbon nanotubes can be carried out by catalytic chemical vapor deposition (CVD) from carbon-containing gaseous compounds which decompose catalytically on transition-metal particles at temperatures lower than 1000 °C. CVD processes are becoming the major way for synthesizing carbon nanotubes in a controlled way (Loiseau et al., 2006). It is possible to modify and control the physicochemical properties of carbon nanotubes by doping processes, introducing either non-carbon atoms or molecules at small concentrations in the plane of the graphene lattice.

Doping carbon nanotubes with nitrogen atoms at lower concentrations than 6.5 wt% induce crystalline disorder in the graphene sheets, as well as an excess of electron donors on the nitrogen-rich areas may result in a more reactive structure compared to pure carbon nanotubes. For nitrogen-doped carbon nanotubes (N-MWCNT) two types of C–N bonds could occur: the first is a three-coordinated N atom within the sp2- hybridized network, with the presemce of additional electrons. The second type is the pyridine type (two-coordinated N) which can be incorporated in the nanotube lattice; an additional carbon atom is removed from the framework. N-MWCNTs have stacked-cone morphology (bamboo-type) and the degree of tubular perfection decreases as a result of the N incorporation into the hexagonal carbon lattice (Terrones et al., 2002). Moreover, properties of CNTs can also be modified by oxidation with thermal treatment or with acidic solutions (HNO_3, KMnO4+, H_2SO4, K2Cr[2]O7, H_2O2). Under acidic oxidation bonds of carbon nanotubes could follow several reactions, for example the formation of quinones which further evolve to different oxygen acidic groups, such as phenolic, lactonic and carboxylic groups. The attached oxygenated groups change the surface chemistry of carbon nanotubes, in particular their wetting behavior and also can be used to attach different chemical groups or to improve their chemical interaction with other substances (Hirsch, 2002; Niyogi et al., 2002). Studies reported previously about oxidation of CNTs in liquid phase suggest that defects at the tips and sidewalls are the most reactive sites to attach oxygen functional groups to the nanotubes structure. These bonded atoms pull the reactive C atom out of the base plane, reducing the curvature strain of nanotubes and creating holes and pores. Long periods of oxidation lead to the shortening and destruction of the nanotubes (Ago et al., 1999; Ovejero et al., 2006; Zhang et al., 2003).

Raman spectroscopy is widely used to evidence the enhancement in defects density along the walls of the carbon nanotubes when oxidation is applied. The main bands of carbon materials are the graphite band (G-band) and the disorder-induced phonon mode (D-band). The relative intensity between the D-band and G-band (ID/IG ratio) indicate a continuously covalent bonding of oxygenated groups along the walls of nanotubes, disrupting the aromatic system of π-electrons (Liu et al., 2004). Complementary information can be obtained with FTIR spectroscopy; it allows determination of functional groups attached to surface of CNTs. This tool was used in some studies carried out by our research group with N-MWCNT, MWCNT and SWCNT, which were oxidized with nitric acid solution (70%) at 80 ± 3 °C for 5 h, as Detailed by Perez-Aguilar et al., (2010). Oxidized carbon nanotubes were identified as ox-N-MWCNT, ox-MWCNT and ox-SWCNT, and their morphology was observed by electronic microscopy (Figure 5). As the oxidation of carbon nanotubes was carried out in liquid phase, oxygen containing functional groups were attached along their entire length. Reactions between carbon materials and nitric acid involve strong electrophilic species in solution, which form nitrogen oxides that eventually are reduced to n_2 through oxidation of carbon. Introduction of a small content of nitrogen has been reported, mainly due to the introduction of nitro (-NO2) groups (Chen and Wu, 2004; Zawadski, 1980).

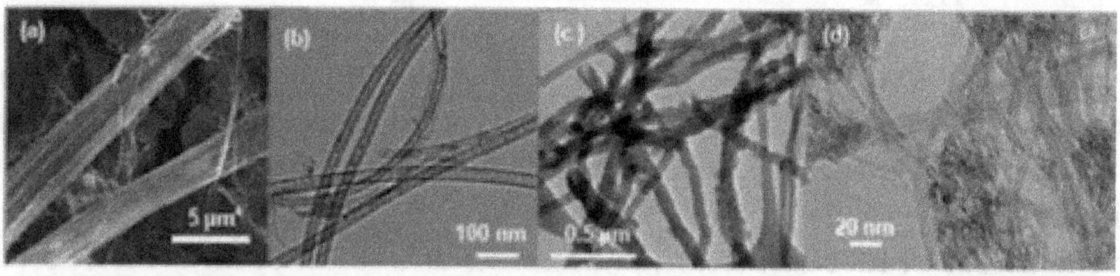

Fig. 5. Micrographs of carbon nanotubes observed by (a) SEM and (b) HRTEM for oxidized N-MWCNTs; HRTEM for (c) oxidized MWCNTs and (d) oxidized SWCNTs. Oxidation with nitric acid at 80 ± 3°C by 5 h

After the oxidation Van der Waals forces that kept CNTs aggregated were overcome by repulsion forces, as result of reducing the π-conjugation of nanotubes and enhancing the surface dipoles by oxygen-containing functional groups. This effect was monitored by determining the elemental composition of pristine nanotubes and partially oxidized CNTs.

An important increment in oxygen content was registered as the carbon content was reduced, caused by the destruction of graphene lattice by electrophilic reaction. The hydrogen content also increased in oxidized nanotubes, maybe as a result of the introduction of carboxylic groups. However, for N-MWCNTs the nitrogen content remained almost constant, probably these atoms might change from pyridinic-type or quaternary-type to an aliphatic form. Equilibrium could be established between the consumption of nitrogen atoms of the lattice and the attachment of nitrogen atoms in some sites of the lattice as nitro groups, as occurs when oxidizing some organic molecules or activated carbons with nitric acid (Zawadski, 1980; Chen and Wu, 2004).

4.1 FTIR characterization of oxidized carbon nanotubes

Functional groups attached to oxidized carbon nanotubes were identified by Fourier- transformed infrared spectroscopy by attenuated total Reflectance, ATR-FTIR in a Nicolet 6700 FT-IR spectrophotometer at 1068 scans, in the frequency interval of 4000 cm^{-1} to 650 cm^{-1} with resolution of 8 cm^{-1}.

The main functional group attached to oxidized carbon nanotubes was carboxylic. Comparison between spectra of pristine nanotubes and oxidized N-MWCNTs showed that the bands at 1444 cm^{-1}, 1373 cm^{-1} and 1251 cm^{-1}, attributed to vibration of MWCNTs and C- N bonding in N-MWCNTs, were overcome by several vibrations in carbonyl and carboxylic functionalities in the range of 1720 to 1250 cm^{-1} (Choi et al., 2004; Misra et al., 2007). A strong band appeared from 3600 cm^{-1} to 3300 cm^{-1} by the stretching of the bonding -OH of carboxylic group in oxidized nanotubes (Zawadski, 1980; Chen and Wu, 2004).

The nitro group (–NO2) is isoelectronic with the carboxylate ion group and their spectra are very similar, but a weak feature related to nitro group was observed at 1535 cm^{-1} for oxidized nanotubes (Zawadski, 1980). By other side, spectra obtained for three types of oxidized CNTs (Figure 6) showed broad features from 3400 cm^{-1} to 3000 cm^{-1} by the stretching of the bonding -OH in carboxylic and hydroxyl groups. Signals near 1710 cm^{-1} and 1685 cm^{-1} were of carbonyl vibrations in carboxyl bonding, a weak band about 1640 cm^{-1} of quinones, and the broad band from 1200 to 1000 cm^{-1} was attributed to single-bonded oxygen atoms such as phenols and lactones.

Two weak signals related to the nitro group were observed at 1538 and 1340 cm^{-1} for ox-N- MWCNTs. Similar spectra have been reported for oxidized multiwall carbon nanotubes; characteristic peaks were assigned to carboxylic, carbonyl, and hydroxyl groups (Wang et al., 2007). These results have probe that spectra carefully acquired by FTIR-ATR are a useful tool for identification of chemical groups attached to surface of carbon nanotubes chemically modified.

5. MICROSTRUCTURE OF STYRENE-BUTADIENE COPOLYMERS AND ITS HYDROGENATION

Elastomers like polybutadiene or thermoplastic elastomers like styrene-butadiene-styrene copolymers are widely used in diverse industries. However, the presemce of double bonds into their structures promotes degradation by heat or light, making them materials with low outdoor resistance. The hydrogenation of these materials is an effective method to improve their performance for outdoor or high temperatures applications (De Sarkar et al., 1997). In this context, the determination of microstructure (relative quantities of isomers), before and after the hydrogenation, is important Since chemical and mechanical resistances are related to it.

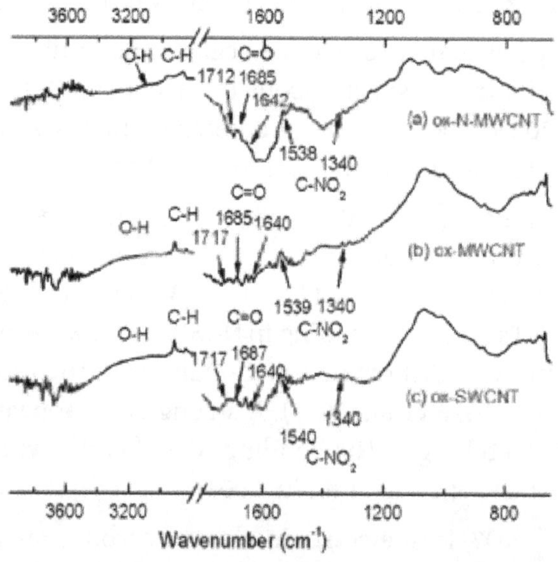

Fig. 6. FTIR spectra of (a) oxidized nitrogen-doped carbon nanotubes, ox-N-MWCNT; (b) oxidized multiwall carbon nanotubes, ox-MWCNT; (c) oxidized single-wall carbon nanotubes, ox-SWCNT, obtained by attenuated total Reflectance (ATR)

In the case of butadiene-containing polymers, the microstructure of the polybutadiene segment (i.e. the relative amount of 1,2 vinyl, 1,4-trans and 1,4-cis bonds) plays an important role in determining their thermal and mechanical properties, as well as their interaction with other materials in the production of composites, such as high-impact polystyrene, modified asphalt, pressure sensitive adhesives, etc. It is also known that anionic polymerization gives copolymers with well-controlled monomer distribution and microstructure, a narrow molecular weight distribution, and the possibility to have end group functionality (Escobar et al., 2000). Hydrogenation of styrene-diene copolymers has received special interest (Bhattacharjee et al., 1993) Since the double bond present in the diene segment of the elastomer is susceptible to thermal and oxidative degradation . Hydrogenation of styrene- diene block copolymers allows the production of thermo-oxidative resistant thermoplastics, and selective saturation of diene units (over the rigid units of styrene) permits to control mechanical and thermal properties. Thus, FTIR is essential to determine quantitatively the microstructure, using a liquid cell. Microstructure can be evaluated with time of hydrogenation by FTIR, allowing the establishment of the mechanism of hydrogenation for styrene-butadiene copolymers.

Most of styrene-butadiene copolymers hydrogenation studies have focused on global saturation of polybutadiene double bonds (i.e. 1,4-trans, 1,4-cis and 1,2-vinyl) and on obtaining high saturation percentages (>90%), however, only a few of them have made a distinction on the saturation of the different types of double bonds present in polybutadiene (Escobar et al., 2000).

The evaluation of microstructure and composition (relative quantity of polystyrene in the copolymer) in styrene-butadiene copolymers by FTIR has been conventionally carried out considering just two of three isomers, and the third isomer percentage has been obtained by difference. The previous characterization by RMN is essential in order to contrasts the results. However, if it is considered that evaluation of microstructure and composition, of such butadiene-based copolymers, by FTIR takes in account the Beer-Lambert law, expressed as:

$$A = K\,C\,L \qquad (1)$$

where:

A represents the absorbance at certain wave number,

K is the extinction coefficient at certain wave number,

C is the solution concentration,

L is the path length for the laser light

Therefore, it is possible to express the Beer-Lambert law (Huang, 1995) for each component as follows:

$$A_i = K_{ij}\,C_j\,L \qquad (2)$$

where:

Ai represents the absorbance at wave number i,

K_{ij} represents the extinction coefficient for j unit at wave number i,

C_j is the solution concentration of j unit,

L is the path length for the laser light

In the case of liquid cell the path length, L, is the spacer thickness between the KBr windows.

5.1 Experimental

The samples of commercial products were weighted (0.25 g) and dissolved in carbon bisulfide (10 ml). A liquid cell with KBr window was used with a spacer thickness of 0.2 mm. The FTIR equipment was a Magna 560 from Nicolet, with 32 scans, in the mid-red: 4000 a 400 cm^{-1}. These samples also were characterized by H^1 NMR.

5.2 Results and discussion

The results for microstructure and composition determination of polybutadiene portion in styrene-butadiene copolymers are presented in the following section. Such characterization was carried out using the indicated spectrometer and contrasted with the results obtained by H^1 NMR.

The structures corresponding to the polybutadiene isomers, which can be evaluated as it is Detailed in the next section, are shown in Figure 7.

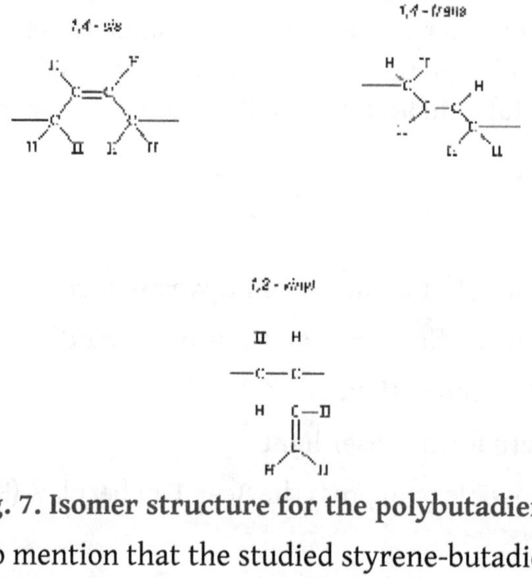

Fig. 7. Isomer structure for the polybutadiene

It is important to mention that the studied styrene-butadiene copolymers and the polybutadiene homopolymer were synthesized by anionic polymerization. Therefore, just three isomers are possible to obtain, the 1,4-cis; 1,4-trans and the 1,2-vinyl.

These isomers are present at different relative quantities, depending on the polymerization conditions. One key factor to modify the microstructure is the presemce of chemical named microstructure modifier, which has an interaction with the counterion of the initiator, modifying the bond length and promotes the ion pair formation.

5.2.1 Microstructure and composition

Considering the Beer-Lambert law, and knowing the extinction coefficients for each isomer at specifics wave numbers: 698, 728, 910 and 966 cm^{-1}, for the case of styrene-butadiene copolymers, it is possible to evaluate microstructure. The Beer-Lambert law can be expressed as:

$$A_i = \Sigma K_{ij} C_j L \quad (3)$$

Thus, the equation system can be expressed as a matrix, which can be solved Since L and K_{ij} are constants. The reported (Huang, 1995) values for the extinction coefficient are shown in Table 1.

Table 1. Extinction coefficients for styrene and isomers of styrene-butadiene copolymers

Extinction coefficient (K)				
Entity	A_{698}	A_{728}	A_{910}	A_{966}
1,4-cis	0.385	0.551	0.037	0.058
1,4-trans	0.005	0.007	0.055	2.542
1,2-vinyl	0.153	0.05	3.193	0.098
styrene	2.703	0.038	0.064	0.05

The path length was constant (0.02 cm). Therefore, the matrix is:

$$A_{698} = 7.7 \times 10^{-3} \; C_{cis} + 3.06 \times 10^{-3} \; C_{vinyl} + 1 \times 10^{-4} \; C_{trans} + 5.406 \times 10^{-2} \; C_{St}$$

$$A_{728} = 1.102 \times 10^{-2} \; C_{cis} + 1 \times 10^{-3} \; C_{vinyl} + 1.4 \times 10^{-4} \; C_{trans} + 7.6 \times 10^{-4} \; C_{St}$$

$$A_{910} = 7.4 \times 10^{-4} \; C_{cis} + 6.386 \times 10^{-2} \; C_{vinyl} + 1.1 \times 10^{-4} \; C_{trans} + 1.28 \times 10^{-3} \; C_{St}$$

$$A_{966} = 1.16 \times 10^{-3} \; C_{cis} + 1.96 \times 10^{-3} \; C_{vinyl} + 5.084 \times 10^{-2} \; C_{trans} + 1 \times 10^{-3} \; C_{St}$$

Once it was established the equations, they were solved. Thus, several commercial samples were evaluated according with matrix. The obtained values are shown in Table 2.

As it can be seen from Table 2, there is a good agreement in the microstructure calculated by FTIR and H^1NMR. Bearing in mind that microstructure and composition calculated by H^1NMR are the real values, the higher deviation, for cis and trans values, was achieved when the styrene content was above 39 % w/w. The last is probably due for overtones for vinyl and trans isomers, which appear with such styrene content.

Table 2. Microstructure and composition of several commercial samples.

Sample	%St (H¹NMR)	%Vinyl (H¹NMR)	%Cis (H¹NMR)	%Trans	%St	%Vinyl	%Cis	%Trans
S-200	0	9.12	40.21	50.67	0	9.51	42.39	48.1
S-1110	16.15	9.12	38.95	51.93	15.43	9.43	40.11	50.44
S-1205	24.39	9.39	40.63	49.98	24.05	9.74	38.5	51.76
S-1322	30.43	8.59	40.18	51.22	29.72	8.87	42.68	48.44
S-1430	39.94	8.82	39.07	52.12	39.14	8.74	43.21	48.05
S-314	69.69	16.89	27.12	55.98	67.35	14.79	33.78	51.43
S-411	30.37	14.56	35.54	49.89	29.7	15.07	35.71	49.23
S-416	29.96	12.69	36.51	50.79	29.37	13.71	36.89	49.39
S-4318	33.17	13.44	36.12	50.44	32.04	14.05	37.29	49.39

Nevertheless, there was a good correlation between the obtained results by both methods, corroborating that FTIR can be used with confidence, in order to evaluate composition and microstructure of styrene-butadiene copolymers. Nevertheless, it is important to mention that the use of the liquid cell is not infallible, Since the path length can be modified when the cell is cleaned.

5.2.2 Hydrogenation

Once the method for microstructure by FTIR was established, the hydrogenation of butadiene-based copolymers can be traced. Disappearance of each isomer of butadiene can be recorded with time. Thus, it is possible to see if hydrogenation is selective toward certain isomer or not, and also it is possible to see if polystyrene portion is saturated or affected by hydrogenation.

The hydrogenation of elastomeric portion (polybutadiene) in the styrene-butadiene copolymers can be visualized as it is shown the Figure 8.

The spectra of styrene-butadiene copolymer and the hydrogenated counterpart are shown in the Figure 9.

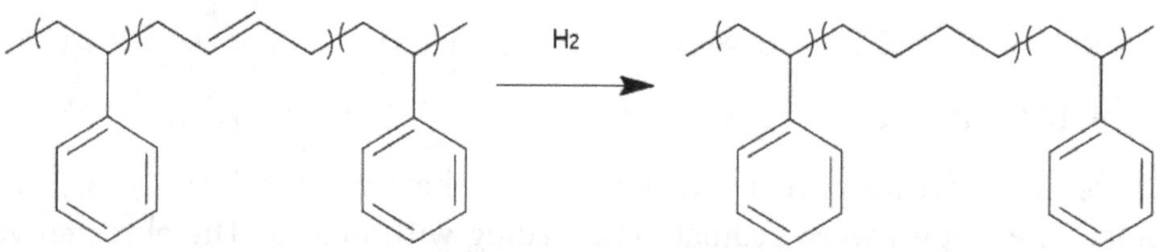

Fig. 8. Hydrogenation of styrene-butadiene copolymer

The spectra shown in Figure 9 indicate that there was a noticeable reduction of the absorbance of 1,2-vinyl, then 1,4-trans followed by 1,4-cis and the absorbance corresponding to styrene remains constant. In previous work (Escobar et al, 2000) it was shown that kinetics of hydrogenation is higher for 1,2-vinyl bonds than for 1,4-trans and 1,4-cis bonds.

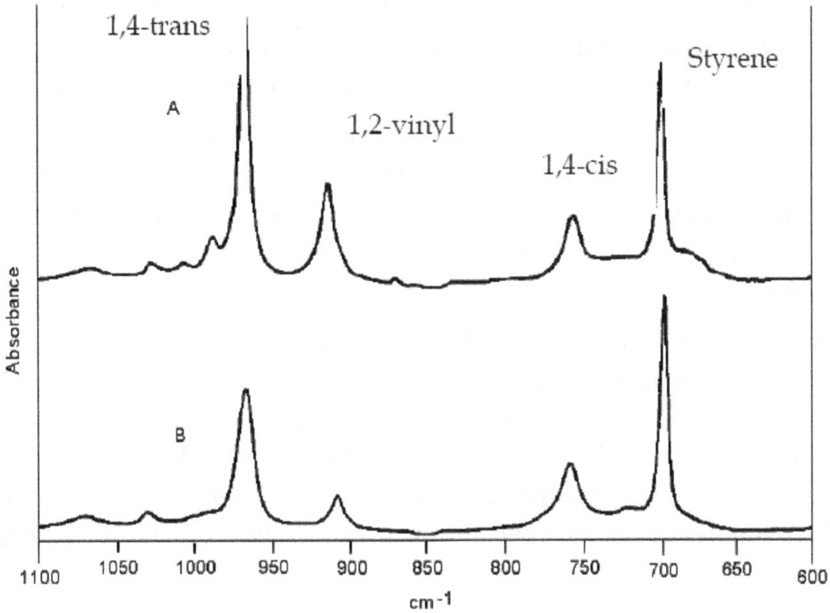

Fig. 9. Spectra of styrene-butadiene copolymer (A) and its hydrogenated product (B)

The 1,2-vinyl bonds are pendant from the backbone and, therefore, they are more accessible to hydrogen. The FTIR is a useful characterization technique in this case, Since it was possible to establish the hydrogenation selectivity toward the 1,2-vinyl bonds and showed that polystyrene remained unsaturated during hydrogenation.

6. CONCLUSIONS

FTIR is a powerful and useful characterization method for polymers, and materials in general. This is an economic, short time characterization that allows to establish the chemical composition, microstructure, chemical interactions and even to follow variation of specific functional groups with time, during reactions.

All this features, makes the FTIR an essential characterization technique for polymers. From hopolymers, copolymers to nanocomposites and biocomposites, can be characterized, qualitative and quantitatively, giving the possibility to understand specific interactions and to establish mechanism of reaction, between materials. Thus, the FTIR has contributed to the development of new materials in the polymer Science.

In the examples presented herein, the FTIR has permitted to follow specific functional groups with time, to establish the end and/or beginning of a particular reaction.

In addition, FTIR was useful to elucidate about interactions between polymeric matrix (polyurethane) and biopolymer (chitin). However, it is necessary to obtain thinner samples in order to see specific interaction in the interface of materials.

Regarding the nanotubes, it was possible to establish carbon functional group present in the surface of them Once these were chemically modified. This is important Since this allows figuring out what kind of interactions with other materials could be taken place, and understanding the mechanism of such interactions.

In the case of the establishment of microstructure of styrene-butadiene copolymers, it was demonstrated the capacity of FTIR to be used with confidence as an alternative technique to primary technique like H^1NMR. The FTIR gives very close values of the relative quantities of the different isomers to those obtained by NMR, besides the composition. Once the methodology was established, it was possible to follow the hydrogenation reaction, in order to see its selectivity toward the isomers.

7. REFERENCES

1. Ago, H., Kugler, T., Cacialli, S., Salaneck, W., Shaffer, M., Windle, A., & Friend, R. (1999). Work functions and surface functional groups of multiwall carbon nanotubes, Journal of Physical Chemistry B, Vol.103, No.38, (September 1999) pp. 8116-8121, ISSn 1520-6106
2. Alhakawati, M. S. & Banks, C. J. (2004). Removal of copper from aqueous solution by Ascophyllum nodosum immobilised in hydrophilic polyurethane foam. Journal of Environmental Management, Vol.72, No.4, (September 2004), pp.195-204, ISSn 0301-4797 Andrade-Espinosa, G, Escobar-Barrios, V.A. & Rangel-Mendez, J.R. (2010). Synthesis and characterization of silica xerogels obtained via fast sol–gel process, Colloid and
3. Polymer Science, Vol. 288, pp. 1697-1704, ISSn 1435-1536
4. Bhattacharjee, S., Bhowmick, A. K., & Avasthi, B,N. (1993). Selective hydrogenation of olefnic bonds in styrene-isoprene-styrene triblock copolymer by palladium acetate catalyst. Polymer, Vol. 49, No. 11, pp. 1971-1977
5. Bhargava, R., Wang, S., & Koening, J. L. (2003). FTIR Microspectroscopy of polymeric systems, Advances in Polymer Science, Vol.163, pp. 137-191, Springer, Verlag Berlin Heidelberg, Germany
6. Bower, D.I., Maddams, W.F. (1989) The Vibrational Spectroscopy of Polymer, Cambridge University Press, ISBN 0-521-24633-4, Cambridge, UK

7. Chen, J., Wu, S. (2004). Acid/base-treated activated carbons: characterization of functional groups and metal adsorptive properties, Langmuir, Vol.20, No.6, (March 2004) pp. 2233-2242, ISSn 0743-7463

8. Choi, H., Bae, S., Park, J., Seo, K., Kim, C., Kim, B., Song, H., & Shin, H. (2004). Experimental and theoretical studies on the structure of N-doped carbon nanotubes: Possibility of intercalated molecular N-2, Applied Physics Letters, Vol.85, No.23, (November 2004) pp. 5742-5744, ISSn 0003-6951

9. Davila-Rodriguez, J. L., Escobar-Barrios, V. A., Shirai, K. & Rangel-Mendez, J. R. (2009). Synthesis of a chitin-based biocomposite for water treatment: Optimization for fluoride removal. Journal of Fluorine Chemistry, Vol.130, (August 2009), No.8, pp.718-726, ISSn 0022-1139

10. De Sarkar, M., De, P.P., & Bhowmick A.K., (1997). Thermoplastic elastomeric hydrogenated styrene-butadiene elastomers: Optimization of reaction conditions, thermodynamiCs and kinetics. Journal of Applied Polymer Science, Vol.66, No. 6, pp. 1151-1162

11. Dresselhaus, M., Dresselhaus, G., & Avouris, Ph. (2001). Carbon Nanotubes. Synthesis, structure, properties and applications, Springer, ISBN 3-540-41086-4, Verlag Berlin Heidelberg, Germany

12. Escobar-Barrios, V. A., Herrera-Najera, R., Petit, A. & Pla, F. (2000) Selective Hydrogenation of butadiene-styrene copolymers using a Ziegler-Natta type catalyst. Part 1. Kinetic study. European Polymer Journal, Vol. 36, pp. 1817-1834

13. Elnashar, M. M., (Ed.). (2010). Biopolymers, Sciyo, ISBN 978-953-307-109-1, Rijeka, Croatia Fidalgo, A., Ilharco L.M. (2005). The influence of the wet gels processing on the structure

14. and properties of silica xerogels. Microporous and Mesoporous Materials Vol. 84, pp. 229–235, ISSn 1387-1811

15. Fowler, P. A., Hughes, J. M. & Elias, R. M. (2006). Biocomposites: technology, environmental credentials and market forces. Journal of the Science of Food and Agriculture, Vol.86, (September 2006), No.12, pp.1781-1789, ISSn 1097-0010

16. Hepburn, C. (1982). Polyurethane Elastomers, Applied Science Publishers, London and New York Hirsch A (2002) functionalization of single-walled carbon nanotubes, Angewandte Chemie

17. International Edition, Vol.41, No.11, (June 2002) pp. 1853-1859, ISSn 1521-3773 Huang, D., Lin, Y. Y., & Tsiang, Ch. (1995). Synthesis of SBS thermoplastic block copolymers

18. in cyclohexane in the presemce of diethylether used as a structure modifier. Journal of Polymer Research, Vol. 2, No. 2, pp. 91-98 ISSn 1022-9760

19. Kane, S. R., Ashby, P. D. & Pruitt, L. A. (2009). ATR-FTIR as a thickness measurement technique for hydrated polymer-on-polymer coatings. Journal of Biomedical Materials Research Part B: Applied Biomaterials, Vol.91B, No.2, (November 2009), pp.613-620, ISSn 1552-4981

20. Koenig, J.L. (2001). Infrared and Raman Spectroscopy of Polymers, Rapra Technology, ISBN-10: 1859572847, Shropshire. UK

21. Lee, L.H., Ed. (1977). Characterization of Metal and Polymer Surfaces: Polymer Surfaces

22. Academic Press , ISBN-10: 0124421016, New York, USA

23. Li, H., Liu, T., Li, Z. & Deng, L. (2008). Low-cost supports used to immobilize fungi and reliable technique for removal hexavalent chromium in wastewater. Bioresource Technology, Vol.99, (May 2008), No.7, pp.2234-2241, ISSn 0960-8524

24. Liu, Y., Pan, C., & Wang, J. (2004) Raman spectra of carbon nanotubes and nanofibers prepared by ethanol flames, Journal of Materials Science, Vol.39, No.3, (February 2004) pp. 1091-1094, ISSn 0022-2461

25. Loiseau, A., Blasé, X., Charlier, J., Gadelle, P., Journet, C., Laurent, Ch., & Peigney, A. (2006). Synthesis methods and growth mechanisms, In: Understanding carbon nanotubes. From basiCs to applications, A. Loiseau, P. Launois, P. Petit, S. Roche, J.P. Salvetat (Eds.), 49-130, Springer, ISBN 103-540-26922-3, Berlin Heidelberg, Germany

26. Mao, J., Won, S. W., Vijayaraghavan, K. & Yun, Y.-S. (2010). Immobilized citric acid-treated bacterial biosorbents for the removal of cationic pollutants. Chemical Engineering Journal, Vol.162, No.2, (August 2010), pp.662-668, ISSn 1385-8947

27. Misra, A., Tyagi, P., Rai, P., & Misra, D. (2007). FTIR spectroscopy of multiwalled carbon nanotubes: A simple approach to study the nitrogen doping, Journal of NanoScience and Nanotechnology, Vol.7, No.12, (December 2007) pp. 1820-1823, ISSn 1533-4880

28. Niyogi, S., Hamon, M., Hu, H., Zhao, B., Bhowmik, P., Sen, R., Itkis, M., & Haddon, R. (2002). Chemistry of single-wall carbon nanotubes, Accounts of Chemical Research, Vol.35, No.12, (October 2002) pp. 1105-1113, ISSn 0001-4842

29. Ovejero, G., Sotelo, J., Romero, M., Rodríguez, A., Ocana, M., Rodríguez, G., & Garcia, J. (2006). Multiwalled carbon nanotubes for liquid-phase oxidation.

functionalization, characterization and catalytic activity. Industrial Engineering Chemistry Research, Vol.45, No.7, (March 2003) pp. 2206-2212, ISSn 0888-5885

30. Pastor, A., (Ed.). (2004). Quitina y Quitosano: obtención, caracterización y aplicaciones, Programa CYTED, CIAD, A.C., Fondo Editorial de la Pontificia Universidad Católica del Perú, ISBN 9972-42-659-9, Lima, Perú

31. Perez-Aguilar NV, Muñoz-Sandoval E, Diaz-Flores PE, & Rangel-Mendez JR (2010) Adsorption of cadmium and lead onto oxidized nitrogen-doped multiwall carbon nanotubes in aqueous solution. Journal of Nanoparticle Research, Vol.12, No.2, (February 2010) pp. 467-480, ISSn 1388-0764

32. Sawyer, L.C., Grubb, D.T., & Meyers, G.F. (2008). Polymer Microscopy (3rd. edition), Springer, ISBN 978-0-387-72627-4, Verlag Berlin Heidelberg, Germany

33. Sudha Bai, R., Abraham, T. E. (2003). Studies on chromium(VI) adsorption-desorption using immobilized fungal biomass. Bioresource Technology, Vol.87, No.1, (March 2003), pp.17-26, ISSn 0960-8524

34. Terrones, M., Ajayan, P.M., Banhart, F., Blasé, X., Carroll, D.L., Charlier, J.C., Czerw, R., Foley, B., Grobert, N., Kamalakaran, R., Kohler-Redlich, P., Ruhle, M., Seeger, T., & Terrones, H. (2002). N-doping and coalescence of carbon nanotubes: synthesis and electronic properties, Applied Physics A Materials Science & Processing, Vol.74, No.3, (December 2001), pp. 355-361, ISSn 0947-8396

35. Vullo, D. L., Ceretti, H. M., Daniel, M. A., Ramirez, S. A. M. & Zalts, A. (2008). Cadmium, zinc and copper biosorption mediated by Pseudomonas veronii 2E. Bioresource Technology, Vol.99, No.13, (September 2008), pp.5574-5581, ISSn 0960-8524

36. Wang, H., Zhou, A., Peng, F., Yu, H., & Yang, J. (2007). Mechanism study on adsorption of acidified multiwalled carbon nanotubes to Pb(II), Journal of Colloid and Interface Science, Vol.316, No.2, (December 2007) pp. 277-283, ISSn 0021-9797

37. Zawadski, J. (1980). IR Spectroscopic investigations of the mechanism of oxidation of carbonaceous films with HNO_3 solution, Carbon, Vol.18, No.4, (April 2003) pp. 281- 285, ISSn 0008-6223

38. Zhang, J., Zou, H., Qing, Q., Yang, Y., Li, Q., Liu, Z., Guo, X., & Du, Z. (2003). Effect of chemical oxidation on the structure of single-walled carbon nanotubes, Journal of Physical Chemistry B, Vol.107, No.16, (March 2003) pp. 3712-3718, ISSn 1520-6106

39. Zhang, P. He, J., & Zhou, X. (2008). An FTIR standard addition method for quantification of bound styrene in its copolymers, Polymer Testing, vol. 27, pp. 153-157 ISSn 0142-9418

40. Zhang, Y., Banks, C. (2006). A comparison of the properties of polyurethane immobilised Sphagnum moss, seaweed, sunflower waste and maize for the biosorption of Cu, Pb, Zn and Ni in continuous flow packed columns. Water Research, Vol.40, No.4, (February 2006), pp.788-798, ISSn 0043-1354

10 Chapter

PREPARATION AND CHARACTERIZATION OF PVDF/PMMA/GRAPHENE POLYMER BLEND NANOCOMPOSITES BY USING ATR-FTIR TECHNIQUE

1. INTRODUCTION

With improvement in the human life, the requisite of the new materials with special properties for many different applications ranging from food packaging and consumer products to use as medical devices and in aerospace technologies can be sensed, strongly. Polymeric materials offer this opportunity to scientists and engineers for designing these new materials. In this regard, precise understanding of Structure and Properties Relationship (SPR) should be very crucial.

Recently, polymeric nanocomposites have opened a new research area and attracted strong attentions. The synthesis of polymer nanocomposites by inserting the nanometric inorganic compounds is an integral aspect of polymer nanotechnology (A. Lagashetty, 2005). These materials, depending upon the inorganic materials present, have particular and improved properties respect to pure polymers that invest their applications.

Fourier transform infrared (FTIR) spectroscopy is a powerful and reliable technique that for many years has been an important tool for investigating chemical processes and structures. In the polymer fields, FTIR data is used in order to study characterization of chemical bonds, polymer microstructure, chain conformation, polymer morphology, crystallinity and etc, consequently is useful in SPR studies.

The combination of infrared spectroscopy with the theories of reflection has made advances in surface analysis possible. Attenuated Total Reflectance (ATR) spectroscopy is an innovative technique for proving chemical information of a sample surface and the ability to quantify newly formed species, based upon Fick's second law. The fundamentals of attenuated total reflection (ATR) spectroscopy date back to the initial work of Jacques Fahrenfort and N.J. Harrick, both of whom independently devised the theories of ATR spectroscopy and suggested a wide range of applications. The schematic showing ATR-FTIR configuration is illustrated in Fig. 1 (KS. Kwan, 1998). The penetration depth, d, can be estimated as:

$$d = \frac{1}{2\pi n_c \sigma (sin^2\theta - n_x^2)^{1/2}} \quad (1)$$

where nc is the refractive index of the ATR crystal and n_x is the ratio between the refractive indexes of the sample and the ATR crystal (s_i which both of them are assumed to be constant in the considered frequency range. Φ is the wave number and Φ is the incident angle. The penetration depth for PVDF as an example by rhe assumption of nc = 2.4, n_x = 1.5 and Φ = 45° from 500 to 4000 cm^{-1} is approximately 0.5-4 μm (Y. Jung Park, 2005).

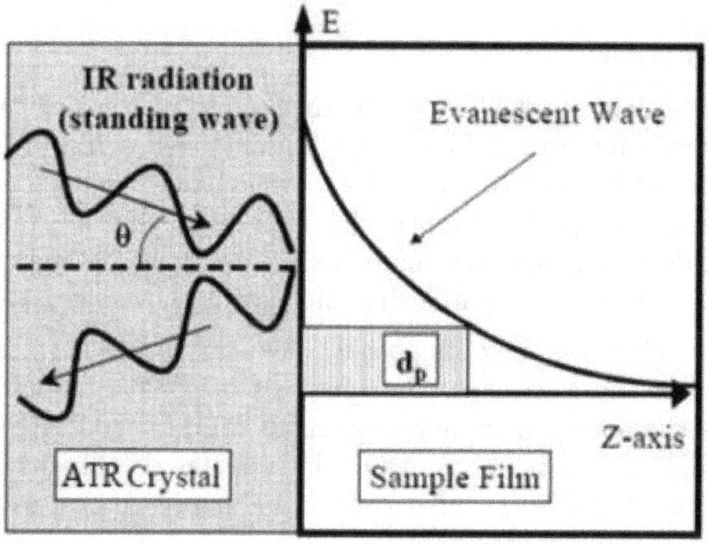

Scheme 1. Schematic illustration of ATR-FTIR configuration (KS. Kwan, 1998)

Graphene and graphite have recently attracted strong attention as versatile, environmentally friendly and available carbon materials which can be used as inexpensive filler in the composite materials (S. Stankovich, 2006, 2007 & L. Al-Mashat, 2010) .Crystalline graphites are used in polymeric systems in order to improve polymer properties such as thermal and electrical conductivity, IR absorption, flame retardancy, barrier resistance, electromagnetic shielding, lubrication and abrasion resistance. When the crystalline graphite is exfoliated to individual graphene sheets, the specific surface would be as large as 2600 m2. g-1 and novel electronic and mechanical properties appeared (Steurer, 2009). Actually graphene sheets are one-atom two-dimensional layers of sp2- network carbon that their fracture strength should be comparable to that of carbon nanotubes with similar types of defects (S. Stankovich, 2007). How to exfoliate the flakes of natural graphite was first described in a US patent in 1891 (Inagaki, 2004). An exfoliation phenomenon was studied mostly and occurs when the graphene layers are forced apart by the sudden vaporization or decomposition of intercalated species (E. H. L. Falcao, 2007).

However, the mentioned properties of graphene-polymer nanocomposites are strongly dependent on the uniformly dispersion in polymeric matrices which is affected by functional groups present on the graphene surface.

The modification of graphene sheets via organic oligomeric and polymeric chains is a favorable way to promote the compatibility of these nanoparticles with polymeric media. So understanding of the functional groups present on the graphene surface should be very vital for designing of modification procedure.

Poly (vinylidene fluoride) (PVDF) is a semicrystalline engineering polymer with very good resistance to chemicals, oxidation, and UV radiation (J. H. Yen, 2006). PVDF is known for its polymorphism crystalline structure and complicated microstructure. It is one of the most widely studied polymers due to its non-linearity, piezo- and pyro-electricity (L. T. Vo, 2007 & K. Pramoda, 2005). PVDF can crystallize in at least five well-known crystalline phases (E. Giannetti, 2001, A. Lovinger, 1982 & N. S. Nalwa, 1995):

α and δ with conformation of the alternating trans-gauche (TG+TG-) which the α is the most common form and is the most thermodynamically stable.

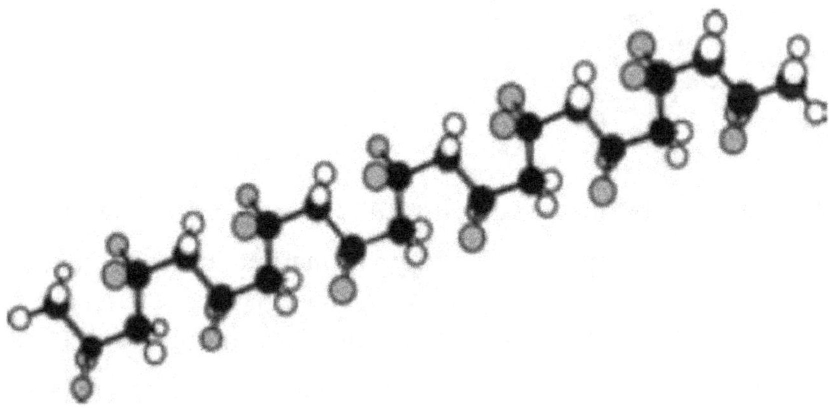

Scheme 2. α and δ form with alternating trans-gauche conformation (J. H. Yen, 2006)

β with all trans (TTT) planer zigzag conformation is polar form and has been extensively studied for its potential applications. This form develops under mechanical deformation (K. Matsushige, 1980) (S. Ramasundaram, 2008) , growth from solution (J. Wang, 2003 & R. L. Miller, 1976), addition of metal salts (X. He, 2006 &W. A. Yee, 2007) ,melt crystallization at high pressures (D. Yang, 1987), application of a strong electric field (J. I. Scheinbeim, 1986), blending with carbonyl-containing polymers (C. Lbonard, 1988 & K. J. Kim, 1995), and recently, addition of nanoparticles (S. Ramasundaram, 2008, T. Ogoshi, 2005 & L. He, 2010). This structure provides some unique properties for PVDF piezo- and pyro-electric activity:

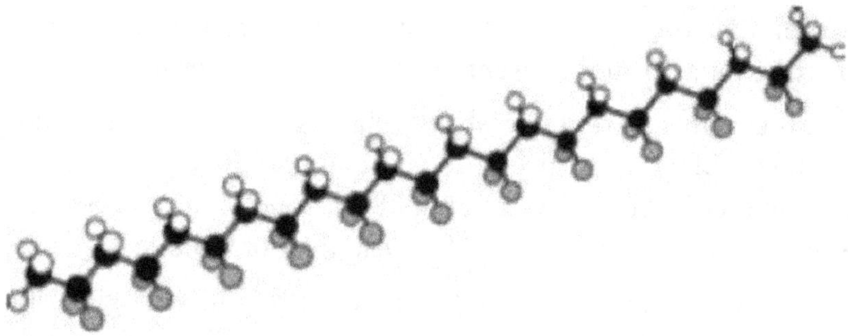

Scheme 3. β form all trans planer zigzag conformation (J. H. Yen, 2006)

γ and ε with T3G+T3G- conformation is also polar but less than β form.

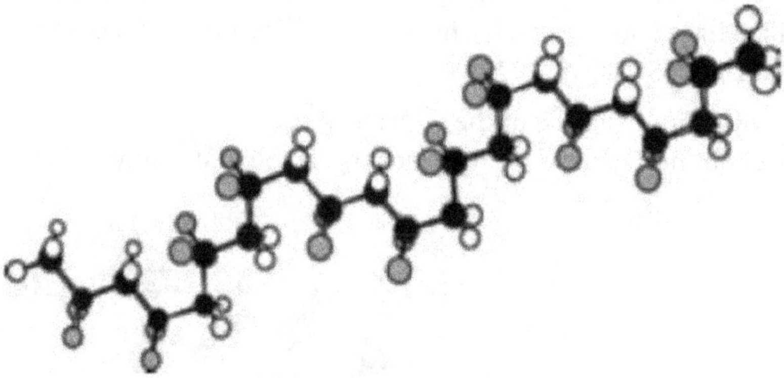

Scheme 4. γ and ε crystalline form with T3G+T3G- conformation (J. H. Yen, 2006)

Subsequently due to specific chain conformation in crystal unit cell and providing the highest remnant polarization, β phase has attracted more attention than the others in pyro- and piezoelectric applications (J.Jungnickel, 1996).

Poly methylmethacrylate (PMMA) can interact with graphene sheets by the interaction of delocalized π-bonds of graphene with π-bonds of PMMA. On the other hand, as reported in articles, PVDF/PMMA blend is a miscible system. Consequently, in attempt to achieve a homogenous dispersion of nanographene layers in PVDF matrix, the use of PMMA chains as a compatiblizer can be useful.

In addition, presemce of PMMA chains in the close touch with PVDF molecules causes the formation of polar crystalline form in PVDF (J. Wang, 2003). Furthermore, previous studies on the CNTs indicated TTT molecular chain prefers to be absorbed on the CNT surface compared with TGTG' molecular chain, and the configuration in which H atoms and CNT surface are face-to-face are more stable than that where F atoms and CNT surface are face- to-face. Since in the PVDF, the negative charge transfer from H to C atom and the negative charge is accumulated around the F atoms, the interaction between the H atoms with positive charge in PVDF and C atoms with π oribital in CNT should be stronger (S. Yu, 2009)

FT-IR is a powerful and reliable technique for description of chemical characterization of graphene and also study of the structure and properties relationships in PVDF/PMMA/graphene polymer blend nanocomposite. The interactions between these three components including: PVDF, PMMA and graphene sheets can be revealed and described by using this technique. As well, exploration of PVDF chains conformations (crystalline structures) are affected by presemce of graphene sheets and PMMA chains can be done which is very important in order to design the new material with special properties.

2. EXPERIMENTAL

2.1 Materials

The graphite used in this study was natural graphite powders with the size of >150 μm, and bulk density of 1.65 gr/cm^3 supplied by Iran Petrochemical Co. The PVDF pelletlet (Kynar® 1000HD) was provided by Atofina Co. The methyl methacrylate (MMA) and Methacyilic anhydride were supplied by Aldrich Co. and MMA used after purification and distillation in order to remove inhibitors. Concentrated sulfuric acid and nitric acid with concentration of 63% were used as chemical oxidizer to prepare expanded graphite. The initiator, Benzoyl peroxide (BPO), dichloromethane (CH_2Cl2), dimethylformamide (DMF), triethylamine (TEA) were bought from Merck Co. and used as received.

2.2 Preparation of exfoliated graphite

The method which was used in this study is as same as general procedure employed in industry for producing exfoliated graphite. Natural graphite powders as starting material were treated with a mixture of 4:1 sulfuric and nitric acid at 80°C for 24h to produce Graphite Intercalated Compounds (GIC)s. Graphite can accept many species into the gallery between graphite layer planes to form GICs.

Sulfuric acid is the most conventional intercalate for achieving a high degree of exfoliation of natural graphite and so it is used most commonly in the industry. Nitric acid was also added as an oxidant to generate some functional groups such as hydroxyl, carboxyl, epoxid groups on the surface and edge of graphene sheets.

The suspension of acid-treated graphite was added to excess distilled water, and this was followed by centrifugation with 5000 rpm and washing with water until the pH of the centrifuge drainage extraction was natural.

These particles dried at oven at 100°C for 12h and vacuumed oven at 80°C for 8h. The oxidized graphite was heat-treated at 1050 °C for 45s to obtain expanded graphite. Actually, In Chung's review, exfoliation phenomena were classified into reversible and irreversible ones. When the oxidized graphite was heated to around 300°C,it expanded to a fractional expansion of about 30 and upon cooling collapsed, again. This expansion–collapse phenomenon was reversible, and was thus called reversible exfoliation.

But if it is heated to a higher temperature (e.g., 1000°C), the intercalating compound and some of functional groups decompose completely and at the same time the host graphite flakes exfoliate up to about 300× in volume, particularly by rapid heating. This exfoliated graphite never returns to the original thickness upon cooling to room temperature, that is, we have irreversible exfoliation. The expanded graphite was immersed in absolute ethanol and sonicated for 30 min in order to exfoliate and break down to individual layers. The dispersion was filtered and dried at oven at 80°C for 8h and vacuumed oven at 80°C for 4h to remove residual moisture in the graphite particles (S. Mohamadi N. S.-S., 2011).

2.3 Modification with methacrylic anhydride

The above functionalized Graphite, which is called FG, was reacted with methacrylic anhydride to provide some organic groups and vinyl moieties on the surface of graphene. Actually, the functional groups which were introduced in the previous process could further react with methacrylic anhydride through hydroxyl groups. A 250-mL flask was charged with 100 mL of DMF, 10 mL TEA and 100 mg FG and after sonication for 30 min, a solution of containing 5 mL methacrylic anhydride and 90 mL DMF was added in to the mixture.

This mixture was allowed to react at 90°C for 24h under magnetic stirring. After that the Modified Graphite, for simplicity it is called MG, were collected by filtration and followed by thorough washing with distilled water to remove any residual TEA and then washed with CH_2Cl2, in order to remove any unreacted methacrylic anhydride. The washed particles were dried in oven and vacuum oven at 80°C for 8h, separately.

2.4 Preparation of PMMA/ graphene masterbatch

The PMMA-graphene masterbatch (PMMA-G master) was prepared via polymerization of MMA with BPO as an initiator. 300 mg of MG was dispersed in 15 mL MMA via ultrasonication for 30 min. The polymerization was initiated by (1.1% wt) of BPO respect to monomer in flask at 90°C, until the polymerization was solidified, completely. Later then, the prepared masterbatch was kept in vacuum oven at 120°C to remove any remaining MMA. The content amount of graphite was kept in 2.5 wt%.

2.5 Preparation of PVDF/PMMA/graphene nanocomposite

PVDF and PMMA-G master solutions were prepared separately by dissolving in DMF at 50°C with stirring for 1 day. Both of the resulting solutions were blended, so that the final concentration of mixture adjusted in 12% wt. on the weight of PVDF/

PMMA-G master respect to solvent. The mixture was sonicated for about 30 min, stirred 2 days and spread on a well-cleaned glass slide of petridish. The prepared sample was carefully evaporated at 60°C in an oven for 20 h and then divided into three parts and annealed at three different temperatures of 50, 90 and 120°C in vacuum oven for 24 h. three samples with varying PVDF/PMMA-G master ratios of 80:20, 70:30 and 60:40 were prepared and analyzed. For convenience the samples were named as 80:20, 70:30, 60:40 and 70:30 No G, respectively (S. Mohamadi, 2011).

2.6 Characterization

The Fourier transform infrared (FT-IR) analysis was performed on a Bruker Equinox55 Analyzer, equipped with a DTGS detector and a golden gate micro ATR from 600 –3500 cm^{-1}. Scanning electron microscopic (SEM) and TEM images were obtained on a Zeiss CEM 902A (Oberkochen, Germany) and Philips-cm^{-1}20, Netherlands at an accelerating voltage of 120 KV, respectively. The thermal behavior was measured with a DSC Q100 from the TA instruments with the heating and cooling ramp of 10°C min-1 from room temperature to 250°C under argon flow. The X-ray diffraction (XRD) patterns of the samples were recorded by Philips, Netherlands advanced diffractometer using Cu($K\alpha$) radiation (wavelength: 1.5405Å) at room temperature in the range of 2θ from 4 to 70° with a scanning rate of 0.04°. S−1.

3. RESULTS AND DISCUSSION

As reported by others in several literatures, presemce of the relatively broad peak at 3427 cm^{-1} and also the peak at 1402 cm -1 in the FTIR spectra of graphite oxide indicate existence of hydroxyl groups. In addition, the peaks at 1720, 1640, 1580 and 1060 cm^{-1} can be assigned to the stretching vibration of carboxyl, carbonyl moiety of quinine, aromatic C=C bonds and epoxide groups, respectively (scheme 1) (C. Hontoria-Lucas, 1995) . To determine which functional groups existed in the graphene sheets after heat treated of graphene oxide (FG) and functionalization with methacrylic anhydride, the related FTIR spectra has been illustrated in Fig. 1.

As can be seen in Fig. 1 (a) most of carboxyl groups in the graphene oxide has decomposed during heat-treatment of graphite oxide so that the peak at 1720 cm^{-1} disappeared. In fact, during the acid treatment various oxygen-functional groups were produced on the graphene surface. While, in the heat-treatment step, the oxygen containing of graphene was reduced but some introduced functional groups remained and still the particles can contain significant amount of oxygen (Steurer, 2009).

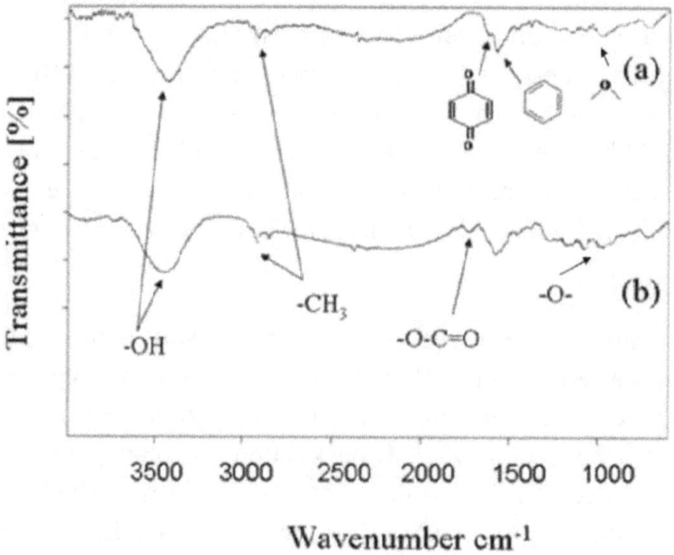

Fig. 1. FTIR spectrums of FG (a) and MG (b).

In addition, the peaks at 2960 and 2930 cm^{-1} are related to the asymmetric and symmetric stretching vibration of pendant methyl groups. Symmetric vibrations are generally weaker than asymmetric vibrations Since the former lead to less of a change in dipole moment. Actually, during sonication it may graphene flakes are broken and this pendant methyl groups came into sight.

In the case of MG, (Fig. 1 (b)) appearance of a new peak at 1740 cm^{-1} which has been assigned to vibration of esteric carbonyl group is due to the formation of esteric linkages between hydroxyl groups of FG and methacrylic anhydride (Scheme 6). Modification of FG by methacrylic anhydride (MG), causes appearance of the new methyl groups which show the peaks at 2960 and 2930 cm^{-1}. Furthermore, C=C band at 1580 cm^{-1} has become more broader in MG rather than FG may due to presemce of aromatic and aliphatic double bonds.

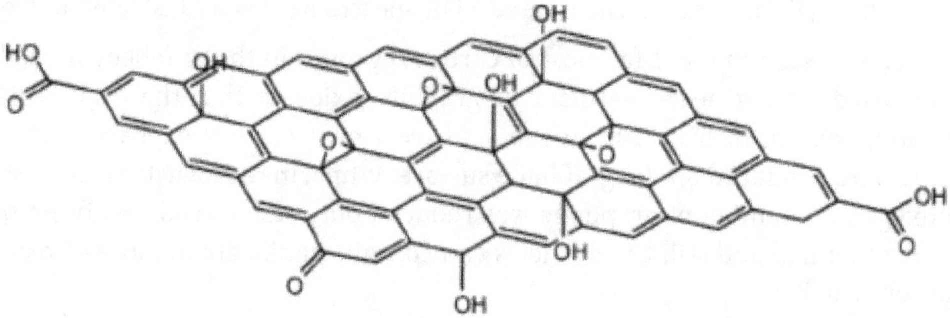

Scheme 5. Graphite oxide (P. Steurer, 2009)

ATR-FTIR spectrum of pure PVDF film which was annealed at 50°C, from 550-3500 cm^{-1} is also shown in Fig.2. The observed pattern originates from oscillations of large parts of the skeleton and/or the skeleton and attached functional groups. Below 1500 cm^{-1}, most single bonds absorb at similar frequencies, and the vibrations couple.

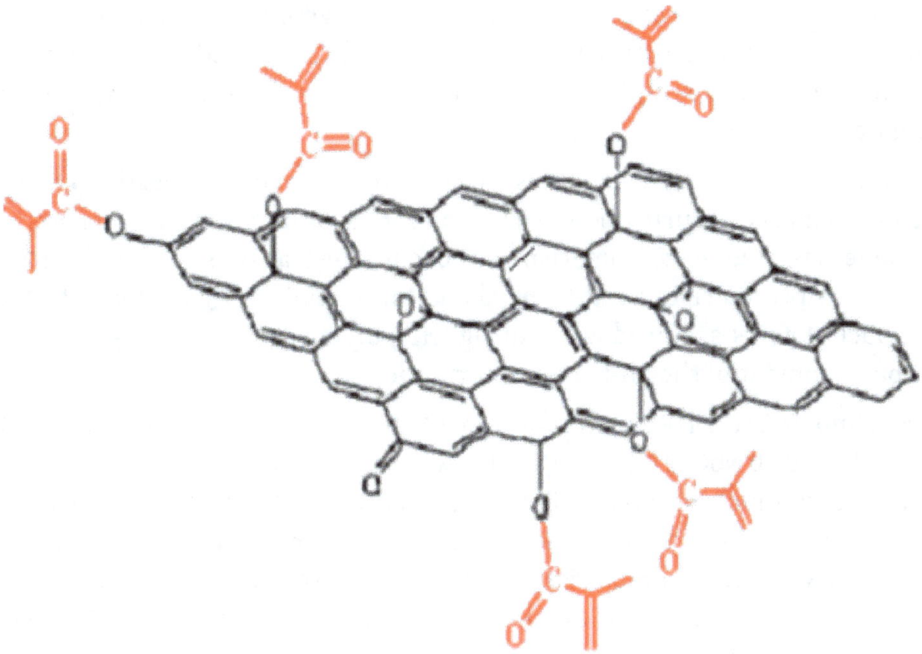

Scheme 6. Modified graphen with methacrylic anhydride (S. Mohamadi, 2011)

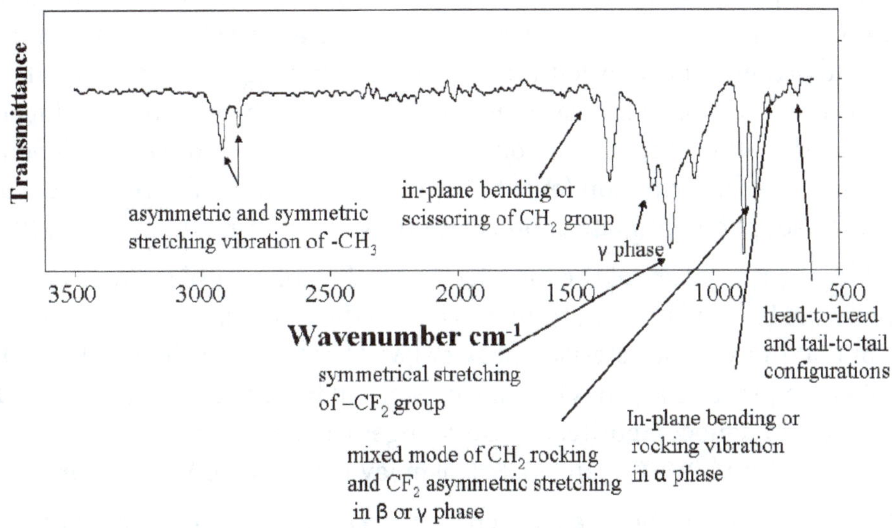

Fig. 2. ATR-FTIR spectrum of pure PVDF film which was annealed at 50°C

It should be pointed that Since the β- and γ-phase are similar to each other in short segmental conformations the identification of crystal phase between β and γ in FTIR spectrum is still in dispute. On the basis of the data from literature the absorption band at 763 cm^{-1} is related to In-plane bending or rocking vibration in α phase (N. Betz, 1994 & S. Lanceros-Me ndez, 2001), the band at 840 cm^{-1} stretching in β or γ phase (G. Chi Chen, 1994 , V. Bharti, 1997, B. Mattsson, 1999 & M. Benz, 2002), the band at 1173 cm^{-1} is associated to the symmetrical stretching of –CF2 group (M. Rajendran, 2010). The band at 1234 cm^{-1} is related to the γ phase and 1453 cm^{-1} is assigned to in-plane bending or scissoring of CH_2 group (M. Rajendran, 2010 & Garton, 1992).

In addition, some irregularities of head-to-tail addition, leading to defect structures, can occur during polymerization for several reasons in polymers of the vinylidene class, sequence isomerism may occur. The transmission band at 677 cm^{-1} points to the presemce of head-to-head and tail-to-tail configurations (B. Hilczer, 1998). Such defects are produced during the polymerization process and reduce the dipole moment of the all-trans conformation.

Fig. 3 shows ATR-FTIR spectra for pure PVDF film (a), G-PMMA master (b), 80:20 (c), 70:30 (d), 70:30 NoG (e), 60:40 (f) which were annealed at 50°C from 550-2000 cm^{-1}. When MG was incorporated in PMMA matrix (G-PMMA master), the absorption originating from MG particles alone were hardly visible. G-PMMA master spectrum clearly shows several characteristic peaks at 1723, 1449 and 1142 cm^{-1} which are related to the presemce of PMMA chains, assigned to ester carbonyl (C=O), O–CH3 and C-O (ester bond) stretching.

Polymer- polymer interactions may also affect PVDF chain conformations. Specific interactions are very sensitive to the distance between the interacting groups and to their relative orientation: hydrogen bonding strength falls off rapidly when the atomic distance increases or when the bond is bent instead of linear. This infers that the efficiency of the contact between two unlike chains depends on their respective conformation (eventually tacticity) and flexibility. The chains are expected to adopt an optimate conformation for interacting (C. Lbonard, 1988).

To identify the nature of the interactions in the polymer blend nanocomposite, we choose the band at 1720 cm^{-1} which is due to the stretching vibration of the C=O group in PMMA as a characteristic peak of PMMA. This band shows a little shift to the higher frequencies and has become broader in both of the polymer blend with no graphen and nanocomposites. These changes can be ascribed by the interaction between PMMA and PVDF chains (I. S. Elashmawi, 2008 & M. M. Colemann, 1995).

The strong absorption peak which appeared at 1173 cm^{-1} is assigned to the symmetrical stretching of –CF2 groups, was choose as characteristic peak of PVDF in polymer blend nanocomposites. The precise consideration reveals shifting of this band to the higher frequencies in the synthesised samples comparing to pure PVDF. Since this shifting is slighter for 70:30 NO G than polymer blend nanocomposites, it can be imagined this change is related to the specific interactions between

PVDF and graphene surface in the nanocomposites. Also, it can be concluded GNPs increase the compatibility of PVDF and PMMA.

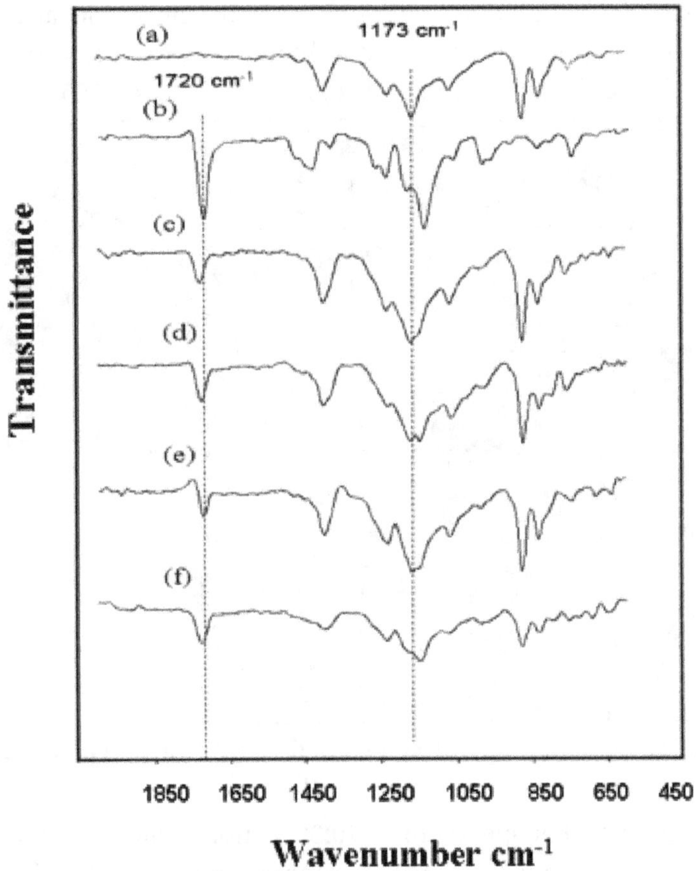

Fig. 3. ATR-FTIR spectra of pure PVDF (a), G-PMMA master (b), 80:20 (c), 70:30 (d), 70:30 NoG (e), 60:40 (f) which were annealed at 50°C.

On the other hand as mentioned above, presemce of polymer-polymer interactions can also affect the polymer conformations. PVDF exhibits short bond sequences tt and tg; the probability of a gg bond is negligible because of steric hindrance. Energy calculations emphasize that PMMA conformation remains close to all trans conformation of PVDF. In our study, presemce of graphene particles along with PMMA chains makes it complex due to PVDF-PMMA, PVDF-graphene and PMMA-graphene interactions.

The surface morphology of the casted films was studied using SEM analysis. In each sample, the surface which is exposed to the Air is examined. The distinguishable spherulitic structure on the top surface of pure PVDF film could be seen which had extended from 9 μm to about 15 μm with increasing the annealing temperature from 50 to 120°C. But in 70:30 some new elongated ribbon like crystals on the spherulitic structures can be observed which became more visible with increasing annealing temperature to 120°C (Fig. 4). Y. J. Park et al reported the oriented align crystalline PVDF lamella which has bamboo-like structure can be achieved when PVDF film casted from polar solvent such as DMF and DMSO and crystallize in the confined spaces of the specific mold at high temperature. In their view, these structures are due to the formation of γ crystals (Y. J. Park, 2005).

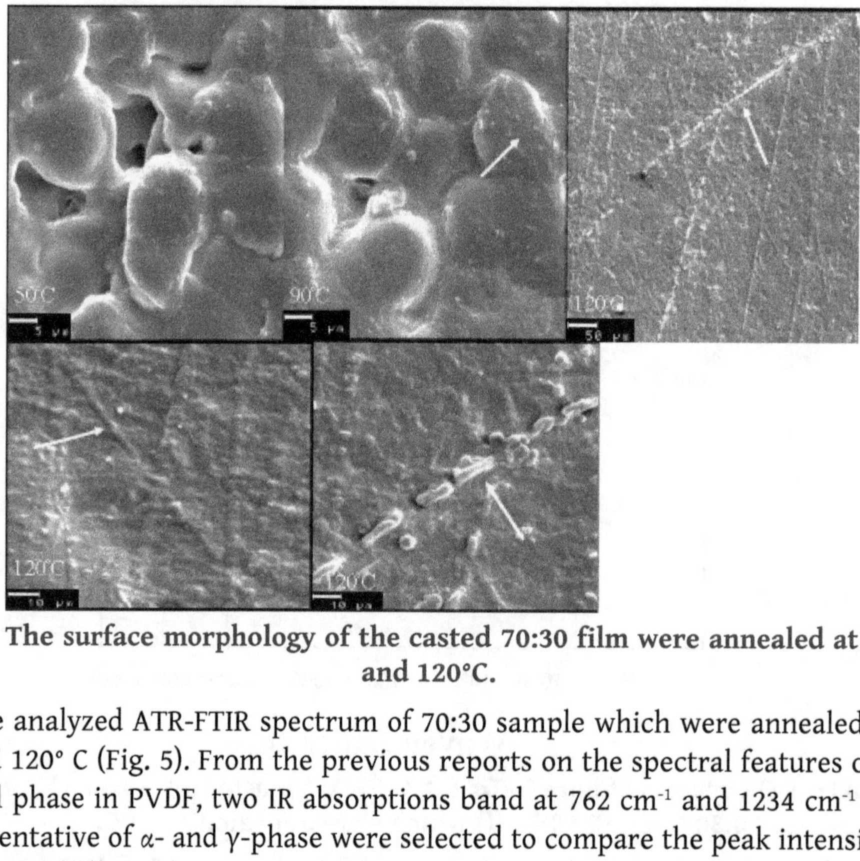

Fig. 4. The surface morphology of the casted 70:30 film were annealed at 50, 90 and 120°C.

We analyzed ATR-FTIR spectrum of 70:30 sample which were annealed at 50, 90 and 120° C (Fig. 5). From the previous reports on the spectral features of each crystal phase in PVDF, two IR absorptions band at 762 cm^{-1} and 1234 cm^{-1} as the representative of α- and γ-phase were selected to compare the peak intensities in each annealed samples.

It indicated with increasing annealing temperature the percentage of γ-phase on the surface increased while α-crystal decreased at 90° C and then didn't change significantly, at 120°C. In this regards, this new droplet like structure identities on the top of the films can be related to γ-phase. Of course, this result is in agreement with X-ray Diffraction (XRD) and Differential Scanning Calorimetery (DSC) results (Fig. 6).

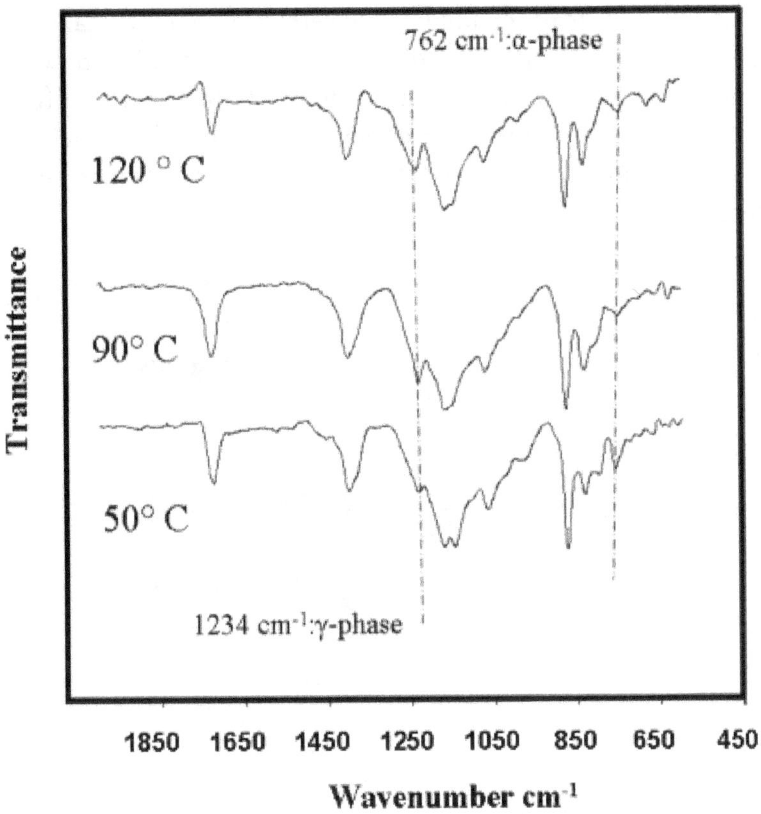

Fig. 5. ATR-FTIR spectra of 70:30 were annealed at 50, 90 and 120° C.

On the basis of the reported data in literatures, WAXD data of neat PVDF contain 100, 020, 110, 021 reflections at 2θ= 17.9°, 18.4°, 20.1°, and 26.7° for α- and only 200/110 reflection at 2θ= 20.8° for β-crystalline form (A. Kaito, 2007, L.Yu, 2009 & G. Guerra, 1986). γ-crystal planes are known to have overlapping reflections with (020), (110) and (021) α-crystal planes and also 200/110 reflection of β-phase. The (100) peak of α- phase at 2θ= 17.9° is the only peak that doesn't have any overlapping with γ-phase. Also the sharp diffraction peak at 26.38° in 70:30 is related to the crystallographic plane of graphite with d-spacing of 3.37Å, indicating that the graphite retained its crystalline structure in nanocomposite but there is a different degree of stacking order and disordered microstructure (K. P. Pramoda, 2010 & T. Ramanathan, 2007). Furthermore, the film which was annealed at 50°C shows the 100, 020, 110 reflections of α-phase and 200/110 reflection of β-crystal In XRD pattern of the films were annealed at 90°C, 200/100 reflection of β-crystal has become stronger, while the 100 peak of α has not changed considerably. It means with increasing temperature to 90°C, the growth of β-crystals is more favorable than α-crystals, increasing the annealed temperature to the 120°C caused the substantial variation in the intensity of reflections. The decrease in the intensity of 200/100 reflection of β-crystal and increase in 020, 110 reflections of α- or γ-phase without any significant change in 100 reflection of α-crystal implies some of β-crystals convert to γ-crystals at 120°C. The DSC results also matched well with those of X-ray diffractions and indicated the existence of the different types of crystal structures. The new appeared endotherm in the film which was annealed at 177° C can be attributed to the melting of the new formed ribbon-like structures on the surface.

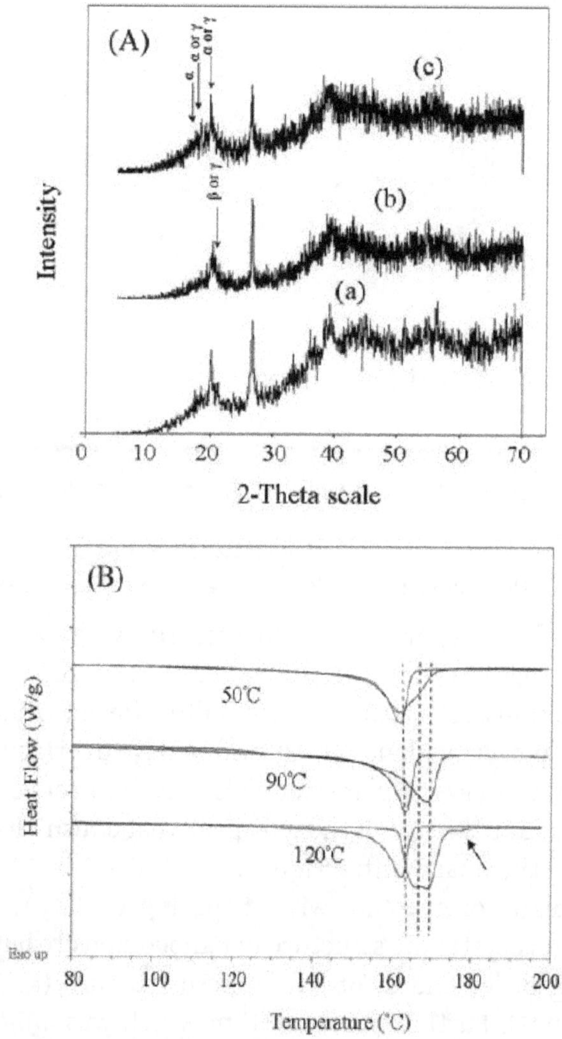

Fig. 6. X-ray diffraction pattern (A) and DSC thermograms of first and second heating (B) of 70:30 films annealed at 50, 90 and 120°C for 24h (S. Mohamadi, 2011)

Actually, in our work, it can be imagined PVDF chains which are trapped between graphen sheets and are in close contact with PMMA chains, crystallize in the confined spaces that are provided by graphen sheets and PMMA chains. It is demonstrated that exfoliated graphene- based materials are often compliant, and when dispersed in a polymer matrix are typically not observed as rigid disks, but rather as bent or crumpled platelets (J. R. Potts, 2011). Consequently, as can be seen in TEM image of 70:30 sample (Fig. 7), the wrinkled surface of the graphene sheets can act as micropattern molds for crystallization of PVDF. In other words, a well-defined feature of micropattern can be provided by the wrinkles on the surface of graphene sheets.

Fig. 7. TEM image and Wrinkled surface of graphene sheets in 70:30 polymer blend nanocomposite (S. Mohamadi, 2011)

Crystallization results basically in the succession of two events: the primary nucleation of a new phase and then the three-dimensional growth of lamellae; these steps can be followed by lamellar thickening, fold surface smoothing, or reorganization into more perfect crystals. As reported by others, with increasing the annealing temperature in PVDF/PMMA blend film, the amount of α phase increases due to easier local internal chain rotation at higher temperature and most of β and γ form convert to α form.

While in our study with increasing the annealing temperature the amount of α phase decreased and γ form increased. It can be said graphene sheets can stabilize the β- and γ- phase at elevated temperature due to restricting effect of graphene sheets on TT conformation chains as is illustrated in scheme:

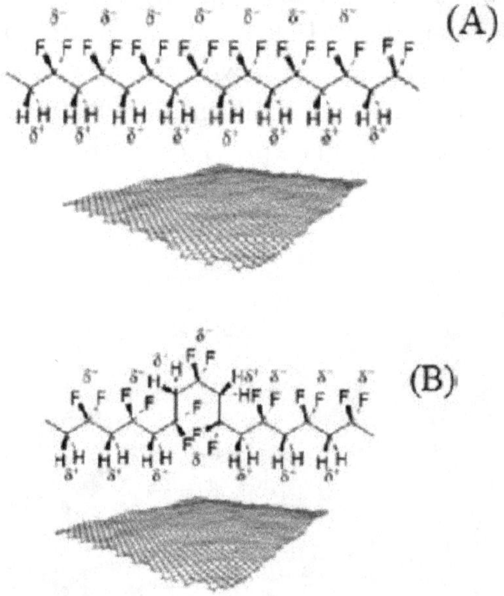

Scheme 7. Schematic illustration of to restricting effect of graphene sheets on TT (A) and TTTG+TTTG- (B) conformation chains

4. CONCLUSION

In this study, modification of graphite was carried out via introduction of hydroxyl groups and then vinyl groups on the surface of graphite by oxidation and estrification reaction respectively. The modified graphene (MG) was used to prepare the PMMA-graphene as a master batch by in-situ polymerization and followed by solution blending with PVDF in different ratios. The series of prepared polymer blend nanocomposite films were annealed at three different temperatures of 50, 90 and 120°C, and characterized.

FTIR results for FG and MG confirmed presemce of different oxygen containing function groups on the surface of FG and vinyl organic moieties on the MG sheets.

ATR-FTIR spectra revealed the specific interaction between PVDF and PMMA chains in polymer blend nanocomposites. The surface morphology of 70:30 film which was annealed at 120°C, some new elongated ribbon like crystals formed. From ATR-FTIR data of this sample the identity of these new structures on the film surface can be related to the formation of γ crystals.

5. REFERENCES

1. Al-Mashat, L., Shin, K., Kalantar-zadeh, K., D. Plessis, J., H. Han, S., W. Kojima, R., B. Kaner, R., Li, D., Gou, X., l J. Ippolito,S. & Wlodarski, W., (2010) Graphene/ Polyaniline Nanocomposite for Hydrogen Sensing, Journal of Physical chemistry C, Vol. 114, pp. 16168–16173, 1932-7447.

2. Benz, M., Euler, W. B., & Gerory, O. J., (2002) The Role of Solution Phase Water on the Deposition of Thin Films of Poly (vinylidene fluoride), Mocromolecules, Vol. 35, No. 7, pp. 2682-2688, ISSn 0024-9297.
3. Betz, N., Le Moel, A., Balanzat, E., Ramillon, J.M., Lamotte, J., & Gallas, J. P., (1994) FTIR Study of PVDF Irradiated by Means of Swift Heavy Ions, Polymer Science Part B: Polymer Physics, Vol. 32, pp.1493–1502, ISSn 1099-0518.
4. Bharti, V., Kaura, & T., Nath, R., (1997) Ferroelectric Hysteresis in Simultaneously Stretched and Corona-Poled PVDF Films, IEEE Transactions on DielectriCs and Electrical Insulation, Vol 4, No. 6, (December 1997) pp. 738–741, ISSn 1070-9878.
5. Chi Chen, G., Su, J., Fina, L. J., (1994) FTIR-ATR Studies of Drawing and Poling in Polymer Bilaminate Films, J. Polymer Science: Part B, Polymer Physics, Vol.32, pp.2065-2075, ISSn 1099-0518
6. Coleman, M. M., Painter. P.C., (1995) Hydrogen bonded polymer blends, Progress in Polymer Science, Vol. 20, No. 1, pp.1- 59, ISSn 0079-6700.
7. Elashmawi, I. E., Hakeem, N.A., (2008), Effect of PMMA Addition on Characterization and Morphology of PVDF, Polymer Engineering Science, Vol. 48, pp. 895-901, ISSn 1548- 2634.
8. Falcao, E. H. L., Blair, R. G., Mack, J. J., Viculis, L. M., Kwon, C., Bendikov, M., Kaner, R. B., Dunn, B. S., & Wudl, F., (2007). Microwave exfoliation of a graphite intercalation compound, Carbon, Vol. 45, No. 6, pp. 1364-1369, ISSn 0008-6223.
9. Garton, A., (1992) Infrared Spectroscopy of Polymer Blends, Composites and Surfaces, Hanser Publisher, Munich, Germany, ISBN 1-56990-034-5
10. Giannetti. E., (2001), Semi-crystalline fluorinated polymers, Polymer International, Vol: 50, No. 1, (February 2000), pp. 10-26, ISSn 1097-0126.
11. Grafting of PMMA chains Through in-situ Polymerization. Macromolecular Science Part A: Pure & Applied Chemistry, Vol. 48, pp. 577–582, ISSn 1060-1325.
12. Guerra, G., Karasz, F. E. , Macknight. W. J., (1986), On blends of poly(vinylidene fluoride) and poly(vinyl fluoride), Macromolecules, Vol. 19, No.7, (March, 1986) pp. 1935- 1938, ISSn 0024-9297
13. He, L., Xu, Q., Hua, C., Song. R., (2010) Effect of multi-walled carbon nanotubes on crystallization, thermal, and mechanical properties of poly(vinylidene fluoride, Polymer Composite, Vol. 31, No. 5, (April 2009), pp. 921-927, ISSn 0272-8397.

14. He, X. & Yao, K., (2006), Crystallization mechanism and piezoelectric properties of solution- derived ferroelectric poly(vinylidene fluoride) thin films, Applied Physics letter. Vol. 89, No. 11, pp. 112909-1 – 112902-3, ISSn 0003-6951.

15. Hilczer, B., & Kulek. J., (1998) The Effect of Dielectric Heterogeneity on Pyroelectric Response of PVDF, IEEE Transactions on DielectriCs and Electrical Insulation, Vol. 5, No. 1, pp. 45–50, ISSn 1070-9878.

16. Hontoria-Lucas, C., Lopez-Peinado, A. J., DE D. Lopez-Gonzalez, J., Rojas-Cervantes, M. L., & .Martin-Aranda, R. M., (1995). Study of oxygen-containing groups in a series of graphite oxides: physical and chemical characterization, Carbon, Vol. 33, No. 11, pp. 1585-1592 ISSn, 0008-6223.

17. Inagaki, M., Kang, F., & Toyoda, M. (2004) Exfoliation of Graphite via Intercalation Compounds in: Chemistry and Physics of Carbon A Series of Advances , Radovic, L. R., pp. (1-65) Marcel Dekker, NewYork, ISBN 0-8247-4088-2.

18. J.Jungnickel, B. (1996). Polymeric Materials Handbook, J. C., Salamone, New York : CRC press Inc, . pp. 7115-7122.

19. Kaito, A., Iwakura, Y., Hatakeyama, K., & Li, Y., (2007) Organization of Oriented Lamellar Structures in a Miscible Crystalline/Crystalline Polymer Blend under Uniaxial Compression Flow near the Melting Temperature, Macromolecules, Vol. 40, (February, 2007), pp. 2751-2759, ISSn 0024-9297

20. Kim, K. J. , Cho, Y. J. , & Kim. Y. H. , (1995) Factors determining the formation of the β crystalline phase of poly(vinylidene fluoride) in poly(vinylidene fluoride)- poly(methyl methacrylate) blends, Vibrational Spectroscopy, Vol. 9, No. 2. (September 1994), pp. 147-159, ISSn 0924-2031.

21. Kwan, KS. (1998) The Role of Penetrant Structure in the Transport and Mechanical Properties of a Thermoset Adhesive, Ph.D. Thesis, Faculty of the Virginia Polytechnic Institute, Blacksburg, 285f.

22. Lagashetty, A., & Venkataraman, A., (2005) Polymer Nanocomposites, Resonance, Vol. 10, Num. 7, (July 2005) pp. 49-57.

23. Lanceros-Me ndez, S., Mano, J. F., Costa. A. M. (2001) FTIR AND DSC STUDIES OF MECHANICALLY DEFORMED B-PVDF FILMS, Macromolecular Science: Physics, Vol. B40, No. 3&4, pp. 517-527, ISSn 0022-2348.

24. Lbonard, C., Halary, J. L., & Monnerie, L., (1988), Crystallization of poly(vinylidene fluoride)-poly(methyl methacrylate) blends: analysis of the molecular parameters controlling the nature of poly(vinylidene fluoride) crystalline phase, Macromolecules, Vol: 21, No. 10, (October 1998) pp. 2988-2994, ISSn 0024-9297.

25. Lovinger, A., (1982) Annealing of poly(vinylidene fluoride) and formation of a fifth phase, Macromolecules, Vol. 15, No. 1, pp. 40-44, ISSn 0024-9297.

26. Matsushige, K., Nagata, K., Imada, S., & Takemura. T., (1980) The II-I crystal transformation of poly(vinylidene fluoride) under tensile and compressional stresses, Polymer, Vol. 21, No. 12, (October 1979), PP. 1391-1397, ISSn 0032-3861.

27. Mattsson, B., EriCson, H., Torell, L. M., Sundholm. F., (1999) Micro-Raman Investigations of PVDF-Based Proton-Conducting Membranes. Polymer Science Part A: Polymer Chemistry, Vol. 37, pp.3317-3327, ISSn 1099-0518.

28. Miller, R. L., Raisoni,. J. J., (1976) Single crystals of poly(vinylidene fluoride), Polymer Science polymer Physics, Vol. 14, No. 12 (March 2003), pp. 2325-2326, ISSn 1099-0518.

29. Mohamadi, S., Sharifi-Sanjani, N. & Mahdavi. H., (2011) functionalization of Graphene Sheets via Chemically

30. Mohamadi. S., & Sharifi-Sanjani, N., (2011) Investigation of the crystalline structure of PVDF in PVDF/PMMA/Graphene polymer blend nanocomposites, Polymer Composites, Vol. 32, pp. 1451–1460, ISSn 0272-8397.

31. Nalwa. N. S., (1995) Ferroelectric Polymers, Marcel Dekker, NewYork , Inc, ISBN 0-8247- 9468-0.

32. Ogoshi, T., & Chujo, Y., (2005) Synthesis of poly(vinylidene fluoride) (PVdF)/ silica hybrids having interpenetrating polymer network structure by using crystallization between PVdF chains, Polymer Science: Part A, Polymer Chemistry, Vol. 43, No. 16, pp. 3543-3550, ISSn 1099-0518.

33. Park, Y. J., Kang , Y. S., Park. C. (2005) Micropatterning of semicrystalline poly(vinylidene fluoride) (PVDF) solutions, European Polymer Journal, Vol. 41, pp. 1002–1012, ISSn 0014-3057.

34. Potts, J. R., Dreyer, D. R., Bielawski, C. W., Ruoff. R. S., (2011), Graphene-based polymer nanocomposites, Polymer, Vol. 53, No. 1, (November 2010), pp. 5-25, ISSn 0032-3861.

35. Pramoda, K. P., Linh, N.T.T. , Tang , P.S. , Tjiu, W.C. , Goh, S. H., He. C.B. , (2010) Thermo- mechanical properties of poly(vinylidene fluoride) modified graphite/poly(methyl methacrylate) nano composites, Composite Science and Technlogy, Vol. 70, No. 4,(December 2009) pp. 578-583, ISSn 0266-3538.

36. Pramoda, K., Mohamed, A., Phang, I. Y., & Liu. T., (2005) Crystal transformation and thermomechanical properties of poly(vinylidene fluoride)/clay nanocomposites, Polymer International, Vol. 54, No. 1, (April 2004)pp. 226-232, ISSn 1097-0126.

37. Ramanathan, T., Stankovich, S., Dikin, D. A., Liu, H., Shen, H., Nguyen, S. T., & Brinson. L. C., (2007), Graphitic nanofillers in PMMA nanocomposites—An investigation of particle size and dispersion and their influence on nanocomposite properties, Polymer Science : Part B: Polymer Physics, Vol. 45, No. 15, pp. 2097-2112, ISSn 1099-0488.

38. Ramasundaram, S., Yoon, S., Kim, K. J. , Park. C., (2008). Preferential Formation of Electroactive Crystalline Phases in Poly(vinylidene fluoride)/Organically Modified Silicate Nanocomposites, Polymer Science: Part B: Polymer Physics, Vol. 46, 2173– 2187, ISSn 1099-0518.

39. Scheinbeim, J. I., Newman, B. A., Sen. A., (1986), Field-induced crystallization in highly plasticized poly(vinylidene fluoride) films, Macromolecules, Vol. 19 (November, 1985), PP. 1454-1458. ISSn 0024-9297.

40. Stankovich, S., Dikin, D. A., Dommett, G. H. B., Kohlhaas, K. M., Zimney, E. J., Stach, E. A., & Piner, R. D. (2006) Graphene-based composite materials, Nature, Vol, 442, pp. 282-286, ISSn 0028-0836.

41. Stankovich, S., Dikin, D. A., Piner, R. D., Kohlhaas, K. A., Kleinhammes, A., Jia, Y., Wu, Y., Nguyen, S. T., & Ruoff, R. S. (2007) Synthesis of graphene-based nanosheets via chemical reduction of exfoliated graphite oxide , Carbon, Vol. 45, pp. 1558-1565, ISSn 0008-6223.

42. Steurer, P., Wissert, R., Thomann, R., & Mülhaupt. R. (2009) functionalized Graphenes and Thermoplastic Nanocomposites Based upon Expanded Graphite Oxide, Macromolecular Rapid Communication, Vol. 30, No. 4-5. pp. 316-327, ISSn 1521-3927.

43. Steurer. P., Wissert, R., Thomann, R., Mülhaupt. R., (2009), functionalized Graphenes and Thermoplastic Nanocomposites Based upon Expanded Graphite Oxide, Macromolecular Rapid Communication. Vol. 30, pp. 316–327, ISNN 1521-3927.

44. Ulaganathan, M., & Rajendran, S. (2010) Preparation and characterizations of PVAc/P(VdF- HFP)-basedpolymer blend electrolytes, IoniCs, Vol. 16, (December 2009) pp. 515- 521, ISSn 0947-7047.

45. Vo, L. T., & Giannelis. E. P., (2007) Compatibilizing Poly(vinylidene fluoride)/ Nylon-6 Blends with Nanoclay. J. Macromolecules, Vol. 41, No. 23(August , 2007) pp. 8271- 8226, ISSn 0024-9297.

46. Wang, J., Li, H., Liu, J., Duan, Y., Jiang, S., Yan. S., (2003), On the $\alpha \rightarrow \beta$ Transition of Carbon-Coated Highly Oriented PVDF Ultrathin Film Induced by Melt Recrystallization, Journal of American chemical society, Vol. 125, No. 6, (November, 2002) pp. 1496-1497, ISSn 0002-7863.

47. Yang, D., & Chen. Y., (1987), β-phase formation of poly(vinylidene fluoride) from the melt induced by quenching, Journal of material Science letter, Vol. 6, No. 5, pp. 599-603, ISSn 0261-8028.

48. Yee, W. A., Kotaki, M., Liu, Y., Lu. X., (2007) Stress-induced structural changes in electrospun polyvinylidene difluoride nanofibers collected using a modified rotating disk, Polymer, Vol. 49, No. 19, pp. 4196-4203, ISSn 0032-3861.

49. Yen, J. H., Amin-Sanayei, R., (2006). Polyvinylidene Fluoride, in: Encyclopedia of Chemical Processing,Vol.1, Lee. S, pp. (2379-2381), Tailor & Francis, ISBN 9780824755003, New York.

50. Yu, L., & Cebe. P., (2009) Effect of nanoclay on relaxation of poly(vinylidene fluoride) nanocomposites, Journal of Polymer Science: Part B: Polymer Physics, Vol. 47, No. 24, (August 2009) pp. 2520–2532, ISSn 1099-0518.

51. Yu, S., Zheng, W., Yu, W., Zhang, Y., Jiang, Q., & Zhao. Z., (2009) Formation Mechanism of β-Phase in PVDF/CNT Composite Prepared by the Sonication Method, Macromolecules, Vol. 42, No. 22, pp. 8870, ISSn 0024-9297.

11 Chapter

REFLECTANCE IR SPECTROSCOPY

1. INTRODUCTION

Infrared spectroscopy is study of the interaction of radiation with molecular vibrations which can be used for a wide range of sample types either in bulk or in microscopic amounts over a wide range of temperatures and physical states. As was discussed in the previous chapters, an infrared spectrum is commonly obtained by passing infrared radiation through a sample and determining what fraction of the incident radiation is absorbed at a particular energy (the energy at which any peak in an absorption spectrum appears corresponds to the frequency of a vibration of a part of a sample molecule).

Aside from the conventional IR spectroscopy of measuring light transmitted from the sample, the reflection IR spectroscopy was developed using combination of IR spectroscopy with reflection theories. In the reflection spectroscopy techniques, the absorption properties of a sample can be extracted from the reflected light.

Reflectance techniques may be used for samples that are difficult to analyze by the conventional Transmittance method. In all, Reflectance techniques can be divided into two categories: internal reflection and external reflection. In internal reflection method, interaction of the electromagnetic radiation on the interface between the sample and a medium with a higher refraction index is studied, while external Reflectance techniques arise from the radiation reflected from the sample surface. External reflection covers two different types of reflection: specular (regular) reflection and diffuse reflection. The former usually associated with reflection from smooth, polished surfaces like mirror, and the latter associated with the reflection from rough surfaces.

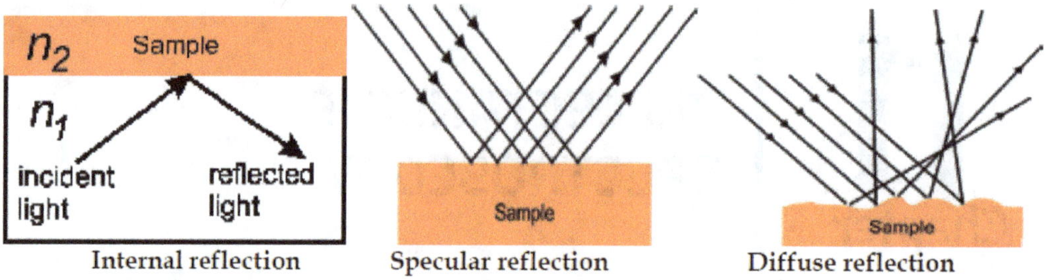

Fig. 1. Illustration of different reflection phenomenon.

2. REFLECTANCE METHODS

Upon interaction of electromagnetic radiation with a sample surface, depending on the characteristic of the surface and its environment, the light may undergo three types of reflection, internal reflection, specular reflection and diffuse reflection. In practice, all three types of reflections can occur at the same time, although with different contributions. Specular reflection is defined as light reflected from a smooth surface such as a mirror (any irregularities in the surface are small compared to λ) at a definite angle whereas diffuse reflection produced by rough surfaces that tend to reflect light in all directions.

2.1 Internal Reflectance spectroscopy (IRS)

Internal Reflectance Spectroscopy (IRS) date back to the initial work of Jacques fahrenfort and N.J.Harrick [1, 2] that independently devised the theories of IRS spectroscopy and suggested a wide range of applications. Internal reflection Spectroscopy is often termed as attenuated total reflection (ATR) spectroscopy. ATR became a popular spectroscopic technique in the early 1960s.

Attenuated total reflection spectroscopy utilizes total internal reflection phenomenon[3]. An internal reflection occurs when a beam of radiation enters from a more dense medium (with a higher refractive index, n_1) into a less-dense medium (with a lower refractive index, n_2), the fraction of the incident beam reflected increases as the angle of incidence rises. When the angle of incidence is greater than the critical angle θc (where is a function of refractive index of two media), all incident radiations are completely reflected at the interface, results in total internal reflection (Figure 2). In ATR spectroscopy a crystal with a high refractive index and excellent IR transmitting properties is used as internal reflection element (IRE, ATR crystal) and is placed in close contact with the sample (Figure 3). The beam of radiation propagating in IRE undergoes total internal reflection at the interface IRE- sample, provided the angle of incidence at the interface exceeds the critical angle θc. Total internal reflection of the light at the interface between two media of different refractive index creates an "evanescent wave" that penetrates into the medium of lower refractive index [3].

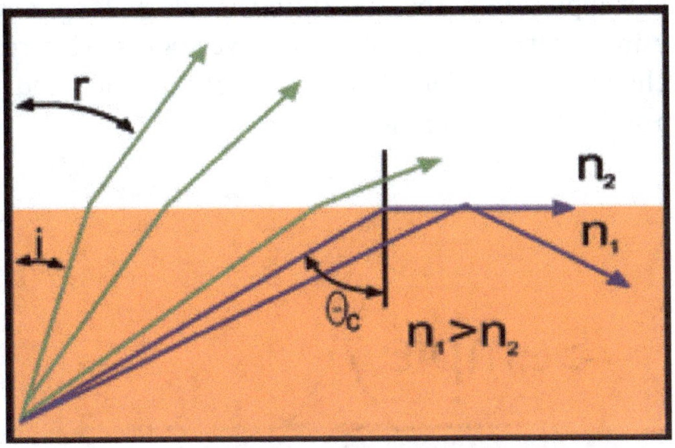

Fig. 2. Illustration of total internal reflection.

The evanescent field is a non-transverse wave along the optical surface, whose intensity decreases with increasing distance into the medium, normal to its surface, therefore, the field exists only the vicinity of the surface. The exponential decay evanescent wave can be expressed by Eq. (1):

$$I_{ev} = I_0 \exp(-Z/P) \quad (1)$$

Where z is the distance normal to the optical interface, d_p is the penetration depth (path length), and I_0 is the intensity at z = 0.

The depth of penetration, d_p, is defined as the distance from the IRE- sample boundary where the intensity of the evanescent wave decays to 1/e (37%) of its original value), is given by Eq. (2):

$$dp = \lambda / \{ 2\pi n_1 [\sin^2 \theta - (n_2/n_1)^2]^{1/2} \} \quad (2)$$

Where λ is the wavelength of the radiation, n_1 is the refractive index of the IRE (ATR crystal), and n_2 refractive index of the sample, and □ is the angle of incidence. Figure 3 illustrates the evanescent wave formed at the internal reflection element-sample interface.

An ATR spectrum can be obtained by measuring the interaction of the evanescent wave with the sample. If an absorbing material is placed in contact with ATR crystal, the evanescent wave will be absorbed by the sample and its intensity is reduced (attenuated) in regions of the IR spectrum where the sample absorbs, thus, less intensity can be reflected (attenuated total reflection). The resultant attenuated radiation as a function of wavelength produces an ATR spectrum which is similar to the conventional absorption spectrum except for the band intensities at longer wavelengths (Figure 4). This difference is due to the dependency of the penetration depth (d_p) on wavelength: at longer wavelength, the evanescent wave penetrates deeper into the sample, thus, the absorption bands at longer wavelengths are relatively more intense than those shorter wavelengths. This results in greater absorption on the longer wavelength side of an absorption band, contributing to band distortion and band broadening.

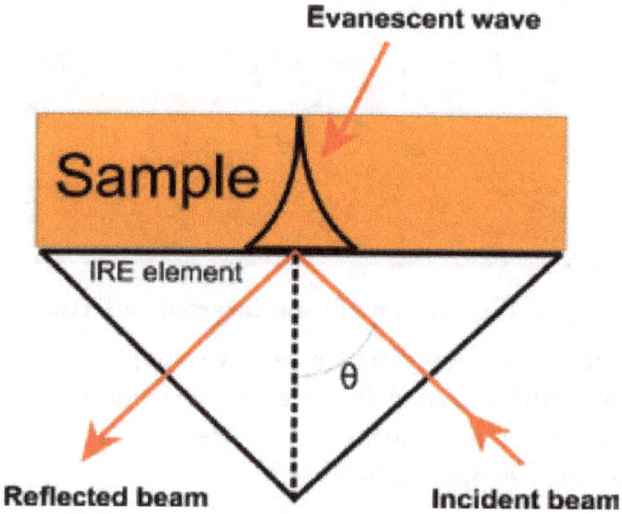

Fig. 3. Penetration of evanescent wave into the sample.

Additionally, compared to the transmission spectrum, small differences may be seen in an ATR spectrum which arises from dispersion effects (variation of refractive index of a material with change of wavelength). An anomalous dispersion causes the refractive index and the penetration depth changes through an absorption band. For instance the effect of dispersion on penetration depth for cocaine is demonstrated in Figure 5. As can be seen, penetration depth changes strongly at wavelength in which the dispersion is the highest.

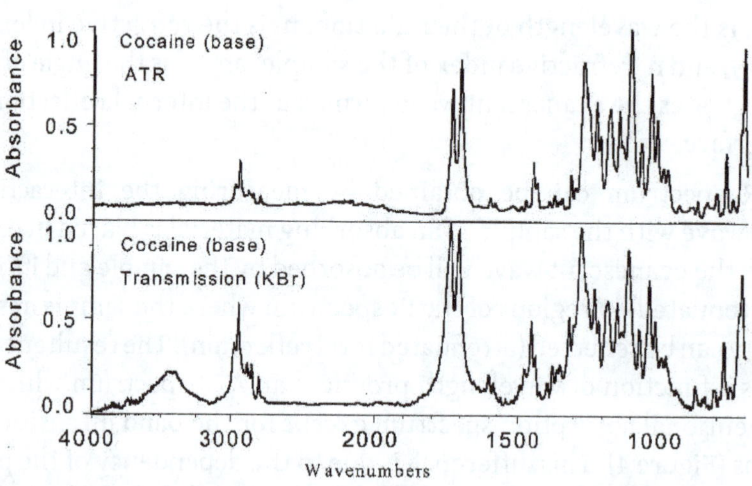

Fig. 4. Representation of ATR and Transmittance IR spectrum of Cocaine [4].

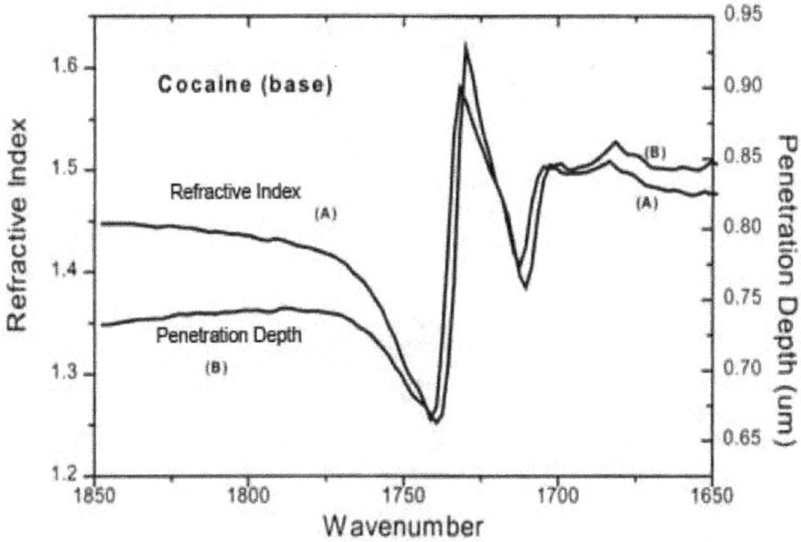

Fig. 5. Penetration depth and refractive index for cocaine at carbonyl absorbance band [4].

Sometimes an empirical so-called ATR correction is applied to compensate across the spectrum for linear wavelength increase, which is termed as Eq.(3):

$$R_{corr} \sim R / \lambda \quad (3)$$

Other differences may occur due to the surface effects between the sample and internal reflection element (IRE crystal). For instance, the degree of physical contact between IRE and the sample influences the sensitivity of an ATR spectrum. Since the evanescent wave only propagates 2-15 μm beyond the surface of the crystal, thus, an intimate contact of the IRE with the sample is essential.

Two configurations of ATR accessory are available. In a single Bounce–ATR, a single internal reflection occurs using a prism whereas Multi-Bounce ATR, undergoes multiple internal reflections (up to 25) using special prisms, as shown in Figure 6. In a multiple internal reflection cell, the effective path length of the sampling surface is the product of the number of reflections at measurement surface and penetration depth. In practice, multiple internal reflections (in Multi-Bounce ATR technique) produce more intensive spectra by multiple reflections and, hence it is useful for weak absorbers, while the single Bounce–ATR is suitable for strong absorbers.

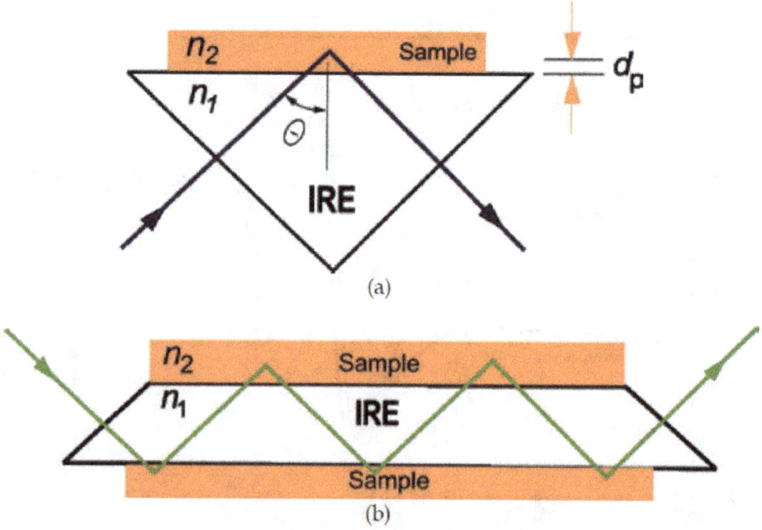

Fig. 6. Schematic representation of total internal reflection with (a) Single reflection, and (b) Multiple reflections.

There are different designs of ATR cell including traditional ATR, horizontal ATR and cylindrical ATR for solids and liquids samples. In the traditional ATR, a thin sample is clamped against the vertical face of the crystal. This design has been replaced by more modern designs, horizontal and cylindrical designs. In horizontal ATR (HATR), the crystal is a parallel-side plate (typically 5 cm by 1 cm) with the upper surface exposed (Figure 6 (b)). The number of reflections at each surface of the crystal depends on length and thickness crystal as well as the angle of incidence (usually between five and ten).

The traditional design is used for continuous surface such as sheets and horizontal ATR (HATR) cells are suitable for liquids and pastes as well as soft powder and sheets films. Application of cylindrical ATR (Figure 7) cell is limited to mobile fluids.

2.1.1 ATR crystals

The ATR crystals are made from materials that have a very high refractive index and low solubility in water. Table 1 summarizes some of the commonly materials used in ATR crystals. Amongst these materials, diamond, zinc selenide (ZnSe) and germanium (Ge) are the most common materials used ATR crystals. Diamond can be used for a wide range of samples, because it is resistant to scratching and abrasion and can tolerate a wide pH range as well as strong oxidants and reductants. Also, due to its ideal properties, it often used as a protective film for ATR crystals such as zinc sulfide (ZnS). However, high cost of diamond causes that its application is limited. Ge with a higher refractive index is suitable for anlayzing of samples with high refractive index. However, ZnSe preferred for all routine applications, but because of its low resistivity towards surface etching, reactivity with complexing agents such as ammonia, and damaging with strong acids and alkalis, its application is limited. AMTIR with similar refractive is a good alternative for ZnSe crystal, when involve strong acids.

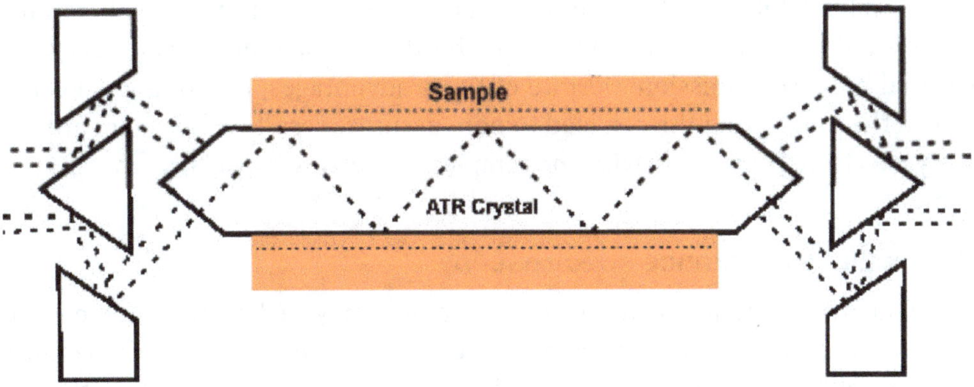

Fig. 7. Schematic representation cylindrical ATR cell.

Table 1. Materials used as ATR crystals

Material	Wave range (cm^{-1})	Refractive index (at 1000 cm^{-1})
Zinc Sulfide (ZnS)	17,000 - 950	2.2
Zinc Selenide (ZnSe)	20,000- 650	2.41
Diamond	45000-2500; 1800 -< 200	2.4
Germanium (Ge)	5,5 00 -870	4.0
Cd telluride (CdTe)	10,000-450	2.65
Saphire (Al_2O_3)	25,000-1800	1.74
Cubic Zirconia (ZrO_2)	25,000-1800	2.15
KRS-5 ($TlI2/TlBr^2$)	20,000-350	2.37
AMTIR (As/Ge/Se glass)	11,000-750	2.5

2.1.2 Application of ATR spectroscopy

ATR technique is used commonly in the near –infrared for obtaining absorption spectra of thin films and opaque materials. However, ATR spectra can be obtained using dispersive IR instruments, but the higher–quality spectra are obtained using FTIR spectrometers.

ATR is one of the most versatile sampling techniques that requires little or no sample preparation for most samples. It only requires that the sample is placed in intimate contact with the IRE crystal, which achieved by pressing the solid onto the crystal with a high pressure clamp. Powders and soft pliable films can be used without any additional preparation and highly crystalline solids should be ground before application. It is an ideal method for liquids and oils, because the contact between the crystal and a liquid is inherently close and hence without requiring the high pressure clamp, liquids are applied directly onto the crystal.

As a conclusion, ATR is a non-destructive technique for a variety of materials including soft solid materials, liquids, powders, gels, pastes, surface layers, polymer films, samples solutions after evaporation of the solvent. It is an ideal technique for thick and dark colored materials which often absorb too much energy to be measured by IR transmission. Despite of these advantages, lack of a good contact between the sample and IRE can lead to non- accurate results. Also, there are a few IRE crystals to be compatible with the samples properties, especially from pH point of view.

2.2 Specular Reflectance spectroscopy

The incident radiation focused onto the sample may be directly reflected by the sample surface, giving rise to specular reflection, and it may also undergo multiple reflections at the sample, resulting in diffuse reflection. In external Reflectance techniques, the radiation reflected from a surface is evaluated (Figure 8).

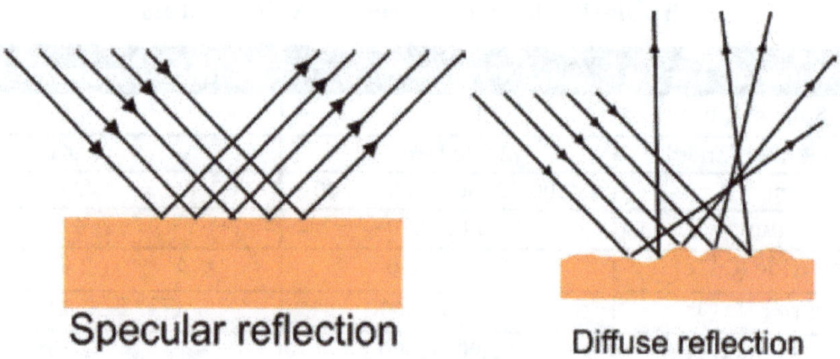

Fig. 8. Illustration of external reflection.

Specular Reflectance techniques basically involve a mirror-like reflection from the sample surface that occurs when the reflection angle equals the angle of incident radiation. It is used for samples that are reflective (smooth surface) or attached to a reflective backing. Thus, specular techniques provide a Reflectance measurement for reflective materials, and a reflection–absorption (*transflectance*) measurement for the surface films deposited on, or pressed against reflective surfaces (Figure 9).

The Reflectance spectra differ from those recorded in transmission, they appear as "derivative-like" bands. These spectra can be converted into absorption one by using of Kramers-Kronig transformation (K-K transformation) that is available in most spectrometer software package. Figure 10 depicts specular Reflectance spectrum of oil on surface of machined steel cylinder.

In absorption-reflection measurement, one fraction of the radiation is reflected on the upper interface and contributes towards the spectrum via specular reflection. Another part of the radiation penetrates the surface film and is reflected by the reflective surface, thus, the light passes through the surface layer twice-to and from the reflective surface, leading to increase the intensity of the Reflectance spectrum as compared to the normal transmission. The effective path length depends on the angle of incidence, therefore, for thin films, a grazing angle of incidence as high as 80°- 85° from normal incidence should be used, and for thick films an angle close to normal incidence is applied.

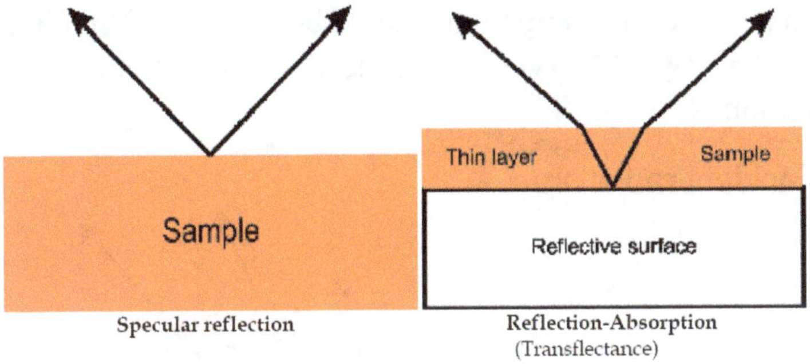

Fig. 9. Illustration of specular Reflectance.

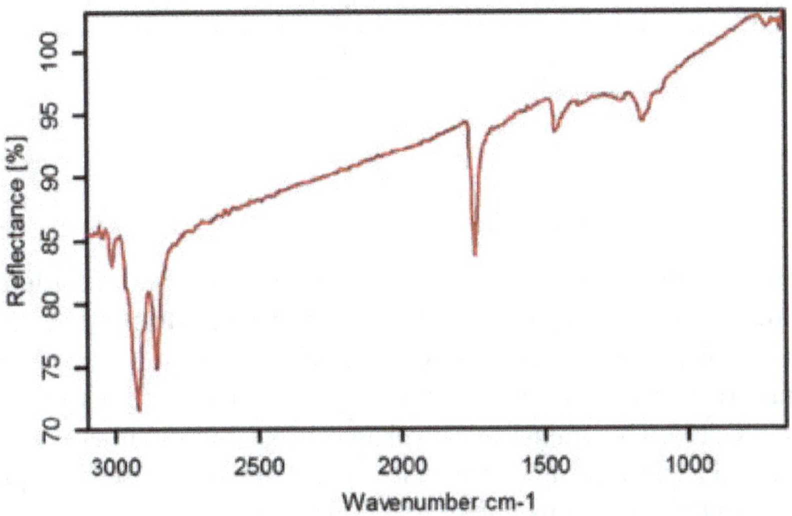

Fig. 10. Specular Reflectance spectrum of oil on surface of machined steel cylinder.

The most common applications of this technique are evaluation of surfaces such as: coating, thin films, contaminated metal surface.

2.3 Diffuse Reflectance Spectroscopy (DRS)

In diffuse Reflectance spectroscopy, the electromagnetic radiation reflected by roughened surfaces is collected and analyzed. When this technique is applied in (FT) IR region, it is termed as diffuse Reflectance infrared Fourier transform spectroscopy (DRIFTS). It is also known "Kubelka–Munk "reflection,because they developed a theory on the radiation transport in scattering media [5].

Light incident onto a solid sample may be partly reflected regularly (specular reflection) by the sample surface, partly scattered diffusely, and partly penetrates into the sample. The latter part may be absorbed within the particles or be diffracted at grain boundaries, giving rise to diffusely scattered light in all directions. Diffuse Reflectance spectroscopy associated with the reflected lights which are produced by diffuse scattering (Figure 11). Since regular reflection distorts the DRS spectra, thus, the regular reflection component should be eliminated in diffuse Reflectance measurement. The DRIFTS accessory is designed to eliminate the specularly reflected radiation.

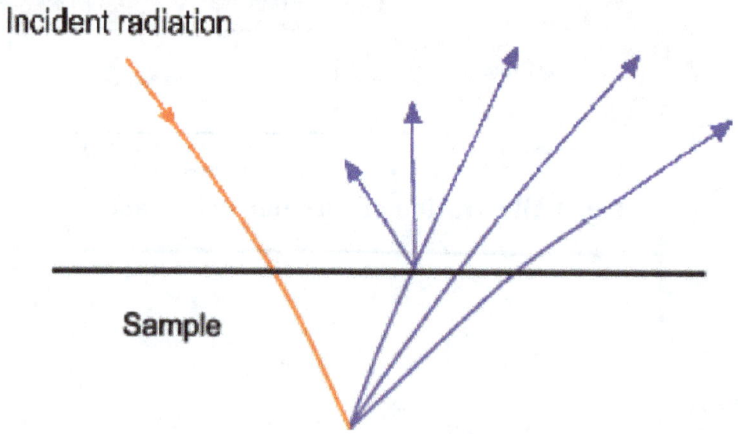

Fig. 11. Representation of diffuse Reflectance.

In diffuse Reflectance spectroscopy, there is no linear relation between the reflected light intensity (band intensity) and concentration, in contrast to traditional transmission spectroscopy in which the band intensity is directly proportional to concentration. Therefore, quantitative analyzes by DRIFTS are rather complicated. The empirical Kubelka - Munk equation relates the intensity of the reflected radiation to the concentration that can be used for quantitative evaluation. The Kubelka-Munk equation is defined as:

$$f(R_\infty) = (1 - R_\infty)^2 / 2R_\infty = c/k = K/s \quad (4)$$

where $f(R\infty)$ is Kubelka-Munk function, $R\infty$ is the absolute Reflectance of layer (the ratio of the sample diffuse Reflectance spectrum and a non-absorbing reference (normally KBr or KCl)), both measured at indefinite depth. K is the absorption coefficient, s is the diffusion (scattering) coefficient which is proportional to the fraction of diffused light, k is molar absorption coefficient (proportional to the fractional of transmitted light) and c is sample concentration. The Reflectance, R∞, indicates the sample is thick enough that no radiation reaches the back surface (indefinite thickness). Appling Kubelka–Munk function on the Reflectance spectrum produces a spectrum (corrected spectrum) that resembling the transmission spectrum (Figure 12). Also the corrected spectrum demonstrates a linear relationship between band intensity and the sample concentration.

In the case of $R_\infty < 0.01$, the simpler function log $1/R_\infty$ is often used for measuring the diffuse Reflectance. Since such small R∞ value is usually found in the near-IR region, therefore, an alternative relationship between concentration and reflected intensity in near–IR is defined as:

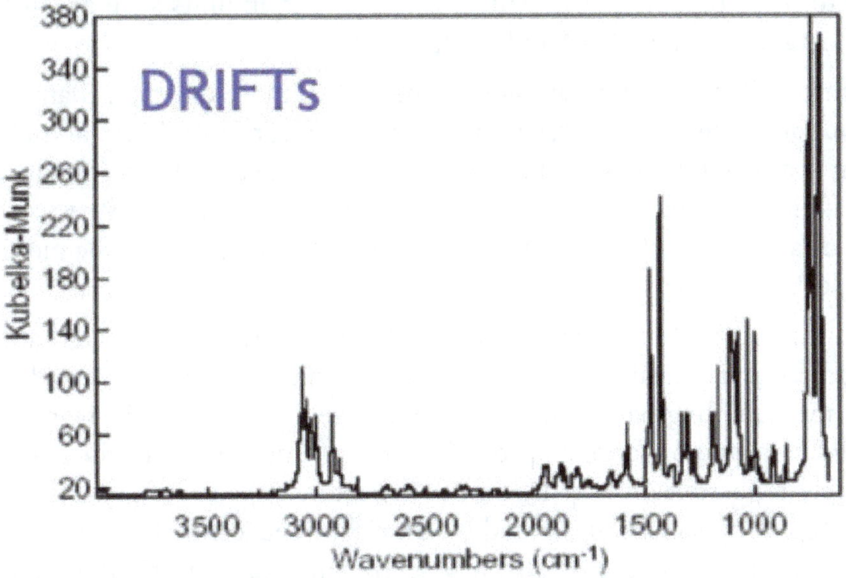

Fig. 12. DRIFT spectrum of 1,2-bis (diphenyl phosphino) ethane, pure powder [6].

$$\log (1/R_\infty) = k'c \quad (5)$$

where k′ is a constant.

In Kubelka-Munk equation it is assumed that s is independent of wavelength and the sample is weakly absorbing. The former condition is achieved by proper sample preparation and the latter by dilution of strong absorbing samples with non-absorbing substrate powder (such as KBr or KCl). Therefore, to obtain reproducible results, particle size, sample packing and dilution should be carefully controlled, especially for quantitative analysis.

2.3.1 Application of diffuse Reflectance spectroscopy

Diffuse Reflectance technique is used for powders and solid samples having rough surface such as paper, cloth. In diffuse Reflectance technique, particles size, homogeneity, and packing density of powdered samples play important role on the quality of spectrum. A sample with smaller particle size having narrow size distribution is preferred. Thus, in order to obtain a qualified spectrum, the sample should be ground into smaller size.

In this method,the sample can be analyzed either directly in bulk form or as dispersions in IR transparent matrices such as KBr and KCl. Sometimes, a thin film of KBr powder placed on the sample surface to improve the quality of the spectrum. Dilution of analyte in a non- absorbing matrix increases the proportion of diffuse Reflectance in the reflected light. Typically, the solid sample is diluted homogeneously to 5 to 10% by weight in KBr. The spectra of diluted samples are similar to those obtained from pellets when plotted in units such as log 1/R (R is the Reflectance) or Kubelka- Munk units.

Diffuse Reflectance measurement in near-IR is more common than in mid-IR. Because non- absorbing scattering substrates are rare in mid-IR, and also more efficient scattering occurs at shorter wavelengths (near-IR). Additionally, due to lower efficiency of the scattering in mid-IR, diffuse Reflectance is very weak in this region, as a consequent, in mid-IR, diffuse Reflectance could only be measured by FT-IR spectrometer (DRIFTS).

In diffuse Reflectance spectroscopy, diffusely scattered light can be directly, collected from material in a sampling cup or, alternatively, collected by using an abrasive sampling pad. In mid-IR, the diffusely reflected light from sample is generally collected by large ellipsoidal mirrors, which cover as much area above the sample as possible. In near-IR, diffuse Reflectance spectra are usually measured by an integrating sphere, described by Ulbrich, (Figure 14). The inner surface of "Ulbrich sphere" is coated by strongly scattering, non- absorbing powder. After repeated reflection, all radiations reach the detector. Thus, with Ulbrich sphere the entire radiation reflected by the sample is integrated.

In mid-IR region, the inner surface of diffuse scattering sphere is treated with gold vapor to guarantee a high degree of reflection, while in near- IR region, Ulbrich sphere consists of spectralon (a thermostatic resin) which is applied as white standard due to its high degree of reflection.

DRIFTS is a versatile technique for analyzing nontransparent samples, powders, roughened surfaces and coating. It offers the advantages of easy sample preparation and applicability to analyze samples at elevated temperature and pressure [6]. Consideration spectrum obtained for an organic material by transmission IR using KBr pellet and DRIFTS reveals the capacity of DRIFTS compared to transmission (Figure 12,13). Both spectra show similar spectrum that corresponds to absorption bands of the sample, but in the spectrum recorded in transmission mode an intensive band appeared at 3500 cm^{-1} which is related to the water present in the

KBr pellet. This example, demonstrates DRIFTS combines the advantages of easy sample preparation with the absence of water band in the spectrum.

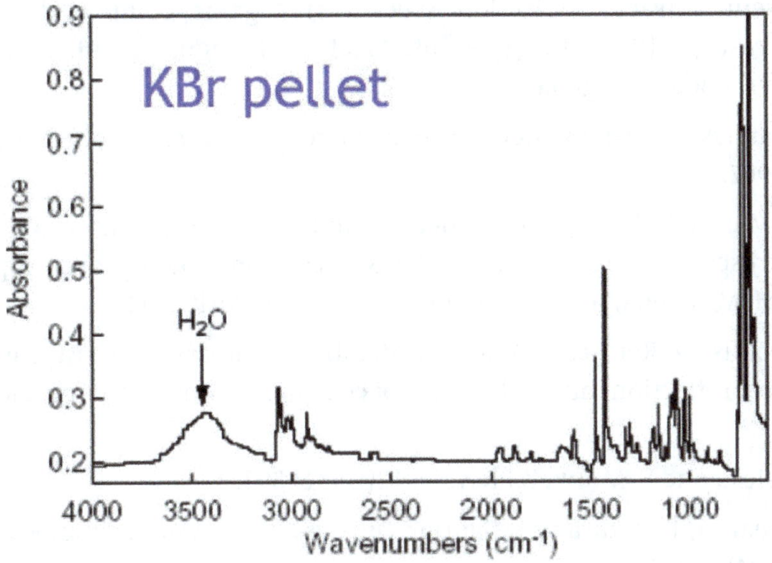

Fig. 13. Transmission spectrum of 1,2-bis (diphenyl phosphino) ethane, KBr pellet [6].

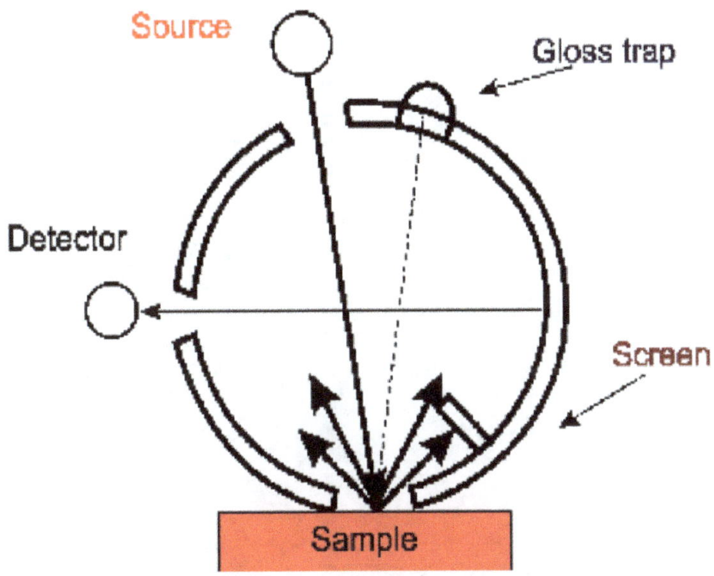

Fig. 14. Integrating Sphere.

The most disadvantage of this technique is difficulty of quantitative analyses. Since the diffusion coefficient strongly depends on sample preparation, thus the Reflectance intensity is influenced by the sample preparation. A same sample may produce different spectra in different experiments.

3. REFERENCES

1. J. Fahrentfort, Attenuated total reflection- A new principles for the production of useful infrared reflection spectra of organic compounds. Molecular Spectroscopy (Proceeding IV Int. Meeting, Bologna, 1959), 2, Mangini A, editor, Londan: Pergamon, 1962, 437.
2. NJ. Harrick, Internal Reflection Spectroscopy. New York, John Wiley & Sons, Int.,1967.
3. FM. Maribeli, "Principles, Theory, and Practice of Internal Reflection Spectroscopy," Internal Reflection Spectroscopy-Theory and Applications, ed. FM. Mirabella, Jr., Marcel Dekker, Inc. New York 1993.
4. CV. Koulis, JA. Reffner, B.A. John, AM. Bibby, Comparison of transmission and internal reflection infrared spectra of cocaine. J. Forensic Science,46(4),(2001) 822.
5. P.Kubleka and F. Munk, Z.Phsysik, 12(1931) 593.
6. T. Armaroli, T.becue and S. Gautier, Oil, Gas Science and technology – Rev.IFP, 59 (2004) 2, 215.

12 Chapter

EVALUATION OF GRAFT COPOLYMERIZATION OF ACRYLIC MONOMERS ONTO NATURAL POLYMERS BY MEANS INFRARED SPECTROSCOPY

1. INTRODUCTION

Infrared spectroscopy (IR spectroscopy) is a technique based on the vibrations of atoms of a molecule. An IR spectrum is commonly obtained by passing IR radiation through a sample and determining what fraction of the incident radiation is absorbed at a particular energy. The energy at which any peak in an absorption spectrum appears corresponds to the frequency of a vibration of a part of a sample molecule (Bickford, 2008).

For a molecule to show IR absorptions, it must possess specific feature: electric dipole moment of molecule must change during the movement that means changes in molecular dipoles which are associated with vibrations and rotations. The atoms in molecules can move relative to one another. This is a description of stretching and bending movements that are collectively referred to as vibrations. Vibrations can involve either change in bond length (stretching) or bond angle (bending). Some bonds can stretch in-phase (symmetrical stretching) or out-of-plane (asymmetric stretching) (Bickford, 2008).

The IR is divided into three regions; the near-, mid- and far- IR. The mid-IR is the most common region to identification and study of organic compounds based on fundamental vibrations and associated rotational-vibrational structure.

IR spectroscopy is a popular method for characterizing polymers. This spectroscopy may used to identify the composition of polymers, to monitor polymerization processes, to characterize polymer structure, to examine polymer surface, and to investigate polymer degradation processes. There are several reports of use of IR spectroscopy to evaluate grafting of acrylic monomers onto natural materials as carboxymethyl cellulose (CMC) and chicken feathers (CF) (Martínez et al, 2003, 2005, Vasile et al, 2004 , Zohuriann-Mehr et al, 2005, Joshi and Sinha, 2006).

Now a days have been an increasing interest in obtation and use of polymeric materials using naturals sources from a diversity of systems, getting atention such biopolymers in research areas as medicine, electronics, textil, corrosion and nanotechnology among others. This rising interest is due variety of properties that offers on depending of chemical structure and source (Martínez et al, 2008).

Use of natural fibers is an research area that allows obtain materials for cotidiane applications, using more resistence materials and with outstanding properties and specially materials that are environmental friendly. Keratin is a natural protein which can be found in wool, hair, claws, horns or nails, and is the main component in birds' feathers, represents from 5% to 7% of the body weight of chickens. Keratin is durable and resistant to organic solvents and chemically unreactive, which give benefits when is exposure to environmental conditions, thinking in industrial applications.

Chicken feathers (CF) are more abundant material of keratin in nature. Birds' feathers characterize by a complex branched structure formed by keratinic filaments that grow in a unique mechanism in cylindric feather follicles. This branched structure is a distinctive characteristic in feathers morphology and its origin is biologic evolution (Xu et al., 2001, 2003).

CF are considered residues of a byproduct of poultry, corresponding to more than 5 million tons around the world (Barone et al 2005). Aminoacids content of CF on depend of breed, feeding and environment of study animals. The aminoacids present in CF are mainly aspartic acid, glutamic acid, arginine, proline, glycine, phenylalanine, alanine, cystine, isoleucine, among others (Schmidt 1998) (see table 1).

Table 1. Aminoacid content in keratin fiber from chicken feather (Martínez et al 2005).

Functional group	Aminoacid	Content (as % mole)
Negatively charged	Aspartic acid	5
	Glutamic acid	7
Positively charged	Arginine	5
Conformationally special	Proline	12
	Glycine	11
Hydrophobic	Phenylalanine	4
	Alanine	4
	Cystine	7
	Valine	9
	Isoleucine	5
	Leucine	6
	Tyrosine	1
Hydrophilic	Threonine	4
	Serine	16

Carboxymethyl cellulose (CMC) is a very important cellulose derivative and it is known by its superabsorbent properties. It is anionic polymer water soluble. It is produced by reaction of alkali cellulose and monochloroacetic sodium salt under strict reaction conditions (Klemm et al, 1998). Byproducts as sodium chloride and sodium glicolate are obtained in reaction whose are removed obtaining CMC sodium salt highly purified. Higher swelling capacity of CMC can be reached and controlled by addition of crosslinking agent, thermal treatment or ionic state conversion. The conversion of some hydroxyl groups from cellulose in hydrophobic substituents diminishes hydrogen bonding decreasing crystallinity and increasing water solubility. Multiple applications of Na-CMC in several areas of food industry made it and almost indispensable material. Other applications are: laundry detergent for instance, mainly as a viscosity modifier or thickener, as a soil suspension to deposit onto cotton and other cellulosic fabriCs (Klemm et al, 1998). Other use of CMC is as a lubricant in non-volatile eye drops. These properties on depend of preparation method, degree of polymerization and degree of substitution (DS), that means how much of hydroxylic groups are displaced for carboxymethyl groups in anhydroglucose (AGU) structure. The more useful DS value is from 0.7 to 1.5. Due these CMC is an interesting material and properties can be modify by means graft reactions.

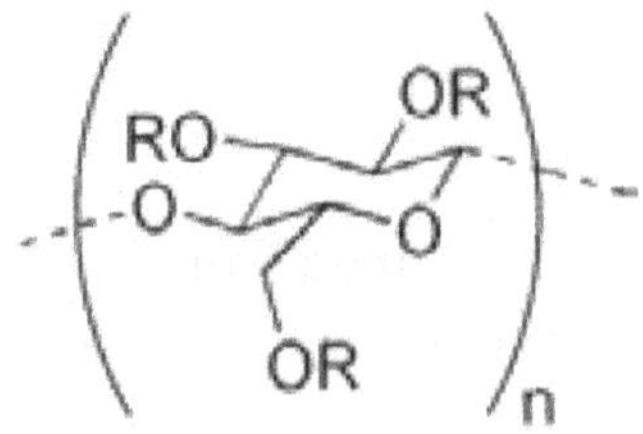

R = H or CH_2CO_2H

Fig. 1. Carboxymethyl cellulose structure.

Modification of natural protein through grafting have become in a widely used path and some works have carry out studies with acrylic monomers as ethylacrylate, acrylic acid, methacrylic acid and methyl methacrylate (Mostafa, 1995; Athawale and Rathi, 1997; Jia and Yong, 2006; Zampano, et al 2009), which carried out graft of natural fibers and proteins. However, just few of works report grafting of hydroxyethyl methacrylate onto naturals fibers (Joshi and Sinha, 2006) (Martínez et al., 2003, 2005).

Graft copolymerization of acrylic monomers onto natural polymers is one of the most useful paths to modify physicochemical properties on order to add new properties to final copolymer with minimal. Acrylic monomers are the most grafted monomers in this kind of research works; being acrylonitrile, acrylic acid, methyl methacrylate more studied species. Redox initiation is an efficient method frequently used to obtain graft copolymers. Polymers with OH groups can react with ceric ion or an oxidant agent to form polymer radical capable to initiate copolymerization. In a redox initiation frequently there is a minimum degree of homopolymerization, due only polymer radicals can be formed. However, limitation of this method is only useful with polymers with functional groups; there are reports of polymers with amides, urethane and nitrile groups by means redox initiation.

Several natural polymers as chitin, cellulose, functionalized cellulose and natural fibers are some of most studied natural polymers in graft copolymerization using redox system as initiator, being cerium ion one with more reports.

Infrared spectroscopy is a useful tool to identify functional groups through vibrational frequencies in polymers to evaluate changes in structure This research was focused in graft copolymerization of Hydroxyethyl methacrylate (HEMA) onto chicken feathers fibers (CFF) and carboxymethyl cellulose (CMC), evaluating effect of reaction conditions (time reaction, monomer concentration, initiator concentrations) on grafting yield and probe presemce of HEMA in copolymers by means Infrared Spectroscopy (IR).

2. METHODOLOGY

2.1 Raw materials

Chicken Feathers (CF) were obtained from a local slaughterhouse. A clean up procedure was carried out before use them for reaction, CF were washed several times with ethanol and dried at room temperature to have them clean white, sanitized and odor-free. Manual procedure by cutting to separate fibers from barbs and barbules was carried out. Chicken feather fiber (CFF) was used in graft copolymerization reaction.

CMC supplied by Sigma-Aldrich with 0.7 degree of substitution was used to graft reactions, without further purification process, same as all other reactives: Hydroxyethyl methacrylate (HEMA) (98%) and malic acid, potassium permanganate (KMnO4), hydrochloric acid (HCl), methanol, ethanol, sulfuric acid (H_2SO4), Cerium ammonium nitrate were from Sigma- Aldrich. Distilled water was used as reaction medium and finally to wash the homopolymer residues.

2.2 CFF graft copolymers

To carry out the grafting reaction procedure proposed by Martinez-Hernandez et al. (2003) was used. In this process the following reagents were used: 0.5 g CFF, distilled water, malic acid (0.005M), sulfuric acid (0.01M), KMnO4 (0.003M) and HEMA monomer in 3 different levels (0.025, 0.05 and 0.075 M).

The substances were mixed at a temperature of 60 ° C, under constant magnetic stirring. The reactions were carried out at three different times: 2, 3 and 4 h. Once the reaction time passed, proceeded to filter and wash the reaction product with hot water and methanol in order to remove residual monomer and homopolymer. Grafting yield was evaluated after a hydrolysis to reaction product, to determinate HEMA amount grafted in CFF. Hydrolysis procedure was: 1g of CFF grafted was swamped in a 6M HCl solution at 130°C for 24 hours in a soxhlet extractor. Once the time was over, hydrolysis residue was dried and weight was determinate, this weight correspond to grafted HEMA, due CFF decompose in hydrolysis process.

Grafted polymer percentage was determinate according with next formulae (Martínez- Hernández et al, 2003; Gupta and Sahoo, 2001):

$$\%\text{Grafted PHEMA} = (X_4 / X_1) * 100 \quad (1)$$

Where:

X_1= CFF without modification

X_2 = container

X_3= container + hydrolyzed sample

X_4= residual polymer after hydrolysis (X3- X2)

2.3 CMC graft copolymers

2.3.1 Initiator preparation

It was necessary preparation of Ce (IV) initiator solution. The required amount of cerium ammonium nitrate salt (0.1, 0.25 and 0.5 M), was dissolved in 100 mL of 1M HNO_3 solution. This solution was stored under refrigeration in an amber bottle.

2.3.2 Grafting reaction

For graft reaction 5 g of CMC were placed in the reactor with distilled water and HEMA was added in selected amount (3 levels: 0.2, 0.3 or 0.4 M), under constant stirring at 70 º C, the initiator was added and reaction was carried out for 3 hours. Once the reaction ends, it was necessary to neutralize the mixture with a NaOH 10% w/w solution, copolymer was precipitated with acetone, and then the material was milled and washed with 90% methanol solution, then dried. After, the material was subjected to a Soxhlet extraction with methanol, in order to extracting the material that has not been in the copolymer. Grafting yield was determinated by weight differences according with next formulae:

$$\text{Grafting yield} = ((W_1 - W_0)/W_0) * 100 \quad (2)$$

Where:

W_0= CMC weight

W_1= graft copolymer weight

2.4 IR Spectroscopy characterization of graft copolymers

IR analysis was carried out to evaluate structural changes of CMC and CFF and its grafted copolymers, by means main functional groups signals. IR spectra were recorded with a Perkin-Elmer Spectrum One Fourier Transform IR spectrophotometer, using an Attenuated Total Reflactance (ATR) accessory, with ZnSe plate, using 12 scans and resolution of 4 cm^{-1}, ranging from 4000 to 600 cm^{-1}.

3. RESULTS AND DISCUSSION

Results will be presented in sections following next order. First, the effect of HEMA concentration and time reaction over grafting yield onto CFF and HEMA and CAN concentration over grafting yield onto CMC. Next, IR spectra are presented for ungrafted and grafted CFF, as well as a comparison of IR spectra for different grafting yield of CFF. Also for CMC grafted and ungrafted are presented the IR spectra and comparison to different grafting yield of HEMA onto CMC.

3.1 CFF-g-HEMA copolymer

Effect of reaction conditions over graft yield were studied. Table 2 shows effect of time reaction and HEMA concentration on grafting yield of CFF, initiators concentration was constant. It can observe that an HEMA concentration increase causes an increase of grafting yield value at 3 reaction times studied. That increase is due the amount of free radicals formed in reaction system, and CFF main component is keratin which posses several pendant functional groups, for like $-NH_2$, -COOH, -SH and -OH, which can form active sites where HEMA can be grafted. This behavior was also reported by Martínez et al (2003) that observed a maximum graft yield value. A higher HEMA concentration value was studied (1 M), but a fiber saturation was observed and also homopolymerization predominated over graft reaction which is not convenient for this study. Scanning Electron Microscopy shows the HEMA cover the CFF structure growing poly HEMA chains (figure 2) on fiber surface of CFF, but CFF reached a saturation of active sites and then homopolymerization happens.

Table 2. Effect of time reaction and HEMA concentration on grafting yield over CFF

Time/hr	HEMA conc./ M	Grafting yield
2	0.075	85%
2	0.05	70%
2	0.025	26%
3	0.075	92%
3	0.05	83%
3	0.025	57%
4	0.075	76%
4	0.05	38%
4	0.025	35%

The increase on graft yield can be also attributed that keratin forms a charge transfer complex with HEMA molecules, so it is possible increase monomer activity at higher concentrations of HEMA which leads homopolymerization.

Other reports (Joshi and Sinha, 2006) indicate an increase of monomer concentration increases graft yield, until a maximum and then decreases the obtained percentage. The difference is attributed to graft yield diminish due the saturation of available active sites in CFF. Also on depend of initiator system.

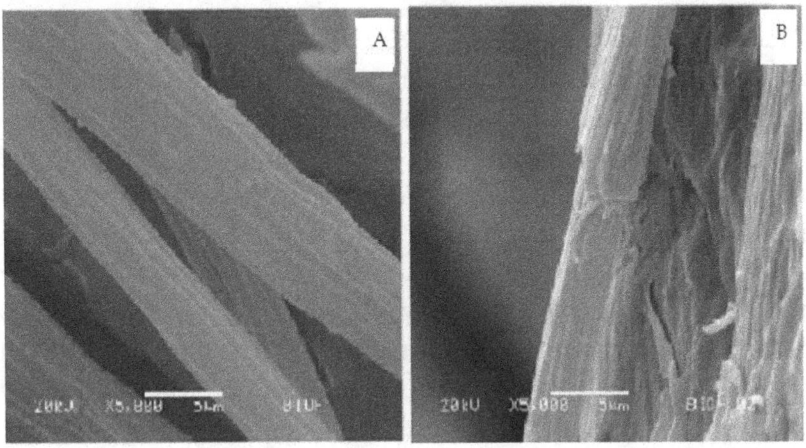

Fig. 2. SEM micrograph of CFF unmodified and CFF-g-HEMA

In the other hand, it can observe that when reaction time increase the graft yield has a maximum value at 3 h and then decrease at 4 h. this behavior is attributed that at higher reaction time the keratin pendant groups open and diffusion of HEMA molecules into the structure occurs, allowing the grafting on keratin structure not only on surface of CFF. However, when reaction time is high, a denaturalization of keratin can occurs, due the acidic medium where reaction is carried out, leading to HEMA homopolymerization. Other reason of this decrease in grafting yield is a reduction in number of free radicals available for grafting as the reaction proceeds, creating a saturation of active sites.

3.2 CMC-g-HEMA copolymer

The graft copolymerization of HEMA onto CMC had differences compared with HEMA-g- CFF: Redox initiator system used was CAN and was conducted at 3 different concentrations; HEMA concentrations values were higher 0.05, 0.1 and 0.15M and reaction time was constant in 3 hours. Table 3 resumes the effect of HEMA and CAN concentrations over grafting yields values of CMC.

Table 3. Effect of CAN and HEMA concentration on grafting yield over CMC

CAN [M]	HEMA [M]	Grafting yield
0.1	0.05	69 %
0.1	0.1	207 %
0.1	0.15	249 %
0.25	0.05	88 %
0.25	0.1	214 %
0.25	0.15	275 %
0.5	0.05	82 %
0.5	0.1	179 %
0.5	0.15	397 %

Graft copolymerization of HEMA onto CMC present a similar behavior as grating of HEMA onto CFF, grafting yield increased continuously with increase in concentration of HEMA when CAN concentration was constant, and reaches a maximum value with 0.15M at 3 CAN concentrations. This behavior could be explained by the fact of that an increase of HEMA concentration lead to the accumulation of monomer molecules in close proximity to CMC backbone. There are reports of modified polysacharides grafted with HEMA that found with CAN concentration of 0.2 M around 200% attributed the primary radicals attack the monomer instead of reacting with backbone polymer (Joshi and Sinha 2006). In this research higher CAN concentration studied was 0.5 M and higher grafting yield was reached with this concentration (397%). Also lower CAN concentrations (0.05 M) were tried in this research work, but not good results in grafting yield was obtained

In the other hand, about effect of CAN concentration over grafting yield, results show the initial increase in grafting yield with increase in initiator concentration levels off and decrease with further increase in CAN concentration. It would seem that termination of graft copolymerization would proceed by the reation of the growing graft polymer chain with ceric ions, the reation point is OH gruops of anhydroglucose, which form a complex with ceric ion. This complex may dissociate, and giving rise to free radical sites onto the polysacharides backbone and these radicals initiate the graft copolymerization (Zohuriaan- Mehr et al 2005).

Figure 3 shows SEM micrographs of CMC unmodified and CMC-g-HEMA. It is clearly seen that CMC morphology was totally modified by grafting of HEMA. CMC shows a porous surface which is covered by grafting of HEMA resulting in a homogeneous surface, which suggests that HEMA chains formed are long.

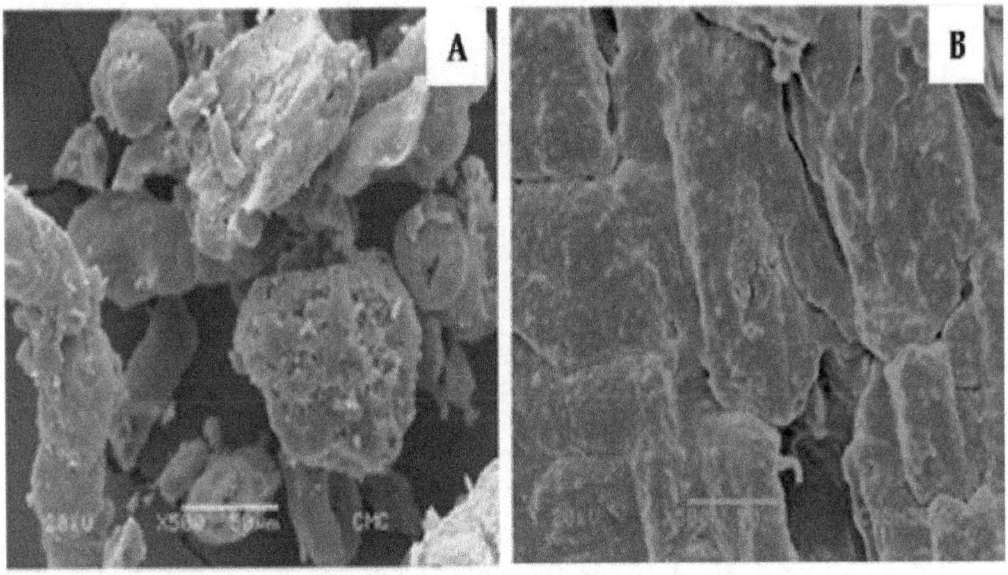

Fig. 3. SEM micrograph of CMC unmodified (A) and CMC-g-HEMA (B).

3.3 IR of CFF-g-HEMA copolymer

IR spectroscopy is a usefull technique to evaluate if a graft reaction is carry out and also to evaluate grafting yields on graft copolymers. Several research works used IR with this purpose for keratin (Martínez et al 2003; Martínez et al 2008; Kavitha et al 2005), starch (Meshram et al 2009), modified starch (Cao et al 2002), chitosan (Mun et al 2008), carboxymehtyl chitosan (Joshi and Sinha, 2006), cellulose (Zampano et al 2009), carboxymethyl cellulose (Okieimen and Ogbeifun, 1996; Vasile et al , 2004), ethyl cellulose (Kang et al 2006), which make it a powerfull tool in graft reactions.

Figure 4 shows IR spectrum of CFF, which main component is keratin, a mixture of aminoacids as serine, proline, glycine, valine, cysteine, and others. The main vibrations attributed to CFF structure were identified according to wavenumber. The region around 3300 cm^{-1} corresponds to range of amide bands, figure shows a peak in 3297 cm^{-1}associated with ordered regions of NH group of amide A a-helix conformation, peak in 2945 cm^{-1} is assigned to the assimetric vibration of CH group od metyl, the strong band at 1715 cm^{-1} is matched with vibration of amide I of b-sheet conformation, band at 1650 cm^{-1} assigned to C=O group of amide I a-helix conformation, peaks at 1520, 1449 and 1243 cm^{-1} attributed to in plane bending of NH group corresponding to b-sheet conformation, bending of CH3 group and C_n group of amide III respectively; vibrations on 1136, 1074 and 1023 cm^{-1} corresponds to asigned to vibrations of C-C group; and finally a peak around 700 cm^{-1} attributed to vibration of C-S group.

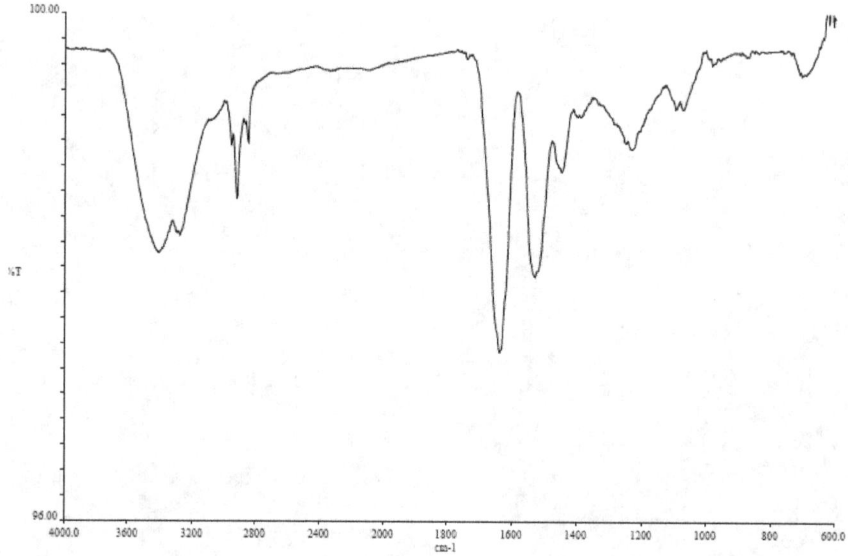

Fig. 4. IR spectrum of CFF ungrafted.

In figure 5 presents IR spectrum of PHEMA which presents signals at 3394 cm^{-1} attributed to vibracion of OH group, 2956 cm^{-1} from antisymetric vibration of CH_2 and CH3, 2925 cm^{-1} symetric vibration of CH_2, CH3, 2855 cm^{-1} symetric vibration of CH_2, 1720 of stretching C=O, a small shoulder around 1652 from stretching C=C, 1369 cm^{-1} deformation of CH_2, CH3, a shoulder at 1260 cm^{-1} from C=O streticing, 1164 vibration of C-O-C, 1048 stretching of CO(H), and 771 CH_2 coupled with skeletal stretching. This is according for reports of IR spectrum of PHEMA (Prachayawarakorn and Boonsawat, 2007).

Figure 6 shows IR spectra of CFF and CFF grafted. It can observe that main differences are IR spectrum of CFF grafted show peaks at 1720, 1160 and a shoulder around 1260 cm^{-1}, attributed to groups C=O and OH respectively, from HEMA structure, and which does not appear in CFF spectrum, which can indicate the graft reaction is carried out. CFF posses functional groups as $-NH_2$, -SH nd -OH where graft reaction can carry out, signals from NH (3290, 3080 and 1525 cm^{-1}) changed which make it suppose that graft reaction is taking place between NH groups and free radical of HEMA.

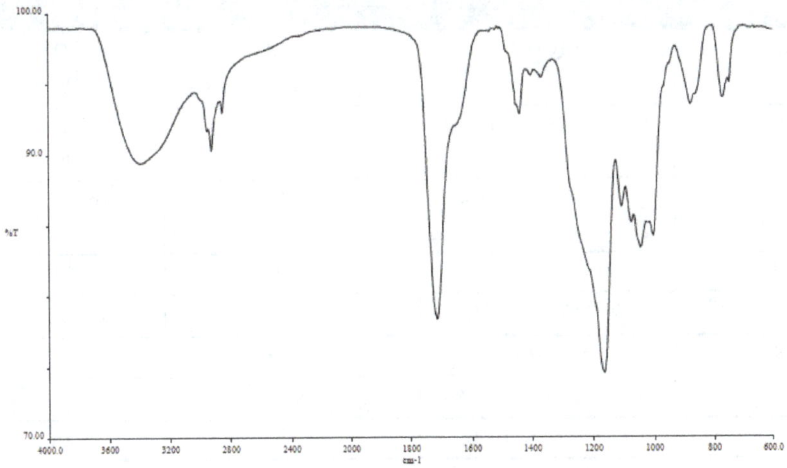

Fig. 5. IR spectrum of PHEMA

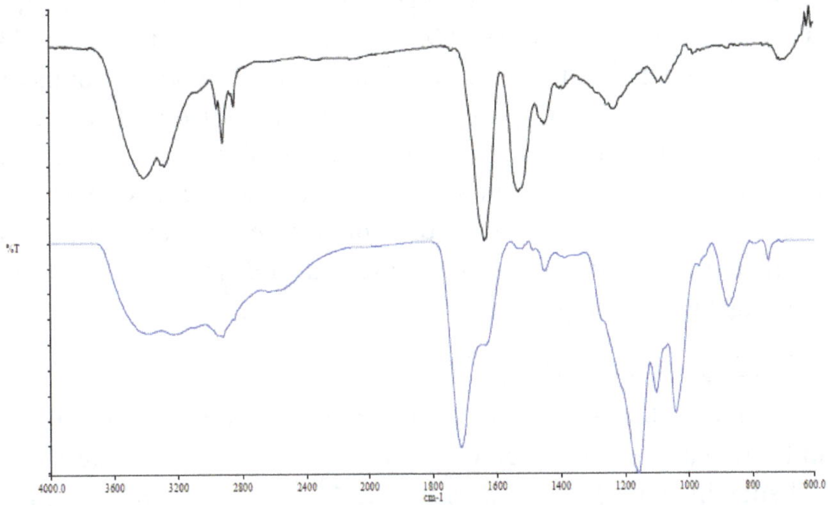

Fig. 6. IR spectra of ungrafted CFF (black) and CFF grafted (blue).

A comparison of main assignments from functional groups of ungrafted CFF and PHEMA are sumarized in table 4.

Table 4. IR assignments of main functional groups of ungrafted CFF and PHEMA.

PHEMA		CFF (Martínez 2005)	
Approximate assignments, cm^{-1}	functional groups	Approximate assignments, cm^{-1}	functional groups
3390	n (OH)	3290	n (NH) Amide A, a- helix
2950	n_a (CH_3, CH_2)	3080	n_a (NH) Amide B
2920	n_s (CH_3, CH_2)	2930	n_a (CH_3)

PHEMA		CFF (Martínez 2005)	
2850	n(CH_2)	1700	n(C=O) Amide I b- sheet
1720	n(C=O) ester group	1650	n(C=O) Amide I a- helix
1650	(C=C) stretching	1525	d(N-H) Amide II b- sheet
1370	d (CH_3, CH_2)	1450	d_a(C-H_3)
1260	C=O stretching	1230	n(CN) Amide III
1160	n (C-O-C)	1170	n(C-C)
1050	(CO-H) stretching	1075	n(C-C)
770	(CH_2) stretching	685	n(C-S)

In figure 7 present IR spectra of CFF grafted with 26 and 92% grafting yield, to evaluate effect of grafting on main assignments. It is evident when graft yield increase assignations attributed to PHEMA are more evident in spectrum (3390 and 1720, 2850 and 1150 cm^{-1}), and signals attributed to NH and OH groups disappear because there are sites where graft reaction is carrying out. There are several reports about use of IR spectroscopy for evaluate graft reactions on keratin from chicken feathers (Martínez et al 2003, Martínez et al 2008 and Kavitha et al 2005). They use IR spectra for identify main functional gruops of keratin and of acrylic monomers in ungrafted and grafted copolymers, in addition to evaluate changes in peaks according with grafting yield.

3.4 IR of CMC-g-HEMA copolymer

IR spectra of the CMC is presented in figure 8. We can notice the characteristic broad band attributed to OH stretching vibration at 3360 cm^{-1} due the CMC has a degree of substitution of 0.7, in average 2.3 OH groups of anhydroglucose ring are present in structure, peak at 2920 cm^{-1} due stretching of C-H, peak at 1620 cm^{-1} a strong absorption band that confirm the presemce of carboxy group (COO-), 1420 and 1320 cm^{-1} are assigned to –CH_2 scissoring and hydroxyl group bending vibration respectively, signal at 1060 cm^{-1} is due to >CH-O-CH_2 stretching vibration and vibrations of the ether groups at 1060, 1110 cm^{-1}. It is worth to remark that CMC used in this work was as sodium salt. The assignments are according with reports of IR spectroscopy studies of CMC (Bono et al 2009, Heydarzadeh et al 2009, Vasile et al 2004).

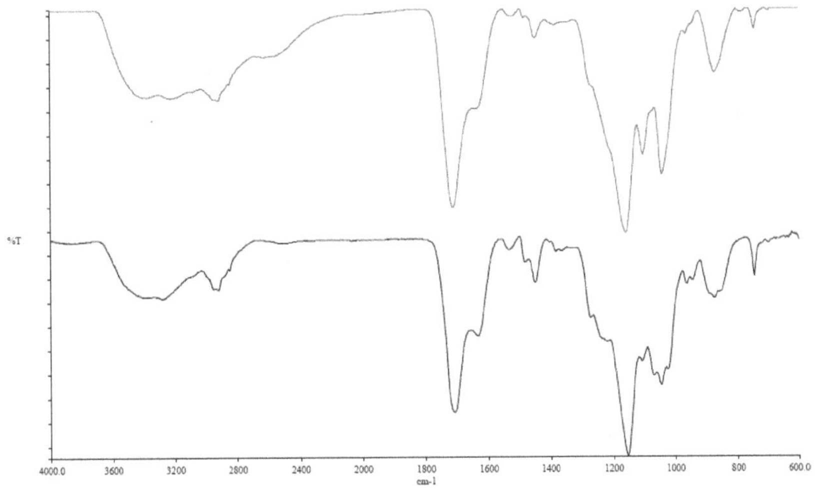

Fig. 7. IR spectra of CFF-g-HEMA with 26 (blue) and 92% (black) graft yield

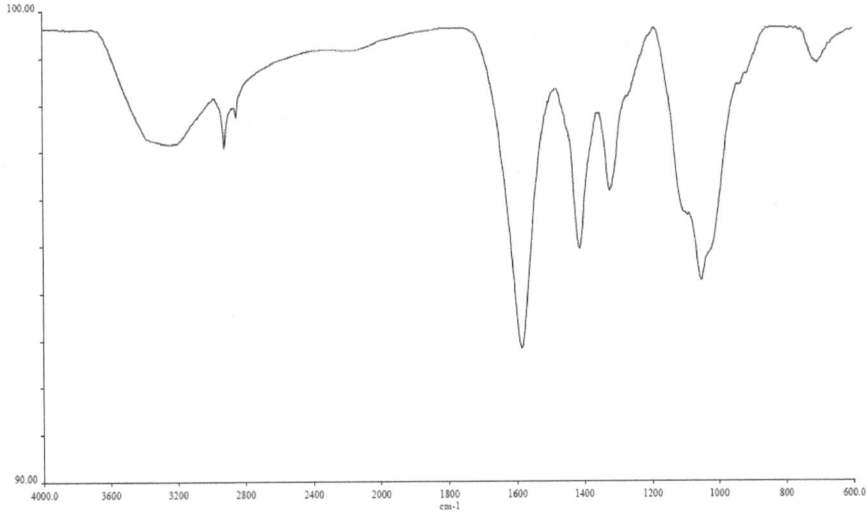

Fig. 8. IR spectrum of CMC

Figure 9 presents a comparative of IR spectra of CMC ungrafted and grafted with HEMA. Before, the main assignments for PHEMA were disscused. The CMC grafted present appearance of peak at 2930 cm^{-1} due antisymetric stretching CH3, 1715 cm^{-1} from stretching vibration of C=O and peaks at 1410, 1250, 1150 and 902 cm^{-1} assigned to bending vibration of CH3, CH_2, stretching of C=O, stretching of CO-H and $=CH_2$ groups respectively, all those signals are from PHEMA structure, and also peaks attributed to CMC structure as 3340, 1650 and broad peak at 1050 cm^{-1}, which are from OH, COO- and C-O-C from CMC, which is indicative the graft reaction was carried out.

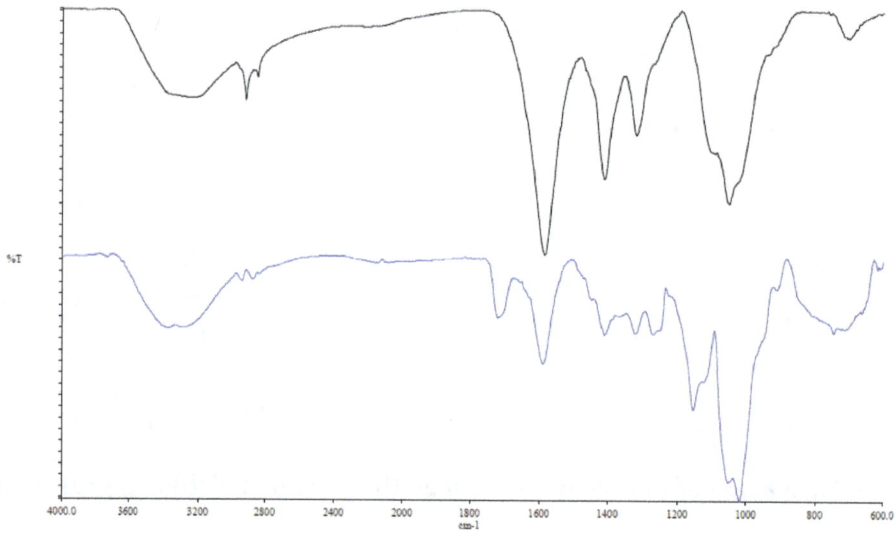

Fig. 9. IR spectra of CMC ungrafted (black) and with 69% grafting yield (blue)

Effect of grafting yield on IR assignments was evaluated in figure 10. It can observe that main peaks attributed to PHEMA (2940, 1720, 1270 and 1060 cm^{-1}) increase according with graft yield, which makes sense due the PHEMA is an higher proportion compared with CMC.

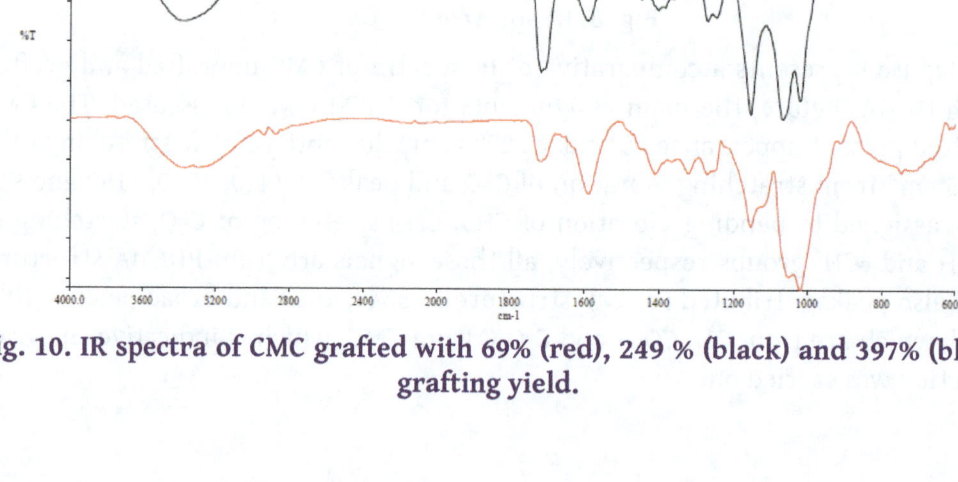

Fig. 10. IR spectra of CMC grafted with 69% (red), 249 % (black) and 397% (blue) grafting yield.

4. CONCLUDING REMARKS

From results presented in this research work it can be conclude that is possible graft HEMA onto CFF and CMC using different initiator system, and that grafting yield obtain on depends of kind of initiator system, being higher with CAN initiator system than KMnO4 malic acid. SEM microgrpah give evidence that the grafting reaction takes place on CFF and CMC surface until a saturation of active sites, and then homopolymerization happens.

One of the most common applications of IR spectroscopy is to the identification of organic compounds. In polymers, may be used to identify the composition, to monitor polymerization process, to characterize polymer structure. In present research, IR spectroscopy showed to be a usefull toll to evaluate if the graft reaction takes place. It was possible identify main functional groups of CFF and CMC ungrafted, and PHEMA. The IR spectra of grafted copolymers presented assignments due to ungrafted materials and PHEMA giving evidence that the graft reaction carried out. Futhermore, it was possible evaluate the changes in peaks of the grafted materials according with grafting yield, increasing signals attributed to PHEMA.

5. REFERENCES

1. Athawale, V.D., Rathi, S.C., (1997), Effect of chain length of the alkyl group of alkyl methacrylates on graft polymerization onto starch using ceric ammonium nitrate as initiator. European Polymer Journal, vol. 33, No. 7, (July 1997), pp 1067-1071, ISSn 0014-3057.
2. Barone, J.R., Schmidt, W.F., Liebner, C.F., (2005). Thermally processed keratin films. Journal of Applied Polymer Science, Vol. 97, No. 4, (august 2005), pp 1644-1651, ISSn 1097-4628.
3. Bickford, M., (2008). Characterization and Analysis of Polymers. John Wiley and Sons. ISBN 978-0-470-23300-9, Hoboken, New Jersey, United States.
4. Bono, A., Ying, P.H., Yan, F.Y., Muei, C.L., Sarbatly, R., Krishnaiah, (2009), Synthesis and Characterization of Carboxymethyl Cellulose from Palm Kernel Cake. Advances in Natural and Applied Sciences, Vol. 3, No. 2, (January 2009), pp 5-11, ISSn 1995-0772.
5. Cao, Y., Qing, X., Sun, J., Zhou, F., Lin, S., (2002), Graft copolymerization of acrylamide onto carboxymethyl starch. European Polymer Journal, vol. 38, No. 9, (September 2002), pp 1921-1924, ISSn 0014-3057.
6. Elizalde Peña, E., Flores Ramírez, N., Luna Barcenas, G., Vásquez García, S. R., Arámbula Villa, G., Garcia Gaitan, B., Rutiaga-Quiñones, J.G., González-Hernández, J. (2007). Syntesis and characterization of chitosan-g-glycidyl

methacrylate with methyl methacrylate. European Polymer Journal, Vol. 43, No. 9, (September 2007) pp 3963-3969, ISSn 0014-3057.

7. Gupta, K.C., Sahoo, S., (2001) Graft copolymerization of 4-vinylpyridine onto cellulose using Co (III) acetylacetonate complex in aqueous medium. Cellulose, Vol. 8, No. 3, (September 2001),pp 233-242, ISSn 0969-0239.

8. Heydarzadeh, H.D., Najafpour, G.D., Nazari-Moghaddam, A.A., (2009), Catalyst-free conversion of alkali cellulose to fine carboxymethyl cellulose at mild conditions. World Applied Science Journal, Vol. 6, No. 4, (2009), pp 564-569, ISSn 1818-4952.

9. Jia, Z.,Yong, Y., (2006); Surface Modification of Poly Acrylic Fibers (PAC) via Grafting of Soybean Protein Isolates (SPI). Iranian Polymer Journal, Vol. 15, No. 10, (October 2006), pp 789-798, ISSn 1026-1265.

10. Joshi, J.M., Sinha, V.K., (2006), Graft copolymerization of 2-hydroxyethylmethacrylate onto carboxymethyl chitosan using CAN as an initiator. Polymer, Vol. 47, No. 6, (March 2006), pp 2198-2204, ISSn 0032-3861.

11. Kang, H., Liu, W., He, B., Shen, D., Ma, L., Huang, Y., (2006), Synthesis of amphiphilic ethyl cellulose grafting poly (acrylic acid) copolymers and their self-assembly morphologies in water. Polymer, Vol. 47, No. 23, (October 2006), pp 7927-7934, ISSn 0032-3861.

12. Klemm, D., Philipp, B., Heinze, T., Heinze, U., Wagenknecht, W., (1998). Comprehensive cellulose chemistry volume 2,Wiley-VCH, ISBN 3-527-29489-9, Weinheim, Federal Republic of Germany.

13. Kavitha, A., Boopalan, K., Radhakrishnan, G., Sankaran, S., Das, B., Sastry, T., (2005), Preparation of Feather Keratin Hydrolyzate-Gelatin Composites and Their Graft Copolymers. Journal of Macromolecular Science Part A: Pure and Applied Chemistry, Vol. 42, No. 12, (December 2005), pp 1703-1713, ISSn 1060-1325.

14. Martínez-Hernández, A.L., Santiago-Valtierra A. L., Alvarez-POnce M. J., (2008), Chemical modification of keratin biofibres by graft polymerisation of methyl methacrylate using redox initiation. Materials Research Innovations, Vol. 12, No. 4, (December 2008), pp 184-191, ISSn 1432-8917.

15. Martínez-Hernández, A.L., Velasco-Santos, C., de Icaza, M., Castaño, V.M., (2005), Microstructural characterization of keratin fibers from chicken feathers. International Journal Environmental and Pollution, Vol. 23, No. 2, (MES 2005), pp 162-178 ISSn 0957-4352.

16. Martínez-Hernández, A.L., Velasco-Santos, C., de Icaza, M., Castaño, V.M., (2003), Grafting of Methyl methacrylate onto natural keratin. e-polymers, No. 016, (April 2003), pp 1- 11, ISSn 1618-7229.

17. Meshram, M.W., Patil, V.V., Mhaske, S.T., Thorat, B.N., (2009) Graft copolymers of starch and its application in textiles. Carbohydrate Polymers, 75, 1, (January 2009), pp 71-78,ISSn 0008-6215.

18. Mostafa, M., Graft polymerization of methacrylic acid on starch and hydrolyzed starches.

19. Polymer Degradation and Stability, (1995), Vol. 50, No. 2, (March 1995), pp 189-194,ISSn 0141-3910.

20. Mun, G.A., Zauresh, S., Nurkeeva, S.A., Dergunov, I.K., Nam, T.P., Maimakov, E.M., Shaikhutdinov, Lee, S.Ch., Park, K., (2008), Studies on graft copolymerization of 2- hydroxyethyl acrylate onto chitosan. Reactive and functional Polymers, Vol. 68, No. 1, (January 2008), pp 389-395, ISSn 1381-5148.

21. Okieimen, F.E., Ogbeifun, D.E., (1996), Graft copolymerization of modified cellulose: grafting of methyl acrylate, ethyl acrylate and ethyl methacrylate on carboxymethyl cellulose. European Polymer Journal, Vol. 32, No. 3, (March 1996), pp 311-315, ISSn 0014-3057.

22. Prachayawarakorn, J., Boonsawat, K., (2007), Physical, chemical and dyeing properties of bomboyxmori silks grafted by 2-hydroxyethyl methacrylate. Journal of Applied Polymer Science, Vol. 106, No. 3., (November 2007), pp 1526-1534, ISSn 1097-4628.

23. Radhakumary, C., Nair, P. D., Mathew, S., Nair, C.P., (2006), HEMA-Grafted Chitosan for Dialysis Membrane Applications. Journal of Applied Polymer Science,Vol. 101, No 5, (September 2006), pp 2960-2966, ISSn 1097-4628.

24. Schmidt, W.F., (1998), Innovative feather utilization strategies. Proceedings of National Poultry Waste Management conference, pp 276-282, Springdale, Ar., October 19-22, 1998.

25. Vasile, C., Bumbu, G.G., Dumitriu, R.P., Staikos, G., (2004), Comparative study of the behavior of carboxymethyl cellulose-g-poly(N-isopropylacrylamide) copolymers and their equivalent physical blends. European polymer Journal, Vol. 40, No. 6, (June 2004) pp 1209-1215, ISSn 0014-3057.

26. Xu, X. Zhou, Z., Prum, R.O., (2001), Branched integumental structures in Sinornithosaurus and the origin of feathers. Nature, Vol. 410, No. 6825, pp 200-204, (March 2001), ISSn 0028-0836.

27. Xu, X., Zhou, Z., Wong, X., Kuang, X. Zhang, F., Du, X., (2003), Four-winged dinosaurs from China. Nature, Vol. 421, No. 6921, pp 335-340, (January 2003), ISSn 0028-0836.

28. Zampano, G., Bertoldo, M., Bronco, S., (2009) Poly(ethyl acrylate) surface-initiated ATRP grafting from Wood pulp cellulose fibers. Carbohydrate Polymers, Vol. 75, No. 1, (January 2009), pp 22-31, ISSn 0008-6215.

29. Zohuriaan-Mehr, J.M., Pourjavadi, A., Sadeghi, M., (2005), Modified CMC: Part1-Optimized Synthesis of Carboxymethyl cellulose-g-Polyacrylonitrile. Iranian Polymer Journal, Vol. 14, No. 2, (February 2005), pp 131-138, ISSn 1026-1265.

www.ingramcontent.com/pod-product-compliance
Lightning Source LLC
LaVergne TN
LVHW062223200926
848484LV00006B/1458